# ATLIN

# ATLIN

# THE STORY OF BRITISH COLUMBIA'S LAST GOLD RUSH

Christine Frances Dickinson
Diane Solie Smith

Atlin Historical Society,
Atlin, British Columbia
1995

Published by the
**Atlin Historical Society**
**P.O. Box 111**
**Atlin, British Columbia**
**V0W 1A0 Canada**

Layout, Design and Prepress by
**Treeline Computing Services**
**P. O. Box 296**
**Atlin, British Columbia**
**V0W 1A0 Canada**

Cover design by Diane S. Smith
Maps by Kate Fisher and Bob McKerihen

Financially assisted by the PROVINCE OF BRITISH COLUMBIA
through the British Columbia Heritage Trust

CANADIAN CATALOGUING IN PUBLICATION DATA

Dickinson, Christine Frances, 1945-
Atlin - The Story of British Columbia's Last Gold Rush
Smith, Diane Solie, 1930-
Atlin - The Story of British Columbia's Last Gold Rush

ISBN 0-9680193-0-7

1. Atlin, British Columbia - History. 2. Gold Rush.
3. Local. 4. Gold miners. 5. Title.

Printed in Canada
First printing 1995
Second printing 1997

Front Cover Photo: First Boat landing at White Pass wharf in the spring of 1903.
Brown Collection - P-2690 AHS
Back Cover Photos: Tlingit Indian youths. BCARS #51386
Steam drill on an Atlin Street. AHS P-1242
A Spruce Creek cabin. AHS P-2049

BCARS - British Columbia Archives and Record Service
AHS - Atlin Historical Society

To the miners, merchants and women
who came to Atlin for gold and
stayed to build the town

and

To our families,
David, Joanne, Rebecca and James Dickinson
and
Bradford, April, and Taylor Smith and J.J. Frost

# Contents

# Maps

## ACKNOWLEDGMENTS

The authors wish to acknowledge the contributions and assistance of the following people: the many Atlin old timers and former residents who so graciously and generously shared their memories with us in conversations, letters and photographs; Shirley Connolly and Norah Smith, whose pioneer work with the Atlin Historical Society gave us the springboard and the inspiration to begin this work; Sue Morhun whose historical sleuthing provided several starting points in the telling of the Atlin story; the very helpful staffs at the British Columbia Provincial Archives and the Yukon Archives; Bob Coutts whose advice on both writing and mining matters was sought and welcomed; Elizabeth Clarke who so generously gave us access to George Clarke's library; Gabrielle Aldeberan who patiently proof read the manuscript; Kate Fisher and Bob McKerihen for their help with the maps; Carrie Knickerbocker who joined Kate with enthusiasm and know how to bring this project to fruition; and the very patient Zachariah Crow whose expertise quite literally brought our words to press. Because the gestation period for this work has been so prolonged there are dozens of others, including our many Atlin friends, to who we owe sincere thanks for their support and encouragement.

# Introduction

Atlin, British Columbia is located in solitary stubbornness a few miles south of the Yukon border on a lake of haunting beauty. A small infusion of new log buildings stand on orderly streets among aged clapboard houses and historic public places as proof of the town's continued regeneration. A major gold rush gave birth to Atlin, but a reason for its present healthy state is not particularly obvious. Many British Columbians have never heard of Atlin, but this was not always the case.

Atlin's nearest provincial neighbour is Telegraph Creek, far away on the Stikine River. They are connected by an overgrown trail 375 km long. This old route, the Telegraph Trail, was once busy with travelers, hunters, and the men who operated and maintained the Dominion Telegraph line. Now it is part of the past as are the other abandoned trails, blazed by stampeders, then used for years by Atlinites, visitors and suppliers on their way in and out of the community. Atlin had passed its half century mark before a road was built linking it with the Alaska Highway. Since 1949 it has been more closely allied with Whitehorse, Yukon than with any place in British Columbia.

In the beginning, though, and for a number of years at the turn of the 20th century, Atlin's name was recognized throughout the province. It had the rapt attention of the mining industry, and was the subject of newspaper columns from Dawson City to San Francisco, New York and beyond. Gold was found on an Atlin creek in 1898, just when the entire world seemed bent on reaching the Yukon. The rousing news of an important strike 1000 km south of the Klondike changed the minds of thousands of stampeders who detoured to the Atlin goldfields instead. Proof of Atlin gold captured crews building the great White Pass railroad, and played serious havoc with construction schedules after 1500 men left their jobs to find gold in Atlin.

In distant Victoria, politicians immediately recognized the implications of a major gold rush within the province. Intent on cornering substantial treasury profits after watching Yukon gold trickle into foreign coffers, and in an urgent attempt to have an all-British Columbian rush, they blundered by meddling with mining regulations. The results caused a serious hindrance to their chances of capitalizing on the stampede into Atlin, and the subsequent development of the mining industry in the area. Many potential investors were turned off, and moved on, but mining and building continued despite the government's botched attempt to manage a gold rush with a life of its own.

When the boom years were over and the Depression ended a glory era of tourism, Atlin's existence depended on the ingenuity of a few hundred people who stayed there because they had no good reason to leave. The Pine Creek town of Discovery eventually disappeared, but Atlin survived because a band of volunteers took charge of town business. They devised clever ways to wheedle services from a suddenly uninterested government, so the town had most common amenities, and functioned with a degree of dignity. While citizens struggled to adjust to a scaled down economy, the propelling spirit instrumental in establishing Atlin never diminished. During the decades following the rush, when the hard business of economic preservation took precedence, memories of the gold rush were pushed aside, more or less forgotten, as was Atlin itself.

The old stories were not lost, but the original players, the tellers, were gone. The human fascination with record keeping had saved most of Atlin's history, although it was scattered as far and wide as the nuggets first mined in Pine Creek. The passing of nearly 100 years had seen to that. As old journals, newspapers, and yellowing ledgers were found, the information they revealed overwhelmed imagination with the clamor of mining, and the voices of thousands of men and women captivated by the search for gold and the lure of the north country. The words wove a vivid backdrop of mighty mountains, seductive and dangerous lakes, and mineral rich creeks, all inseparable from the narratives of those first citizens.

Allowing such exciting tales to vanish, as was nearly the case, would have constituted the theft of an important part of British Columbia's pioneering history. To ensure against such loss, the Atlin stories were retrieved and told in this book.

The Location

CHAPTER 1

# Before Gold

Throughout the eighteenth century, the British, Spanish and Russians jockeyed for position and influence on the southeast Alaskan coast. Preoccupied with the baffling search for the Northwest Passage, and trade with China for luxurious silks, tea and other exotic treasures, they had little interest in what lay beyond the shorelines. The sweeping forests brimming with wildlife, the cascading rivers, icy clear lakes and buried minerals, remained undiscovered by European adventurers. Also unknown to those early coastal explorers were the small bands of aboriginal people, who, for centuries, had cleverly sustained themselves and lived in reasonable harmony with the wild northern land and its rigorous climate. It is with these stalwart people the human history of the Atlin region begins.

In the early years of the next century, the sea otter, a vital exchange item for the far east trade, had been harvested close to extinction. The Russians and British, desperate for other furs, were at last stirred to consider the potential of the interior lands. The coastal Tlingits, who had access to abundant fur supplies from beyond the Coast Mountains, became their trading partners. Had the Europeans been motivated to venture inland themselves, their plans would soon have been put to rest by the Tlingits. The astute aboriginal businessmen had no intention of allowing their trading position to be usurped. They became steadfast guards of all mountain passes and river routes to the interior. Until the 1880's they alone had contact with the inland people, the trappers, who furnished the luxurious furs, so coveted by the Europeans. The Tlingits often traveled far to the south to obtain the tools, buttons, beads, tea, tobacco and guns for exchange with their inland neighbours. As middle men they profited enormously from their trading arrangement.

The coastal Tlingits who lived in the vicinity of the Taku River, traded with two small bands of Tlingits who maintained camps inland on some of the tributaries up river and at Teslin Lake. They were not acknowledged as kin by the coastal Indians who considered them to be foreigners or "gunana".

George Emmons,[1] a late 19th century anthropologist, believed that the inland Tlingits had originally lived in the upper Taku Basin. He also believed the Tlingits from the Taku delta were a relatively recent people and the result of extensive intermarriage between Athabaskans from the Stikine and Taku Rivers, and coastal Tlingits.

The language, customs and much of the social organization of the interior Tlingits was similar to that of the coastal people. Their way of life, though, was quite different. The restless search for food made them semi-nomadic. They moved through the country seeking supplies of freshwater fish, birds, small game, caribou and moose, and furs for trade and clothing. Anthropologist, Catherine McClellan pointed out, that they lived in a much harsher climate, where time and energy was spent on the mere task of survival, and little was left for the development of a sophisticated material culture, or the amassing of wealth[2] which characterized the coastal peoples.

Apart from obvious feelings of superiority on the part of the coastal Tlingits, the precise nature of the relationship between the two groups is unclear. McClellan believed, however, that the Tlingits of the interior came originally from the coast.[3] The valley of the Taku River was a crossing to the Yukon long used by the coastal people.[4] A migration up the Taku River and inland into already known country, may have resulted from some kind of social dislocation of groups or individuals among the coastal people. Certainly McClellan offered evidence that the interior was a haven for outcasts from the coast. It is also clear that both the Tlingits and the Athabaskans of the interior occasionally stole each other's women, and that there was a certain amount of intermarriage between the two groups. Sometimes marriages were arranged to promote trading advantages. These mixed families may well have chosen, or been forced to live in the interior. James Ward of the Atlin Indian Band believed settlements on the Taku River became overcrowded, and some of the people moved into the interior as well as to other points on the coast.[5] This theory may well have applied to some early time, but in 1880 coastal Taku Tlingits numbered only 269 people living in four settlements on the lower Taku River and its mouth.[6]

The presence of non-Tlingit influences among the interior people is substantiated by their rapid and nasal speech, characteristic of the Athabaskans, but not the coastal Tlingits. Physical signs of Athabaskan influences are evident among Atlin and Teslin Tlingits today. As well as the short, broad-bodied lighter-skinned coastal-types, there are some with the tall, lean darker Athabaskan characteristics.[7]

By the coming of each spring, the inland Tlingits had cached large supplies of furs, for trading. An account from the 1840's said that when the Taku River was ice-free, the coastal Indians would travel by canoe, "100 English miles up the river and proceed about the same distance on foot to the trading places of the interior."[8] Later traders rigged their canoes with square moose hide sails, and with the wind at their backs reached the head of navigation at Canoe Landing in about a week. There they would meet the people of the lakes for the exchange of goods, or they would take another three or four days to pack their goods to the Indians in the lake country.[9]

The coastal Taku Tlingits were willing to trek across the mountains to meet the inland Indians and, in so doing, prevent them from dealing directly with the European traders. This position became particularly difficult to maintain, when, in 1840 James Douglas opened a Hudson Bay Company trading post near the mouth of the Taku River. The outpost, officially named Fort Durham, but commonly called Fort Taku, was not particularly successful. According to Bancroft, the native people were "so absorbed in dealing with slaves, who were brought from a distance and used in commerce as a sort of currency, that they had but few skins left to buy whiskey with."[10] In 1843 Fort Durham and a number of other coastal forts were replaced by the steamer *Beaver*, a mobile trading vessel.[11]

Emmons noted that, even forty years later, the coastal Tlingits remained anxious to prevent extensive contact between the inland Tlingits, and other interior people, and the white traders. After the establishment of Juneau (in 1880), in the early summer when the river was high, the up-river (Tlingits) came down with their furs and lived with their coast relatives. In trading with the stores, they were carefully supervised by their hosts who took part in every transaction to their own advantage.[12]

It appears it was relatively easy for the coastal Tlingits to maintain their monopoly at that time since the inland Tlingits were spending less time on the river. They had settled in more permanent homes on Teslin and Atlin Lakes.[13] This was supported by Johnny Jack, second son of Chief Taku Jack, who believed that his people maintained a regular summer camp at the same site as they established themselves after the discovery of gold in the Atlin valley.[14]

If the visits to the coast became less frequent, they remained well supervised. Trips were sometimes made to obtain supplies in Juneau. John Bone, one of the last Tlingits who remembered pre-gold rush days, recalled going there for shoe-packs, like those worn by the first white man he ever saw, a man whose pale skin frightened him. They also purchased black powder by the cake, and lead to make shot for their muzzle loaders.[15] Through these trips some limited knowledge of European culture was gained.

Although there is no record of Russian missionaries venturing inland from the coast, a small number of Atlin and Teslin Tlingits were first converted to Christianity by the Russian Orthodox priests, whom they apparently encountered on trips to Juneau. Only a few were baptized, but they clung to their faith although isolated from the Russian church. They were ignored by the missionaries in Atlin until 1907, and the arrival of a French priest with allegiances to the Church of Rome.

The Hudson Bay Company was a shadowy presence in the area throughout the nineteenth century. Fort Durham was abandoned after two years, but as late as 1892 the Company had a trading post on the Sheslay River called Egnel Post.[16] There was a post at MacDame Creek, northwest of Dease Lake and others in northeastern British Columbia. Whether their knowledge of the head water regions of the Yukon River was extensive is not clear. Obviously they had an understanding of the fur potential of the area before establishing trading posts, but it was not in their interest to share their knowledge of the country with interlopers.

In 1930, Ruth and Bill Albee,[17] a newly married couple, keen for an unusual adventure, decided to walk from Prince George, B.C. through the Yukon and Alaska to the Bering Sea. Before departure they searched in vain for a map of the northern interior of the province. When they reached the Hudson's Bay post at Fort Liard, they spent several days there during which they amused themselves looking through old company records. Among the papers they discovered detailed maps of the territory through which they had just passed. Apparently these had been kept secret for the Company's own reasons. It is possible similar mapping was done in the Atlin Lake and surrounding regions. Future research in the Company Archives may reveal such evidence.

Records suggest that the first white man to see Atlin Lake was a Cariboo miner named Michael Byrnes. He was hired by the Western Union Telegraph Company to help scout a path for an overland telegraph line that would connect the United States with Europe. The idea was Perry McDonough Collins', who had made a fortune in the California gold rush. He pursued and promoted his scheme to run an overland line through northern British Columbia, the Yukon, Alaska and Siberia. Finally Western Union was convinced of its

AHS Noland Collection P-194

*Atlin Lake was named by the Tlingit Indians. A Tlen is their word for great water.*

AHS Henning Collection P-723

*Llewllyn Glacier originates in the ancient ice of the Juneau Ice Cap. Its slow retreat from the south end of Atlin Lake exposes a maze of rocky islands and deep inlets.*

merits and in 1864 work began on the 27,000 km communication link.[18] Exploration and construction started in British Columbia in May 1865, after permission was granted by the Legislative Council. At the end of the working season in 1866, the line had been installed from Hazelton, and was operating to a point at the confluence of the Kispiox and Skeena Rivers. It would go no farther. The Trans-Atlantic cable had been successfully laid in July of that year, and intercontinental communications firmly established. Many months would pass before labourers and surveyors on remote sections of the route could be notified to abandon their camps.

Michael Byrnes was a prospector of merit, and a first rate man in the bush. With Vital La Force he is credited with the discovery of gold on the Omineca River during the winter of 1868- 69,[19] and Burns Lake was named for him.[20] Because he was not a professional surveyor or explorer, Byrnes left neither survey notes nor any journal accounts of the country he traveled while scouting for Western Union. It is uncertain just which tracts of the interior he covered. Nonetheless, two later explorers, Frederick Schwatka and George Dawson were well aware of his work, and credit him with investigating a huge area of previously unknown British Columbia territory.[21] Both believed that he reached Atlin Lake, although the evidence is unclear. Bilsland[22] casts some doubt on Byrnes' ability to progress further north than Fort Fraser in 1866, since the correspondence of Colonel Bulkley, the American military man in charge of the telegraph installation in British Columbia, places him for the most part between Quesnel and Fort Fraser between April and October that year.

Dawson, however, suggests that Byrne's exploration of northern British Columbia took place in 1867.[23] He may have proceeded with the advance parties dispatched towards the Skeena, Taku, and Chilcat Rivers in October, 1866.[24] Schwatka thought he came into the region by way of the Tahk-o Pass (the Taku River), "got on one of the sources of the Yukon, and . . . descended it to the vicinity of the lake" referred to as Tako Lake. An early map shows Atlin Lake as Tako Lake. It was in this area a courier, dispatched by Western Union, caught up and told him of the Atlantic cable.[25]

In the absence of written records, it cannot be stated positively that Byrnes reached Atlin Lake. However, three other distinguished explorers, Schwatka, Dawson and Gauvreau, did pass by the largest lake in British Columbia, were aware of its existence, if not its size, and did not veer from their routes to investigate.

From the time of the Alaska purchase in 1867, the United States Army had planned a reconnaissance expedition to the territory. For years there were delays and wrangling over funding. It was not until May, 1883, that a small, not particularly well equipped party led by Lieutenant Frederick Schwatka left Portland, Oregon for Alaska Territory. Their purpose was to gather information and map the routes to interior Alaska for future reference, should the military need it.[26] Approaching from Chilkoot Inlet, they proceeded up the Taiya River, crossed the Chilkoot Pass, and rafted down Lindeman, Bennett, Tagish and Marsh Lakes to the Yukon River.[27]

Schwatka returned north in 1891, intending to enter via the Chilkoot Pass, as he had done previously. In Juneau, however, he was persuaded by local businessmen to travel up the Taku River instead, to investigate the possibility of building a pack trail along the old Indian route to Teslin Lake. He proceeded up the river with six Indian guides to just above the confluence with the Inklin. From there he went overland to Teslin Lake. Although Schwatka was close to Atlin Lake again, he did not investigate.

In 1887 the Canadian Government organized its own expedition under the auspices of the Canadian Geological Survey, led by George Mercer Dawson. Unlike Schwatka's expedition the Canadians' purposes were not explicitly military, but they were apparently seeking similar information. Dawson, who believed an important discovery of gold in the Yukon District was imminent,[28] said, his purpose was to gain "information on the vast and hitherto almost unknown tract of country which forms the extreme north westerly portion of the Northwest Territory . . ."[29]

Dawson passed along Tagish Lake but did not explore the area of it called Taku Arm. On the authority of local Indians he described a "considerable river" which entered the Arm from the east. His description of the Atlintoo River as "tranquil" is hardly accurate, but his information appears correct in most other respects. He continued, "It flows out of the west side of another very long lake which lies nearly parallel to Tako Arm."[30] His Indian informants offered three possible names for this body of water, one of which was Atlin, and this Dawson adopted for his map.

The expedition was highly successful, and the vast fund of knowledge it collected was a valuable resource for the Canadian Government and the many adventurers who would soon strike out for the Klondike goldfields. Dawson too, however, missed Atlin Lake, but he did select the name which would be adopted by posterity. Astutely he also identified Atlin Lake as the headwaters of the Yukon River system.

In 1892 the British Columbia government finally decided it was time to investigate the province's most northwesterly corner. Narcisse Gauvreau and three colleagues were dispatched to explore and report on the unmapped territory which lay between the Stikine River and the northern border, and west to the Alaska border. Gauvreau, a land surveyor, originally from Quebec, had practised his profession in New Westminster since 1885.

The Gauvreau party proceeded from Telegraph Creek to the south end of Teslin Lake. A portion of the route was over a rough trail hacked out by the Hudson Bay Company employees from Egnel Post on the Sheslay River. It was not a well traveled route and progress was difficult. From Teslin the party turned southwest as far as the Inklin River, and explored a route which led them to Sloko Lake, located just a short walk from the most southerly finger of Atlin Lake at Sloko Inlet. From his own account Gauvreau saw Atlin Lake, although, because of "bad weather and the lateness of the season,"[31] he did not proceed far, and his estimate that the lake was 45 miles (72 km) long was short by almost half its length.

Of all of the explorers in the region of the Atlin valley, it seems most certain that Gauvreau did reach Atlin Lake, but it is unlikely that he was the first white man to do so. There are persistent stories that early Atlin miners discovered the remnants of successful Russian mine workings in the area. One such tale tells of a broken down cabin which contained the remains of a man, who was presumed to be Russian, a 30 year old rifle, and about 30 kg of gold.[32] Boulder Creek was first named Musket Creek, presumably because an old musket was discovered there.

The assumption that earlier activity was Russian has not been substantiated, and in fact evidence[33] supports the notion that, like the Hudson Bay Company, the Russians were not interested in mining. Mineral exploration was counter productive to their prime interests which were to preserve the habitat of the fur bearing animals on which their trade depended.

The Russian pottery, and various relics, such as tea chests, discovered by early miners in the Atlin area, probably belonged to the Tlingit people, since discoveries of such items have been made as far east as Great Slave lake.[34]

Other evidence suggests that, before recorded settlement began on Atlin Lake in 1898, there were other miners in the area, not the suspected early Russians, but men from the Cassiar country where gold was discovered on Thibert Creek in 1873. There was a rush in 1874 and 1875, and thereafter hundreds of prospectors fanned out over the vast northern landscape, following the Liard River and investigating the Frances Lake country. Others moved west to the islands of southeastern Alaska, using Sitka as a base.[35] In 1877 Hibbens estimated the Cassiar district to be "a tract of country at least 300 miles (500 km) square."[36] For the next twenty years rumours, tall tales, and minor successes kept an army of hopefuls roaming the far flung creeks and valleys of the northwest in search of gold in paying quantities.

The *Alaska Searchlight*,[37] a Juneau newspaper, reported in the summer of 1896 that a party of five miners, T. R. Brown, F. A. Limbocker, P. J. Packard, A. E. Johnson, and W. H. Griffin, had spent more than four months prospecting the bars of the Tulsequah and Silver Salmon Rivers, and creeks flowing into Atlin Lake. They set out on April 6, carrying over 900 kg of provisions, but their long search revealed no gold in significant amounts. The prospectors were not alone in the area. They met at least four other parties, including the "Bryant party of 10 coming out from the Lakes".[38] They also encountered a large group from the Cassiar - ". . . eleven white men, and two Indians - a guide and a packer, in the party, and they had 13 pack horses with them. They had a supply of grub sufficient to last them 8 months."[39]

Rosalind Watson Young[40] suggested that the rotten sluice boxes and old caches discovered on Atlin creeks, during the early days of the gold rush, were the work of the Cassiar miners who had not overlooked the lakes which formed the headwaters of the Yukon River. Her suggestion seems valid as many of those wanderers were veterans of rushes to California, Colorado, Idaho, the Fraser River, Cariboo, and Cassiar. Small successes kept the allure alive for those who knew nothing but the addictive restlessness of the true prospector and explorer. Clearly the mountains, valleys, and waters of the Atlin district were known to some white men before the gold discovery, but little of their knowledge was documented.

Meanwhile as these men made cursory searches and a few planned investigations of the region, the Tlingit people continued to live in harmony with the renewable resources of the land, trapping, hunting, gathering, harnessing them to meet their needs. They were barely in touch with the newcomers or their society, and unconcerned for the potential that lay beneath the ground. They remained unaware of the impending dramatic change to their way of life.

CHAPTER 2

# Discovery on Pine Creek

Good prospectors were solitary men who ghosted through the deep wilderness testing creeks and chipping samples from remote mountains. Tight lipped, and wary of misinformation, claim jumping, and the general skulduggery that was a hallmark of their trade, they quietly staked and recorded promising ground, before leaking news to a tense mining world. Often likely locations were filed in memory for later inspection.

The secretive nature of prospectors, enthusiastic, but careless newspaper reporting, and the flood of rumour that accompanies a major mineral find have all contributed to the fact that the true discoverer of Atlin gold has never been positively identified. At least a dozen accounts of the finding are known, some are romantic, and most are believable. Although the records show that Frederick (Fritz) Miller and Kenneth McLaren were the first to stake claims in the Atlin district,[41] most accounts do not credit them with the initial discovery.

The *Victoria Colonist* printed the story of a butcher named Murphy, originally from Tacoma, who was driving a flock of sheep through the country on his way to Dawson City. One of the animals wandering along a creek unearthed a gold nugget. Murphy investigated and stayed to mine until he ran short of food. Taking some of his newly mined gold, he went to Bennett City for supplies. After two more supply trips, Murphy's gold aroused the suspicions of the Bennett storekeeper who reckoned it must be coming from somewhere nearby. Murphy was followed to his workings, and the secret was out.[42] Or so it was reported.

At the time of the Atlin strike, George Miller, older brother of Fritz, lived in Juneau, Alaska. Juneau and its smaller neighbour, Douglas, both founded on hard rock mining, had become a hub of mining activity in southeastern Alaska in the 1880's. Many of the men who prospected the Yukon River, its headwaters and tributaries, made the Alaska mining centre their home base. Jobs were readily available in the mines, and by workir winter, they could grubstake themselves for each new season.

Where mining men gathered, gold was a lively topic of conver... Living in that whirlpool of talk, George, a packer and hotel keeper, might easily have been privy to any number of leads to potential gold properties. And the common element in most accounts of the Atlin discovery is George Miller.

Just how George knew there was gold in the Atlin valley is the point of departure for another set of variations to the story. There is a persistent theme that he was given some nuggets by a dying prospector, who told him clearly where they had been found.[43] Another version identifies the informant as a French Canadian trapper, who had been told by an Indian that he had seen gun powder in the streams of Atlin Lake. The trapper determined the gun powder was actually the magnetic black sand found with placer gold.[44]

More likely, George Miller himself was in the Atlin valley, and a number of accounts agree on this point.[45] The dates suggested for his trip vary with the sources, from 1890 to 1896, as does his purpose. He may have been driving livestock through the country on his way to one of the mining camps on the Yukon River. He may have been specifically prospecting for gold,[46] or he could have made a purely accidental discovery.[47]

If Miller did find gold, he left the area without staking claims, and without mining. That omission could be explained by the extreme difficulty of the trip. He met with accident after accident, and even suffered broken bones. He was pursued by Indians, who were aroused by his presence or his activities. Apparently the experiences so haunted him that he had little incentive to return to the region.[48] It has also been suggested that as a newly wed, he was unwilling to stray from the comforts of his home.

Word was eventually passed to Fritz. Grubstaked by George, and his partner, Lockie McKinnon, he headed north to investigate. It is unclear just when Fritz and Kenneth McLaren became partners, but Fritz wrote in his diary[49], how in February, 1898,[50] he and one other made a trip along frozen lakes to a creek that emptied into Atlin Lake. His companion must have been Kenneth McLaren, since it was on this expedition that they found gold, and were given official credit for the discovery.

The partners followed one of the trails used by the Klondike stampeders to cross the Coast Mountains. They climbed over White Pass with loaded sled and dogs. In the mountains they left the northbound traffic, swung east to Tutshi Lake, and travelled its frozen surface and outlet to Tagish Lake.[51] Bitter

*Frederick "Fritz" Miller came to Juneau, Alaska at the age of seventeen. Seven years later he was heralded as one of the discoverers of the Atlin gold fields and named the founder of Atlin City.*

*Kenneth McLaren, from Moose River, Nova Scotia, travelled west as a young man and became recognized as a first rate prospector. He and Fritz Miller made the historic gold strike on Pine Creek in the winter of 1898.*

cold and fierce blizzards accompanied them as they moved down the lake. One long night was spent holding fast to their tent, to keep gale force winds from sweeping it away. On the move they slogged on snowshoes ahead of the team, breaking trail, as the dogs strained against leather harness to move the heavy sled. They were not, however, alone in the land.

Two surveyors and four others were also abroad, battling their way along the winter trail. Somewhere on this route they met a Tlingit hunter, and gratefully bought moose meat and fish from him. Toward the lower end of Taku Arm,the partners turned east into a long narrow inlet, and followed it to the Atlintoo River, the only outlet of Atlin Lake.

The gold searchers' information may have led them cross country, directly to the upper waters of the creek flowing into Atlin Lake from the east. If, instead, they prospected from its mouth, they would have recovered mere traces of gold in the first several kilometres. In geological history the level of Atlin Lake was 150 metres higher, and the rich gold deposits were laid down before the lake found its present level. Mining below the ancient shoreline never proved profitable.

The men made camp in a narrow valley filled with evergreens, and the winter skeletons of black cottonwoods and aspen. Fritz Miller named the creek flowing sluggishly under ice ledges for the predominant pines. To the south a long low mountain blocked the morning sun, still travelling low in the winter sky. Flashy magpies and Canada Jays scrounged scraps and ravens squawked at the intruders.

They began immediately to sink test pits into the frozen ground. Fires were built, and as they died out, the thawed ground could be shovelled away with the ashes. This slow process was repeated as holes deepened a few centimetres at a time. The first few pits revealed black sand, signalling the presence of gold.

On March 25th they struck pay dirt on bedrock, at about 1.7 metres down. The next two holes yielded good pay at only 1.5 and,0.6 metres.[52] The partners were jubilant to have found such good showings so near to the surface.

All too soon food supplies for both men and dogs had dwindled. With only 10 kg of provisions left it was imperative they leave Pine Creek. Inspired by their success, and with the short winter days behind them, they travelled quickly back to Juneau to report to George and Lockie McKinnon, who had been waiting anxiously for news. Clearly time was precious, and on April 5, just eleven days after making their strike, Miller and McLaren were re-outfitted and headed back to the creek.[53]

At the head of Lynn Canal the partners opted again to travel over White Pass. Several tramlines had been operating on Chilkoot Pass where their goods could have been transported to the summit with relative ease, except that two days earlier a devastating avalanche buried the upper part of the Chilkoot Trail. Eventually the bodies of more than 60 luckless travellers would be dug from the smothering load of wet snow. With rescue work still in progress, the partners joined the stream of men moving steadily out of Skagway.

Climbing slowly through the mountains with a motley procession of men and animals, Miller and McLaren became nearly invisible in the throng. Most men were clothed in woollen pants and knee length canvas parkas, with hoods pulled forward against the driving wind and snow that had plagued the passes for weeks. They trod slowly forward in dreary anonymity. Bowed nearly double under huge packs, urging overloaded horses and panting dogs, tugging at hand sleds, pushing toboggans, the stampeders were a surging human force moving relentlessly along a messy trail. For Miller and McLaren, experienced young men in their prime, the White Pass trail was not a serious obstacle, but both must have pondered how they had arrived so far from their origins, to take a place in that unbelievable parade.

Fritz Miller, tall and confident and only 24 years old, was born in Linden, Hanover, Germany. He had a rugged square face with a hank of dark hair that fell above eyes sparked with good humour. In 1891, after both of his parents had died, he set out to join his brother, George in Juneau. He attended school there, and later made a trip to London. In the Spring of 1894 he went to Circle City, Alaska, where George, ever alert to financial possibilities, had already established a business and mining interests in the new gold town. Circle City, with 500 people, was the largest community on the Yukon River, and George's investments must have looked solid. Then, unexpectedly, the town emptied when news of a major strike in the Klondike broke, and the restless population headed up river.

With little reason to remain in Circle, Fritz returned to Juneau in the summer of 1897,[54] and it is likely George returned about the same time. Soon George and Lockie McKinnon became owners of the Circle City Hotel. Perhaps the name reflected George's success in the Yukon River town, despite the short-lived glory of that camp. Although Fritz visited Dawson, and it is possible George was there too, neither was seduced by the Klondike. They may have decided it would be another flash in the pan as Circle had been.

Kenneth McLaren was 28 years old the year he and Miller first ventured into Atlin Lake. He was from Moose River, Pictou County, Nova Scotia.[55] By coincidence, George Henderson, one of the discoverers of Klondike gold, also came from Pictou, not far from Moose River. McLaren was a quiet gentle person with remarkable blue eyes. He wore a bushy moustache, and was often seen with a cigar clamped between his lips. It is unknown when McLaren headed west, but his travels earned him extensive prospecting experience. He was well respected among mining men. He was a boyhood friend of Lockie McKinnon,[56] and may have met the Miller brothers through him.

Along the icy edges of Bennett Lake and Lindeman, toilers who had moved their awesome loads across the passes, were camped in haphazard cities of tents and rough hewn buildings. The prevailing atmosphere was of hope and boundless impatience. Building boats and rafts was the immediate job. Scraping whipsaws, pounding hammers, and shouting men made a din that rose to the snowy peaks as stampeders prepared for the last dangerous leg of the Klondike journey.

On May 28 dark ice candled by the spring sun, released its grip on the lakes. As it floated away, so too did hundreds of rustic crafts. In the midst of the bobbing armada were Fritz Miller and Kenneth McLaren. At Fort Sifton or Tagish Post, as it was commonly known, the North West Mounted Policemen inspected and recorded each vessel. Goods were checked by Her Majesty's Customs Officers stationed nearby. Once through this check point, Miller and McLaren turned south into Tagish Lake, on course for their camp at Pine Creek, where they arrived on June 4.[57]

The men worked for a month to set up their mining operation, whipsawing lumber for a sluice box, and stock piling gravel for testing. The sluice was located at the edge of the creek where a swift flow of water could be easily diverted through it. Nuggets too large to hang up in the riffles landed in a pit at the end of the trough. With equipment in place, the only job left was shovelling. Miller claimed that on July 3 they commenced operations, and on that first day recovered $120.00 worth of gold.[58]

The partners were well aware of the astounding changes that took place at the site of an important gold strike. They were, after all, walking on the very heels of North America's greatest gold rush, which was in boisterous full swing only a few mountains and rivers away. They may also have believed that news of their strike would soon be known beyond the quiet Atlin valley. Traffic in the area was not uncommon, as they discovered on their initial trip to Pine

Creek, and other accounts reveal that there were travellers at their camp early that summer.

In June 1898, E. G. Tennant, who would later operate a sawmill in Atlin, was returning with partners to the coast from Dawson City when their boat capsized on Atlin Lake, spilling their entire outfit. The hungry survivors eventually stumbled into Miller and McLaren's camp.[59] It seems unlikely the shipwrecked party would have turned east up Pine Creek unless they had good reason to think they would find someone there.

When Miller and McLaren returned to Juneau for supplies, indications are that some others were told of the Atlin gold. One may have been Jim Greenslate, whose name appeared as one of the group whose claims were first recorded on Pine Creek.

Greenslate, an experienced miner, was lured north in 1894, to join the rush to the Unuk River, and he had been at the mines on the Shumagin Islands. By 1897 he was operating a roadhouse in a large tent just above the Scales on the Chilkoot Trail.

One day Greenslate was visited by three of Soapy Smith's men. Smith, a confidence man from Colorado, had a network of hoodlums working every angle of the vast movement of gold seekers through Skagway, Dyea, and as far down the coast as Wrangel. The gangsters wanted Greenslate to install gambling tables in his place, and offered him various considerations and handsome percentages in return. Although few men had the courage to resist Smith, Greenslate flatly refused, and was duly threatened. Fortunately the former soldier was fast with his gun, and by drawing first, gained the upper hand. The men left his establishment, but Greenslate was no fool, and he knew this was not the end of the matter. Wisely, he packed up his possessions and left the Pass. He must have known about the mining on Pine Creek, because he disappeared into the Atlin valley where Smith's gang would have been hard pressed to find him.[60]

During July vague rumours began circulating about a big strike somewhere not far from Tagish Lake.[61] A party of Indians had come upon Miller and McLaren's camp, and reported the information at Bennett.[62] One of their routes to Teslin Lake from the Taku River passed along Pine Creek, and it is not surprising they observed carefully any outsiders in the region, so long their exclusive domain.

With growing apprehension that their secret was no longer safe, the miners felt the pressure to protect their interests. Official government records show that Fritz Miller was elected to act as Mining Recorder at a traditional miners' meeting on July 3. No record was made of the attendance at the meeting. However, it appears that friends and associates, including Jim Greenslate, who had been invited to join Miller and McLaren at Pine Creek before the larger mining community knew of their new strike, were at the gathering.[63]

Later in July Miller returned to the coast, leaving McLaren at the camp. In Juneau, he organized another party of miners, prospectors and businessmen, who left immediately for Atlin Lake. Soon a number of claims were staked on Pine Creek.[64]

The historic Discovery Claims were staked on 27 July 1898, and recorded by Fritz Miller at Tagish Post before Inspector D'Arcy Strickland of the North West Mounted Police on July 30. Nine claims were signed by "F.Miller." The Discovery Claim was recorded jointly for Miller and McLaren. Eight others were registered for James Greenslate, Amelia Greenslate, Al Short, Almyra Short, John Eisenburg, Linneous Shank and Fred McClosky, all of Juneau, and Walter R. Weeks of Almeda, California. Another group of claims, also staked on July 27th, were registered on August 4th, and signed by their respective owners.[65]

The rights of discovery granted Miller and McLaren 370 metres of creek bed. All additional claims were 77 metres long. This was the law of the North West Territories, which included the Yukon Territory. Inspector Strickland had acted under the assumption that the gold strike was made north of the 60th parallel. Strickland's error was to prove disastrous in more than one respect, but, oblivious to his mistake, he left for Pine Creek immediately with eleven men. His entire party staked below the Discovery Claim.[66] While at the gold camp, in his capacity as acting Gold Commissioner, Strickland ratified the earlier election of Fritz Miller as acting Mining Recorder.[67]

Although the Atlin Lake region was not formally mapped, many men along the coast and inland had knowledge of its terrain. Captain W. J. Rant, the Government Agent and Gold Commissioner stationed at Bennett, was certainly one of them. Soon after the Discovery Claims were filed, he began having serious doubts about the true location of the new strike. As a result he consulted H. H. Eggerton, who was a civil engineer from New York, employed by the builders of the new White Pass Railway. Eggerton took observations and concluded the strike had been made 50 km within the British Columbia boundary.[68] His estimate was to prove remarkably accurate. Rant reported his misgivings and Eggerton's findings to Superintendent Sam Steele, and together they left Bennett on July 31 aboard the *S. S. Kilbourn* heading for the new workings. Near the mouth of the Tutshi River on Tagish Lake they met the returning Strickland[69] who was still convinced that Pine Creek was in the Yukon.[70]

Once at Pine Creek, Rant and Steele soon confirmed Eggerton's calculations. The gold strike was indeed in British Columbia, and so it was officially declared. A grave error had been corrected, but not before 40 miners had registered their claims according to the wrong set of regulations. They had been granted 77 metres of working ground, rather than the 30 metre claims allowed by British Columbia law. This blunder would have serious consequences for the miners and for the gold camp as a whole.

When Fritz Miller recorded the first claims, the mining community, always strung taut, reacted like fighters primed for battle. The stampede was immediate. Rant reported 30 miners working on Pine Creek, just days after the first recordings, and others were arriving every hour.[71] He realised he could not cope with the situation from as far away as Bennett City, where he was already short staffed.[72] Taking matters into his own hands, he appointed his son, Norman W. F. Rant, as acting Mining Recorder and Constable on August 5.[73] It was at Rant's request that Steele later sent in Sergeant Davis and Constable Woodhouse. In his Annual Report Steele confidently reported that this action had prevented "claim jumping and its attendant evils".[74]

Meanwhile at Bennett Lake, Strickland began spreading the news and the excitement was contagious. A party of employees and customers, quickly organized by the manager of the Bennett Lake and Klondyke Navigation Company, embarked on a trip to an unknown place for unknown purpose. Their mysterious travels would, of course, be to Pine Creek, and the destination was dramatically announced while enroute aboard the *S. S. Nora*. In spite of some mishaps and bungling along the way, the excited party made it to Miller and McLaren's camp. The miners were surprised to see the Bennett group, but soon reconciled to the fact that news of their discovery was spreading. Miller was even persuaded to sell $100 worth of gold to the manager, who planned to send it to Victoria for display to promote the goldfields, and the fortunes of the Navigation Company.[75]

Each day more miners arrived in the Pine Creek valley. Some were returning from the Klondike, either beaten by the game there, or floating on success, and willing to try for one more strike. Miners still struggling across mountain passes were hearing disturbing rumours that the Klondike was plugged and all the good ground already staked. Going to Atlin instead would cut 1000 km of the journey, and for many it became an attractive alternative.

The resulting Pine Creek fever struck hard in places like Dyea, Skagway and Juneau, as gold hungry men quickly changed plans, and prepared to head for Atlin Lake. As most of the male population exited the towns, the running of stores and businesses was left to the relatively few women who were left behind. At least one steamer landing in Dyea had its mooring lines handled by women.[76] The Bennett Lake and Klondyke Navigation Company lost about half of its Dawson passengers to the Atlin stampede.[77] There was a severe shortage of cooks, and anyone willing to work at makeshift hotels on the trail to Atlin could demand an unheard of $15.00 a day.[78]

An unexpected and startling consequence of the rush was the sudden evaporation of the huge labour force working on the construction of the White Pass Railway. Very few of these workers actually came north for the purpose of building a railroad. They were gold hunters who needed the work to finance their way to, or retreat from, the Klondike. So when three prospectors on the way to Skagway stopped at White Pass City, and showed labourers a dazzling five and a half kilograms of gold dust and nuggets from the gravels of Pine Creek, railroad jobs with their poor wages were easily sacrificed.[79] Of the 2000 men working on the enormous rail building project, 1500 quit within days. Some left with the company's picks and shovels in lieu of waiting for pay cheques. Others hung around to harass the paymasters for their wages.[80] Later in the season, one disgruntled Atlin storekeeper complained that there were "too many useless short handled railroad shovels" in the camp.[81]

The Bennett Lake and Klondyke Navigation Company was ready to serve the Atlin region, particularly with the decline of passengers on their Dawson City run. The voyage from Bennett to Taku City was unbelievably beautiful, and many thought the trip well worth taking for that reason alone, and sometimes unusual encounters made the passage truly unforgettable. On the initial voyage of the *S. S. Nora* passengers spotted three swimming moose in Tagish Lake. The captain circled and herded the animals until two could be shot for meat for the ship's galley. Just as the third moose made a break for shore, a lanky passenger baled over the railing and landed on its back. He clung to the animal until it reached the beach, where he jumped to the ground fully expecting to be attacked. Instead the moose stood still looking at him. They remained eye to eye for some time, then the man reached out and patted the animal's side. The *Nora* had by then pulled ashore nearby, and Captain Woodman called for the passenger to come aboard. As the man walked to the boat the moose followed meekly beside him. That moose was spared from the ship's menu.[82]

The *S. S. Nora* continued doing unusually good business taking miners and their equipment to the mouth of the Atlintoo River. She sailed from Bennett every three days, and carried hundreds on the trip at $30.00 one way.[83] While the *Nora* provided safe and comfortable, if sometimes exciting passage, this was not the case on all vessels plying the lakes. Other small steamers, some of dubious design and seaworthiness, were pressed into service for the run to the river, where a tiny collection of tents and buildings provided some services, and soon became known as Taku City.

Hamlin Garland, an American writer, made the journey as a tourist on one such vessel. Fully loaded it carried 8 - 10 passengers and their outfits. Hastily built with no thought to comfort, it was described by Garland as:

> ... a mere scow with an upright engine in the center and a sternwheel tacked on the outside. There were no staterooms of course, and almost no bunks. The interior resembled a lumberman's shanty.[84]

Somehow this floating shack delivered Garland to Taku City, and later when he had hiked the four kilometre portage to Atlin Lake, he met an enterprising fellow waiting to ferry passengers across the lake in his naptha launch. Unfortunately engine problems caused a lengthy delay. The next morning Garland's group hired a primitive skiff. Although they paid an exorbitant fare, it was still necessary for all hands to pull their own weight during the 8 km crossing.[85]

Water travel in the region was always hazardous. The immense ice cap crowning the mountains created unusual turbulence. Prevailing winds rushing down the broken ice falls gathered force as they met the warmer air over the open lakes, and made a nightmare of waves and flying spindrift. These dangerous storms materialized in minutes with no warning to unfortunate boaters needing a safe retreat.

The Atlintoo River harboured its own obstacles. In its short 3km length, it dropped more than 15 metres in elevation. Its fast moving waters masked an uneven and treacherous bed. Boulders hidden just beneath the surface upset boats or tore at their planking. One party of eight travellers was surprised to come down the river in just 21 minutes, after they had spent nearly ten hours polling and towing their boat up.[86]

With enthusiasm surpassing technical skill, many stampeders built row boats and scows in order to reach the alluring new gold fields, and the consequences were often disastrous. It was estimated that poorly built and over loaded boats allowed the unpredictable lakes, and dangerous Atlintoo River to claim at least 20 lives during the 1898 rush.[87]

News of the gold strike first reached Victoria on the 9th of August and throughout that month and the next the press heralded it with unbounded enthusiasm. Pine Creek was billed as a "poor man's paradise".[88] British Columbia had its very own gold rush, and expectations were high that it would be as rich as the Klondike.

Gold rushes were big business and the province still chaffed from missing much of the financial spin-offs of the Klondike strike. The stampeders had chosen Seattle and San Francisco as major supply centres, and up to this point they had largely ignored the "all Canadian" routes to Dawson City. Most miners opted for the faster easier water routes which skirted the British Columbia border. They sailed through its waters, but rarely stopped at her ports.

There were warnings that the strike on Pine Creek might not equal the Klondike in sheer quantities of gold, but still there was great optimism that with proper promotion this gold strike would serve commercial interests of Victoria and Vancouver, and fill provincial coffers as the Klondike rush had failed to do.

The many advantages of mining in British Columbia were extolled. A great irony in the light of future events, was that provincial mining laws were deemed to be "intelligent", and a "model for the world."[89] The climate was promoted as being more hospitable, and the area better served by transportation systems. There was good timber available, ample water supplies, and best of all, gold bearing bedrock close to the surface of the ground, and easily reached with a minimum of digging.[90] Surely all of this would answer any miner's dreams. These advantages were of course, a thinly veiled attempt to detract from the lure of the Klondike.[91]

To further enhance the appeal of British Columbia's own gold rush, the *Victoria Colonist* published numerous articles about rich strikes in the region, the ease with which the gold fields could be reached, and the rapid development there. The *Colonist* printed accounts of claims that yielded $6.00 worth of gold per pan, and more. It reported that five men working on the Miller and McLaren's Discovery Claim were being paid $12.00 a day each, but the claims' actual yields were $50.00 - $60.00 a day per working miner.[92] To make quite sure that it was clearly understood that the country was easy to travel, various accounts of women travelling in the region were published.[93] So that the readers of the *Colonist* might not miss their opportunity, there was a report that five townsites had been laid out in the area.[94]

In fact for some months nothing resembling permanent towns existed on Atlin Lake or Pine Creek, although loose collections of tents and shacks at both sites had been given names, Atlin City and Pine City. In October the insistent *Colonist* announced that a sawmill was expected to commence operations, and there were two large hotels, five restaurants, six general stores, two barber shops, and about 100 tents in Atlin.[95] Basically this was correct, except that most of these establishments were operating in haphazard disorder in drafty tents. There was no sawmill yet, and regular means of transportation had not been established. There were, however, numbers of interested merchants and speculators settling into the area, and engaging in preparations for the business opportunities which would accompany the mining in 1899.

Although summer was near its end when the Pine Creek strike became widely known, 3000 people came to the area during the last weeks of the 1898 mining season.[96] Most came for brief visits during the initial frenzy of staking to assess the business potential in preparation for the expected boom. They came, as Garland wrote, " . . . with a cigarette and a sandwich", and were certainly not planning to spend the winter.[97] Nevertheless, the men who stayed to mine the waning season recovered 3,750 ounces of gold, valued at $75,000.[98] They also prospected widely and before winter set in, other creeks tributary to Pine, and several of those feeding Surprise Lake had been investigated and staked. The last discovery claims recorded that year were on McKee Creek. It was clear that, in spite of runaway promotion and distortion of fact surrounding the strike, there *was* gold to be mined in Atlin. Enough had already been found to give proof, and there was confidence that a viable mining camp had been born.

A tantalising question remains- just how much gold was taken from Miller and McLaren's Discovery Claims on Pine Creek that first exciting season. There are many stories. It is likely the young discoverers lost accurate count themselves, but years later an old miner who knew McLaren said that before the claims had ever been officially registered, the partners took out $42,000 to $43,000 worth of gold. That was approximately 2,350 troy ounces or 66 kilograms.[99]

CHAPTER 3

# The Trails to Atlin Lake

The winter of 1898-1899 saw a trickling migration of gold-hungry men, inching along each of the difficult routes to the Klondike. They were scattered through the dark forests of British Columbia, and along its rivers. They were camped on frozen lake shores, and huddled aboard crowded ships off the bleak coasts of Canada and Alaska. The routes were rougher and longer than ever imagined. Optimism withered in the face of doubt and despair. Many feared the curtain would fall on the Klondike scene, leaving them without a part in the big show. When gold was struck on Pine Creek, that winter army of stampeders could visualize a new Eldorado almost within their reach. Thousands changed course to Atlin Lake.

At the first whisper of gold in the Klondike, adventurers had flocked from across the continent, and from across oceans to gather at ports in the western states and British Columbia. Seaport cities competed fiercely for the business of outfitting the stampeders. San Francisco, Portland, Tacoma, Seattle, Victoria, Vancouver vied for a part of the action, and from them came a bewildering fleet of vessels. Ships were thrown together, resurrected from bone yards or quickly refitted to provide transportation to obscure Alaskan ports like Wrangell, Skagway, Dyea, or St. Michael. Since the summer of 1897 this flotilla had hauled hundreds of passengers and tonnes of goods. While most of these vessels were sound, and mastered by men of integrity, too many were death traps, overloaded with passengers, pack animals and mountains of supplies. Many amenities were advertized by unscrupulous operators, but few existed. Tickets guaranteed berths where there were none. Food was often inedible, and the filth and stench from animals jammed below decks was unbearable. Rampant overcrowding and the notorious weather complicated many precarious voyages.

The sidewheeler, *North Pacific*, which was actually a river boat, left Tacoma in February, 1898. Over-booked, and with scant accommodations, passengers were left to sleep on the floor of the dining room, in passageways and toilet rooms. She was so overloaded that every ocean swell posed a danger. The cargo was so poorly distributed that passengers were forced to take shifts standing on one side of the vessel to prevent a catastrophe. The Captain spent most of the journey lying drunk in his cabin, while the dreadful responsibility for all lives aboard rested with a young pilot, who had never before sailed the northern coast. That they reached Wrangel in just five days was sheer good fortune.[100]

By the time gold was struck in the Atlin region, George A. Brackett had built a passable road across the White Pass, and hundreds of travellers gladly paid a toll to use it.[101] The White Pass Railway was well under construction, and that was an even better proposition. On the Chilkoot Trail, aerial tramways were transporting a steady stream of freight to the summit.[102] Stampeders off loaded with their goods at the head of the Lynn Canal, had a choice of following either of these routes. These were the most practical and direct passages to the interior, and over them passed the greatest numbers of people bound for the Klondike and Atlin.

Local Chilkat Indians, taking advantage of opportunities presented by the initial rush to the Klondike, sold their services as packers on the Chilkoot Trail. Entrepreneurs among the newcomers organized pack trains and vast inventories of supplies moved steadily toward the interior. Huge fortunes were made in just a few months by men involved with moving goods.

Stampeders challenged the steep passes with an astonishing variety of pack animals, horses, mules and dogs. Some brought steers and goats, and many were more trouble than they were worth. Thousands of horses plodded the uneven trails at the hands of ruthless owners. They were goaded on, even when covered with saddle sores and crushed under loads weighing up to 120 kg, and kept on the move for long painful hours each day. Suffering horses, injured or completely worn out, were left to die in agony beside the trail. Edward Curtis witnessed the horror: "many animals died from exhaustion; but by far the greatest number were destroyed by falling amongst the boulders, the heavy packs nearly always causing broken limbs".[103] Scores of animals were underfed since hay ranged in price from $45-$150 a tonne.[104] Most abhorrent were the men who drove their horses or oxen to the point of fatal exhaustion, then slaughtered them and hung out signs on their tents advertising, "Meals $2.50."[105]

AHSP-1243

*At the Chilkoot Pass summit, stampeders had to thread their way through a chaotic area of supplies and hand sleds. An aerial tramline is visable at the upper right, its empty carriers travelling back to the lower terminal.*

AHSP-566

*Lindeman City, on Lake Lindeman, was a sprawling tent town where north bound stampeders stopped to build boats and rafts to carry them along the water route to Dawson City and Atlin.*

AHSP-444

*Log Cabin, located a few miles north of the White Pass Summit, developed as a trailside town of tents, rough log buildings and a customs station in the charge of P.F. Scharschmidt. 1898*

AHSP-446

*At Log Cabin miners bound for the Atlin gold fields turned east on the Fantail Trail, which started at this point.*

The cost of using the aerial tramway or hiring professional packers was beyond the financial reach of most stampeders. Yet each man's outfit was composed of heavy winter clothing, tents, tools, mining and boat building equipment, and a year's supply of food; a dead load of 450 kg of essential goods.[106] His alternative was to move the supplies one load at a time. If a man had no animals, he used his own muscles and brawn, and faced the reality of travelling the trail countless times over. Hard working, sweaty bodies were vulnerable to fierce biting winds that chilled flesh and spread frostbite. Snow obscured trails, hid the landscape and buried cached supplies. Meals needed to be thawed on open fires or portable stoves, but many were wolfed down ice cold. Runaway sleds, ailing animals, tent fires and the crowded trails were needling frustrations added to the back breaking work. Little wonder then that thousands of Klondike stampeders were glad to modify their plans and shorten the distance, by heading to Atlin when they learned gold was discovered there.

The North West Mounted Police assumed responsibility for the welfare of the stampeders. Their decree that all parties entering Canada come equipped with supplies to last a year was sternly enforced. They aimed to avoid a winter of possible starvation, a reality that had haunted the Klondike region when miners flocked there, unprepared, for the winter of 1896-97. It was reckoned, though, that a high proportion of miners would have sacrificed food for speed. The police, located in cramped quarters deep in the snows at the Chilkoot summit, monitored stampeders and collected duty. As well, in manning that mountain post, they established once and for all, one of the points on the nebulous boundary between British Columbia and Alaska.

The builders of the White Pass railroad, recovering somewhat from the paralysing loss of their work force to the Atlin gold strike, recognized the potential of a new destination. They began opening a trail toward Atlin from Log Cabin. From the congested mountain camp, a stop over for packers, the route lead in a south easterly direction to Fantail Lake (known then as Otter Lake). From there it followed the drainage river to the west shore of Taku Arm. After that it was a matter of travelling frozen Taku Arm to Taku City, then going on 15 km to Atlin. This route, the Fantail Trail, was a direct 110 km. It was only feasible during the winter months, but was particularly suitable for dogs and handsledding.[107]

Pulling a loaded sled in stages, a man could cover the distance in 11 or 12 days, if conditions were good.[108] Often, though, they were far from favourable. The beginning of the trail was extremely steep, and sleds frequently escaped and overturned, scattering a carefully packed load in all directions. Travellers, exhausted by the trip, sometimes just turned their sleds loose, but had to face the chore of gathering up smashed goods, buried deep in the snow at the bottom of the hill.

There were several primitive stopping places, on the Fantail Trail such as Teepee, Hale, Golden Gate and others where food and a place to sleep could be bought. At the lower end of Fantail Lake, an old Cockney served coffee and ham sandwiches in a small tent equipped with only a small stove and a table. At Golden Gate, where travellers entered the narrow waterway[109] leading to the mouth of the Atlintoo River, another tent establishment sold beans. Taku City, rising to meet the demand, offered a Japanese restaurant where a complete meal could be bought for 75¢.[110]

Although the food was poorly cooked, often greasy, and the comforts were minimal, trail stops were welcome sights to exhausted travellers. Just the same, the comforts promised by the prospect of finding vast riches were never far from mind. Roger Hayes Hicks, an English doctor, travelled the Fantail Trail with a partner in March, 1899. He spoke for many when he lamented:

> (We) both feel rather weary and stiff and say that we want a good dinner, a cigar, with an easy chair and a book, a good glass of grog and a comfortable bed afterwards - none of which are to be obtained.[111]

By early spring, the Fantail Trail was no longer safe. It was reported that as many as 20 lost their lives on the uncertain ice of Fantail Lake.[112] Heavily loaded travellers were advised to use a safer, if slightly longer, route known as the Too-chi Trail.[113] This alternate route crossed Tutshi Lake to the north and served stampeders for a few additional weeks in the spring. It was 142 km from Log Cabin to Atlin via this trail. After break up, men who crossed White Pass went on to Bennett City, and made the rest of the trip to Atlin by boat.[114]

The first train from Skagway reached Bennett Lake on July 6, 1899,[115] and from that day on summer travel to Atlin could be accomplished in comfort, and supplies transported with ease. At the opening of navigation that year, there was lively competition for the Atlin trade. The *Olive May*,[116] the *William Ogilvie*,[117] and the *Gleaner*[118] all made regular runs from Bennett to Taku City.[119] The portage between Taku City and the wharf at Scotia Bay, on Atlin Lake, had been groomed to a good wagon road. The steamship companies provided horse teams to carry freight and passengers on the three and a half kilometre wagon road.[120] Later Captain John Irving operated horse drawn

*Tagish Post is where the North West Mounted Police and customs agents checked stampeders into Canada. They inspected supplies and made safety checks on crafts. (1898)*

*Taku City, at the mouth of the Atlintoo River, was the western end of the portage connecting Graham Inlet and Atlin Lake. (1899)*

Yukon Archives #248, Vogee Collection

*At Taku City, lodging houses and restaurants were quickly established to accomodate the thousands of people going to Atlin.*

AHS P-939

*Huge loads of food, clothing and equipment, the essentials for life in the gold fields, made for slow progress on the trails.*

rail cars on the portage. In 1900, a railroad was built and horse conveyances disappeared from the scene.

The unbroken chain of rail and steamships linking the Klondike and Atlin with southern ports effectively closed down all other routes. Trails once teeming with travellers grew still. The few settlements along them either disappeared or returned to their previous state. In essence most trails had been the advertising schemes of cities which stood to profit from their use, rather than practical routes for travellers.

The Taku Trail was the dream of Juneau businessmen who wanted to cash in on the Klondike rush by providing an attractive route to Teslin Lake. From Juneau, the trail followed the Taku and Nakina Rivers to the south end of Teslin Lake. They claimed it had been used for years.[121] This was substantially correct as both the Taku Tlingits, and Juneau-based prospectors had travelled extensively in the area. The distance from tide water to Teslin Lake was about 192 km.[122] It was not, however, the "leisurely" route,[123] "suitable for horses"[124] touted in the promotional literature.

In 1894, Captain Charles Constantine of the North West Mounted Police was delayed in Juneau for three days when going north.[125] Persuaded by Juneau people, who extolled the merits of the Taku River route, he sent a written report to Ottawa strongly advocating it as an easy route to the upper Yukon River. As a result, William Ogilvie, a Canadian government surveyor, was instructed to make a reconnaissance of the area the following spring. During their exploration, Ogilvie and his party were besieged by raging storms and hampered by treacherous glaciers. He returned on March 1, 1895 to report, "a most rigid search revealed no practical wagon road."[126] Juneau people continued to promote it vigorously, but the Taku Trail was never used extensively by Atlin-bound miners.

Merchants in Victoria, Vancouver and Spokane[127] promoted a trail they touted as the "all-Canadian route". They distributed persuasive literature that played on economy, British patriotism, and the fear of American customs agents in Dyea and Skagway. This tortuously long route, the Ashcroft Trail, was chosen by many Americans and British subjects after the April avalanche on the Chilkoot Trail was dramatically reported.

The first 250 km of the trail was the old wagon road from Ashcroft to Barkerville and Soda Creek, lifeline of the Cariboo goldfields in the 1860's. About 80 km east of Quesnel, the Ashcroft Trail reached the nearly forgotten survey trail made by the men of the Western Union Company. Remnants of their line crossed the Fraser and Nechako Rivers, passed near Fort Fraser, Burns, and Decker Lakes, and followed the Bulkley Valley to Hazelton. At this point the trail turned north, into the little known northern interior of the province, to Telegraph Creek on the Stikine River.

When Western Union crews withdrew from their remote survey camps, they left little sign of their project except for some wire and insulators, and a thin trail blazed through the wilderness. Thirty years later a ragged parade of men and animals pushed uncertainly, but steadily north, along those vanishing survey lines deep in the unforgiving forests of northern British Columbia.

In the spring of 1898 more than 1500 men and 3,000 horses left Ashcroft, B.C.[128] bound for the Klondike. Misled by colourful advertising, and ignorant of the land, the stampeders were unprepared for the hazards ahead of them. Dozens of dangerous rivers, and immense stretches of tangled timber were daily barriers. Constant drenching rain, rapacious mosquitoes and black flies tormented men and animals. Neither had the eager gold hunters anticipated the dangers of driving burdened horses through areas beset by forest fires, endless swamps, or miles of land already stripped of natural feed by the hungry animals who had passed before. Horses died in the quick sands, floundered in the muskeg and starved. Men suffered from the unrelenting physical toil on a monotonous diets of desiccated potatoes, beans and mouldy bacon.

Young Albert Tervo, a Finn, left North Dakota in 1897 with a group called the Alaska Mercantile and Mining Company, a diverse group of 15 men representing a variety of occupations. They travelled together for economic reasons. When they left Ashcroft, they had 39 horses, a large wagon, and an ambitious scheme. The company was capitalized at $100,000, and they aimed to mine and do business in Dawson City. Their plans had not dealt with the trials of just getting to the Klondike. One by one the horses died; they drowned, ate poisonous weeds, or became hopelessly stranded in swamps and had to be shot. One hundred and sixty-five kilometres south of Telegraph Creek the company disbanded, and split up the remaining goods. Each man was on his own. Some retraced their steps, but Tervo and a few others continued to Glenora,[129] 20 km down river from Telegraph Creek.

Northern access to the all-Canadian route was the Stikine Trail. Stampeders travelled from Fort Wrangell, Alaska, by boat or on the ice of the broad Stikine River to Glenora, 245 km away. This was the staging ground for miners going to Teslin Lake.

Glenora's first boom had faded with the end of the Cassiar gold rush in the 1870's, but in the winter of 1898-99 it was again a bustling centre. Several stores, two hotels and four or five restaurants were in business. Cabins and

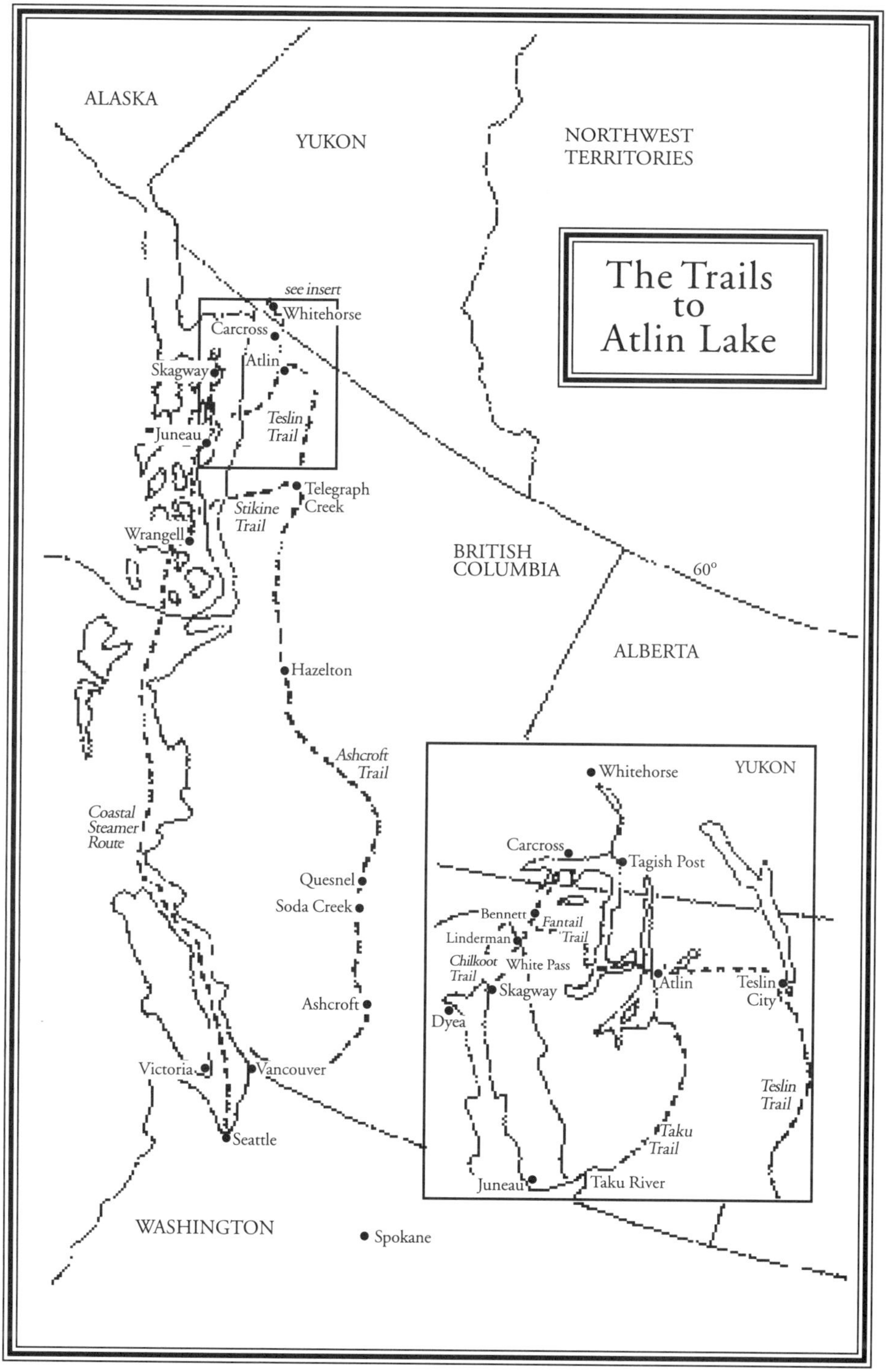
The Trails to Atlin Lake
ALASKA
YUKON
NORTHWEST TERRITORIES
see insert
Whitehorse
Carcross
Atlin
Skagway
Teslin Trail
Juneau
Telegraph Creek
Stikine Trail
Wrangell
BRITISH COLUMBIA
60°
ALBERTA
Hazelton
Ashcroft Trail
Coastal Steamer Route
Quesnel
Soda Creek
Ashcroft
Victoria
Vancouver
Seattle
WASHINGTON
Spokane
Whitehorse
YUKON
Carcross
Tagish Post
Bennett
Fantail Trail
Linderman
Chilkoot Trail
White Pass
Skagway
Atlin
Teslin City
Dyea
Teslin Trail
Taku Trail
Juneau
Taku River

tents stretched out along the broad bench, a former farm, at the edge of the Stikine. It was "a most orderly community" thanks to the detachment of North West Mounted Police stationed 48 km down river at the border. They kept watch on the "transient population of 4,000-5,000, some of them women, moving through the country."[130]

Telegraph Creek, on the other hand, was little more than a camp. It was described by one Ashcroft Trail veteran as a "a miserable one horse place with one or two stores, the inevitable whiskey house, and a few tents".[131] It marked the start of the Teslin Trail.

It was at Glenora that Hulet M. Wells and his father, having travelled together from Agassiz, British Columbia, parted company. Wells Senior was exhausted by long months on the Ashcroft Trail; his enthusiasm for gold had disappeared. He boarded the last boat in the fall of 1898, and headed down the Stikine. Hulet had a teacher's certificate but the search for gold overshadowed his interest in a country classroom, where he might have found work, so he decided to press on. He was 19 years old and had $15 in his pocket.[132]

Another father and son team stuck together. Seventeen year old Guy Lawrence and his father, an English portrait painter, arrived at Glenora by steamer in July, 1898. Guy was intrigued with the busy town, and spent his days making friends with people from around the world, who had gathered in that unlikely place. He met Dominic Burns, who with his brother would later build a meat packing empire, but in Glenora he was busy buying up outfits of men who had abandoned the trail. The young stampeder listened to exciting stories of past gold rushes, and caught and sold fish to earn extra money. All the while, his father was wasting valuable time investigating rumours of a strike on the Tuya River.[133]

The demand for an all-Canadian route to the Klondike had prompted the government to contract with the company of McKenzie and Mann to build a wagon road from Telegraph Creek to Teslin Lake. Only 20 km had been built, some of the survey work had been done, and a good deal of equipment delivered to the site, when the scheme was abandoned. The Senate refused to sanction the project, since they believed that the construction of White Pass Railway made the expenditure unwarranted.[134]

The shock to hundreds of stampeders who had expected to find a good wagon road to Teslin Lake reverberated through the area. Rallying together they held indignation meetings, but nothing changed. Men who had negotiated the treacherous Ashcroft Trail were staggered to learn the promised 260 km road did not exist. Plans to move supplies with relative ease in large wagons on the last overland lap to Teslin Lake were crushed. The existing 1.25 metre wide trail would accommodate nothing larger than a hand cart. Bitterly they realized there was no easy leg of the all-Canadian route.

Hard decisions were made at Glenora and Telegraph Creek. Broken men turned down the Stikine and headed home in defeat. Others sailed north to the Passes. The rest, still determined to try for Klondike gold, waited in Glenora for the right conditions to push through on the Teslin Trail.

Rousing news of gold in the Atlin Lake country reached Glenora in August 1898. The bright prospect of lopping hundreds of miles off their journeys by going to Atlin instead of the Klondike, kindled hope and renewed enthusiasm. The Kendall Brothers were among the scores who moved out that fall. They had started their northern argosy in Malheur, Oregon in May. At Ashcroft, they bought 10 horses and 50 head of cattle, and hired a hand. By September they had negotiated all the dismal, dangerous and soul straining hazards of the Ashcroft Trail, and landed pretty much intact in Telegraph Creek. News of the Atlin strike was too compelling to ignore. They decided to take a chance on the weather and attempt to drive their herd through to Atlin before the first snowfall.[135]

The Lawrences bought an axle and two wheels, and by shortening the axle, made a cart that would just skim the edges of the narrow trail. In August, 1898, they set out for Teslin Lake with one horse pulling 200 kg loads. Still they had to relay ten to twelve times to move their supplies ahead. They reached the Little Tahltan River in mid-October. They had travelled only 60 km, but reckoned they had covered "five hundred and sixty miles, at the very least, to say nothing of the miles we had to hunt the damn horse every morning." They decided they must spend the winter at the river, and used their failing horse to pull in the logs needed for a cabin. With that done, they put him out of his misery.[136]

Stirring reports of the new diggings continued to arrive in Telegraph Creek and Glenora throughout the fall. The remaining residents idled away the weeks and months, waiting for snow. "Dawson or Atlin Lake was the question in every mind."[137] The Klondike lost favour as they worried about finding open ground there, as their food supplies and finances diminished. If Atlin was the logical destination, then the difficulties of the trail were by no means over. The Teslin section was comparatively short, but the winter of 1898-99 was severe and food was scarce.

Young Tervo of the defunct Alaska Mercantile and Mining Company, and four of his former partners decided on Atlin. Hulet Wells also succumbed to the lure of the nearer gold creeks. He left Glenora alone on February 13, pulling a hand sled packed with everything he owned. He faced 500 km of trail and particularly vicious weather. He had youth on his side, though, and stamina teamed with determination pulled him through several close calls. He learned immediately the mistake of omitting a stove from his outfit. Making camp at -47° C was hard enough, but cooking over an open fire was impossible.[138] He plunged through the ice on the Inklin River. Soaked to the armpits, he desperately needed a fire in order to survive. The only wood around was green brush near the river. Knowing his hands would soon freeze, but that keeping on the move might save him, he ran back along his trail until he located his fire pit from the previous night. Live coals still smouldered there.[139]

Guy Lawrence was enjoying the winter, the company and the entertainment provided by the men living in nine other cabins on the Little Tahltan River. Then in early February, 1899, conditions seemed fair and they all decided to move on. The summer trail had been an ordeal, but it was "mere child's play (compared with what) they now faced." The extreme cold of the winter of 1899 with temperatures ranging at -40° and -45° C. meant that every step was a challenge. Moving supplies was the main job, but keeping warm was a conscious task. Guy and his father spent one night on the ice of a frozen creek, with only a rubber ground sheet beneath them. At -42° C., and desperate for heat, they overloaded the stove, and burned the tent, their only shelter.

Several days later they hauled their load to Long Lake, where they headed for an empty cabin, eagerly anticipating a warm, comfortable night. The door opened to a chilling sight almost hidden by the darkness within the cabin. Two bodies lay on the bunks, the frozen remains of two men dead from scurvy.[140]

Martin Leinweber and his partner overtook two young men suffering so badly from scurvy they were barely able to drag their sled. Still they would not give up the dogged attempt to reach their goal. Perhaps they knew that their only chance was to survive the trail itself. Little could be done to help the desperate pair, but Leinweber gave them some potatoes and dried vegetables in exchange for pork. Much chastened, Leinweber and his companion moved on, taking care to cook whatever greens they could find, such as fireweed and lambs quarter that grew near the edge of the melting snow in the spring. They regularly brewed up quantities of foul-tasting spruce tea.[141] Later they learned that the ailing pair had been rescued. The vegetables they had traded may well have saved their lives.[142]

The fortunes of four French Canadians were not so happy. Like Guy Lawrence and his father, they had built a cabin to wait out the winter. Their food cache contained only beans and pork; apparently they had no knowledge of the dangers of travelling without vegetables or some source of ascorbic acid. One by one they died, until there was only one survivor, who expended valuable energy burying his last companion. He then waged a losing battle against his own inevitable death.[143]

At the southern end of Teslin Lake, within British Columbia, a small settlement developed where about 100 people waited out the winter of 1898-99. Barren little Teslin City was a scattering of tents and cabins along the lake shore. There were also two restaurants, a hotel, a sawmill, a city jail, and a detachment of the British Columbia Provincial Police, in the charge of Constable A. D. Drummond. Despite a population of impatiently stalled and bored would-be miners, Drummond had to deal with little crime. He was, however, sincerely concerned for the plight of those at Teslin Lake, and others wintering on the trail, who suffered from scurvy. A few were being well cared for by their partners, but others were alone and destitute.

If there were any heroes on the gold trails, then surely Constable Drummond was one of them. As the only representative of government, he heard the pathetic pleas for help. However, he had strict orders from his superior, Government Agent James Porter, in Telegraph Creek, to incur no expenses without consultation. With the northward movement of traffic on the trail, and no regular mail service, permission to expend government money to rescue the sick, could take as long as eight weeks to reach Teslin City.

Ultimately Drummond, unable to accept the indifference of the bureaucracy, decided to act on his own. As winter deepened and scurvy ravaged the waiting miners, he organized a group of friends and a dog team to rescue the dying men. Drummond was personally willing to finance the rescue operation and any subsequent nursing fees. Whether or not the Provincial Government ever reimbursed him is unknown.[144]

While in Teslin City, Hulet Wells gained two partners. On this last leg of the trip to Atlin, and so near their destination, both of his companions were suffering from the destructive symptoms of scurvy. Regular doses of spruce tea kept them on their feet and moving, and they stoically refused any help with their share of the load.[145]

From Teslin the route to Atlin was across frozen lakes and a range of mountains, a trail frequently travelled by the Tlingit people, but few others. Because of the extreme cold and unknown terrain, this part of the trail presented its own problems.

In Teslin City on Christmas Eve in 1898, the Teslin Tlingits were holding a potlatch for the Peel River Indians, who arrived driving teams of huskies with jingling bells. Two Tlingit Indians from Atlin were also in town for the celebration. Walter Hamilton, and his partner Will Harrigan, were worried about their dwindling supply of horse fodder. They turned down a hospitable invitation to stay for the festivities. Anxious to be on their way, they asked the Atlin Indians for directions. They were told there was no trail, but if snow did not obscure their tracks, the partners might follow the path made on their journey to the potlatch. First they were directed to a cabin about 25 km down Teslin Lake, and from there to turn west, and follow a small creek into the hills. Hamilton and Harrigan set out on snow shoes; their two horses pulling double-ended sleds. The rough descriptions given by the Indians were their only guide for a trip of over 225 km.

The first night they camped in freezing rain that turned to heavy snow as temperatures began to drop. On Christmas morning they discovered their sleds frozen fast in the ice. Before they could move, they had to chop the runners loose. Block and tackle were needed to assist the sleighs and horses to the top of the first ridge as the temperature continued to drop. The trail was becoming hard to follow. Finally, when the horses could only flounder helplessly in the snow, they were shot. Then they relayed their supplies above timberline, and hauled along enough firewood to last until they were again among the trees. Camped one night in a clump of willows, the only available shelter, the mercury hit the lowest point on their thermometer, -54° C.

Still searching for the Indian trail along the mountain, the men separated for a day to cover more ground. Harrigan's hands had started to freeze by the time he returned to their tent, but he managed to fumble a match between stiff, senseless fingers, and to strike it. When it flared, he knew he was safe. The terrible exertion, thin high air and intense cold caused Hamilton's left lung to haemorrhage. When he coughed, blood splattered in the snow. Descending from the mountain pass eased his frozen lung, but the last assault from the Teslin Trail was still ahead. Painful snow blindness inflamed Hamilton's eyes as the pair crossed the glaring ice of Surprise Lake.

When the partners met Robert Little at the mouth of Boulder Creek, they knew their journey was over. They were at their first real gold mining camp. On the evening of 5 January 1899, in a large tent illuminated by candles, the travellers enjoyed their first restaurant meal since leaving Ashcroft in the early spring of 1898. They were at Pine City, the place where Fritz Miller and Kenneth McLaren had discovered gold one year earlier.[146]

The welcome was less auspicious for Wells and his two scurvy-stricken companions. They made their painful way into Atlin in the spring of 1899, and Wells took them immediately to the hospital for a cure that was simply a diet of raw potatoes.[147]

Disappointment surrounded the arrival of the Kendall brothers, in October 1898. They had branched off from the Teslin Trail, found the Taku Trail, and driven their herd to the Pike River. Making good time, they followed it to the eastern shore of Atlin Lake. At the mouth of the Pike, 43 km south of Atlin, they slaughtered the herd, most of which had survived the long trek. They took their beef to town only to find unexpected competition in the meat business. Their enterprise was reduced to financial failure by the Tlingit Indians, who true to their nature as traders, were already doing a brisk business in caribou and moose meat. The defeated brothers left for home early in 1899.[148]

The Edmonton Trail to the Klondike is regarded as the most difficult, and certainly it claimed its casualties. Those who perished or turned back far outnumber the few who reached their destination. Of the thousands who stampeded to Atlin, perhaps only two came via that route.

On 11 March 1898 a party of eight left Edmonton with 48 head of cattle. Arduous and long, the trail took its toll. Two brothers turned back at Hay River when one became ill. One member of the party drowned when a canoe overturned. Finally they lost the trail entirely, and had to backtrack a discouraging 170 km to find Fort Nelson.

The remaining five members of the party made their way into the Cassiar country just as winter set in. All the cattle were in surprisingly good condition, but the winter played havoc, and the entire herd died of black leg. Three of the party returned to the States. One man, whose name was Harris, made his way to Atlin independently. He sent word back to his former partners, who were still waiting out the winter in the Cassiar district, to join him. S. R. Geddis and H. Brown accepted his invitation. They arrived in Atlin on August 4, 1899, just as the initial rush of stampeders had thinned. One of them had been on the trail since leaving Edmonton, 17 months earlier. Harris successfully established the Vendome Hotel on First Street in Atlin.[149]

The story of the trails that led to Atlin would not be complete without the account of the young man who chose his own path. He was Frank Neill, a student from Philadelphia, who was truly hell bent on his quest for gold. When the schooner he was aboard ran aground near Juneau, most of his companions wasted no time returning home. Not Frank. Once in Juneau, he listened closely to news of the strike on Pine Creek. His money was nearly gone, so he studied maps and decided his best course was the shortest possible route to Atlin, disregarding all obstacles. Equipped with a hand sled, sleeping bag, some food and gear, he started along the Taku Trail. Then, with unbelievable determination, he took a direct route up the Taku Glacier and climbed to the vast icefields at the summit. He crossed the Coast Mountains over a death trap of bottomless crevasses, and made his way down the dangerous face of the Llewellyn Glacier. From these dazzling heights he could see a hazy town, his mecca, far down Atlin Lake. It was March, 1899 when he pulled his sled off the lake and onto the streets of Atlin. By some miracle this wilful stampeder made the only known glacier crossing to the Atlin gold rush. He was too late to stake a good claim, but he did venture into the saw mill and lumber business instead. Later he pursued similar ventures in Dawson City and New York.[150]

The men who came to Atlin via the Chilkoot Trail, White Pass, the Ashcroft, Stikine, Teslin, Fantail and Taku Trails experienced adventures rich with bravery, resourcefulness, compassion, and ordinary luck. The promise of gold was motivation enough for thousands of men who struck out to find fortunes, but if the gold was free for the finding, then the price of getting it was often too great. Too many lives were lost; too many wills were broken.

Still in every important human episode there lurks an element of pure fun, and many hit the trails for sheer adventure. For them the gold rush was a roaring good time. Unrelenting weather, disgusting food, brutal work, and needling homesickness would eventually take second place in their memories. The real tales were of loyal partners, prickly scrapes with danger, and the joy of reaching a lofty summit with a dead load of supplies. Outstanding was the electrifying excitement of starting out, and the foggy euphoria of finally reaching the end. The wild trails themselves would never be forgotten.

CHAPTER 4

# 1899: The Boom Year

*Our young city has had such wide advertisement that the word Atlin is in the English speaking world's mouth today.*

- The *Atlin Claim*, April 29, 1899

A mere twelve months after the day Fritz Miller and Kenneth McLaren found the gold in Pine Creek, a part of the wilderness was transformed, perhaps beyond the miners' wildest expectations. A great lake of transcendent beauty, bound on the west by sentinel mountains, and fed from the east by dozens of spirited creeks and rivers, became the focus for a clamorous growing mining camp. The hope and energy fueled by runaway gold fever had found sudden, unanticipated direction. No man was immune to the atmosphere of expectancy that vibrated around him. Experienced miners and prospectors, aspiring businessmen full of schemes, and zealous chechakos gathered on the shores of Atlin Lake. They were engulfed in a sea of pervading confidence. There was animated speculation about the camp's rosy future and anticipation of the thousands who would arrive to secure those prospects. Ever in each mind was the gambler's dream of striking it rich.

The ebullient mood in Atlin was given substance by dozens of optimistic articles emanating from coastal newspapers. They carried headlines such as *Rich as the Klondike,*[151] *More Famous Than Klondike,*[152] *Atlin City Will be the Wonder of 1899.*[153] Accounts of gold nuggets which would "choke a mule,"[154] and a single nugget which tipped the scales at over 3 kg[155] made seductive reading. While many of the reports fell somewhere short of the truth, and had been fed by investors to enhance personal ventures, they stoked the fires of the rampant fever, and within Atlin, grandiose plans were laid while stampeders converged on the Eldorado of British Columbia.

Although most of the initial influx of claim stakers and curious speculators had disappeared before the winter of 1898-99, a few hundred men stayed to gain early advantage.[156] If the booming Atlin City of newspaper frenzy did not yet exist, there is no doubt that a town was taking shape. The ubiquitous tent, the starter kit for all instant frontier settlements, sprouted like mushrooms along the lake shore. Some of the first businesses were impressive false fronts, but behind the facades there usually lurked a tent. They billowed and snapped in the wind until log or lumber structures could replace them. In the interest of haste and to conquer the cold, buildings were sometimes simply erected around tents, which became interior walls. Often the luxury of windows was foregone, since there was a great shortage of window glass.[157]

E. G. Tennant's new sawmill was one of two producing lumber at a rapid pace, and this combined with the need to heat drafty businesses and cabins created a pressing demand for wood. Trees fell in waves that originated at the townsite, and spread toward the hills. The islands off shore stood in nude relief against the mountains as logs were skidded across the ice to stoves and whining saws. Throughout the autumn, forests were also being leveled by "immense brush fires . . . that were raging all the time."[158] Notices were put up by authorities ordering prospectors to put out fires when breaking camp.

In December, the 20 metre wide clearing for Discovery Avenue, often called Dominion Road[159] stretched toward Pine Creek. A reporter for the *Skagway News* was obviously overwhelmed by the stunning view on all sides when he abandoned straight forward reporting and raved,

> When the moon rises from behind the surrounding mountains, turning night into day, the scene is one of glorious splendour; making one feel that this is indeed one of the most glorious spots he could find, even in our famed north, celebrated as it is for the grandeur of its scenery.[160]

As the pattern of trade, commerce and settlement in the Atlin district began to emerge, Discovery, known simultaneously as Pine City,[161] was being established at the site of the gold strike. About 10 km away at the west end of Surprise Lake, approximately 100 people lived under canvas and in a collection of small cabins clustered around a large tent bearing a sketch of Queen Victoria over the wooden doorway. This was the Victoria Hotel in Surprise City, a welcome sight for the stragglers coming off the Teslin Trail.[162] At first many thought the strategic location of Taku City on the major route to Pine Creek from the coast would make it the business centre for the area, but that distinction was Atlin's from the start.

The impressive development of the area during the first winter at the hands of the pioneer residents was supplemented in the spring when both Dominion and Provincial Governments began building roads. A new winter trail suitable for horses, was cut from the west shore of Tagish Lake to Log Cabin. It skirted the unpredictable Fantail (Otter) Lake, and thus provided a safer and shorter alternative to the Fantail Trail. Pack trains from Atlin could now reach the railroad station at Log Cabin in just a day and a half, while a dog team could usually do it in less.[163] Roads were linking the townsites and gold creeks, and by summer it was possible to take a wagon into almost all mining areas.[164]

Late one evening in June, shortly after the ice went out on the lake, a pretty 25 metre stern-wheeler, the S. S. *Scotia*, arrived at the wharf in Atlin.[165] Several other large boats also began plying the waters to bring hundreds of passengers and mountains of freight across the 6.5 km stretch from Scotia Bay on the west side of the lake. The new arrivals scattered to the creeks or joined the work of building the towns and creating a society which made few concessions to the isolation or the wilderness which surrounded it.

Undoubtedly the most profitable concerns operating in the spring of 1899 were the two sawmills. Yet production failed to meet demands, even with the saws howling 24 hours a day.[166] Customers arrived at dawn and waited in line or dozed in chairs. Hotel owner, Richard Anzer, was told that to buy lumber, he should have money in his hand, because at the mill the rule was: "No check, by heck, but we dash for cash."[167]

To satisfy the demand for local building materials, a number of other mills were soon in operation. Rough lumber sold for as much as $150 per 1000 feet, and a bank official was heard to say that it was better to own a sawmill in Atlin than a bank.[168]

If lumber prices were sky high, the cost of labour was not far behind. Dr. Roger Hayes Hicks complained when he had to pay a carpenter $1.00 an hour, for two and a half days to erect a small cabin at Discovery. The building was little more than a roof over four walls. By the time the doctor added a tarpaulin to cover the floor and furnished a table, bunk, stools, and cheese cloth curtains, his total cost had soared to $64.00, a price he thought exorbitant for a building he considered "prison-like".[169]

When the Provincial Government auctioned land on the newly surveyed townsite of Atlin in April, there was stiff competition for the best commercial lots. Most measured only 8x35 metres, and many were narrower, but they sold for as much as $500.00 each.[170] Not surprisingly there was more than a little resentment about this and other instances of price gouging by Victoria.

The land sale heated up the dispute between the merchants on Discovery and Pearl Avenues as to which area constituted the business district of Atlin. The lake boats docked and loaded from the wharf at the foot of Pearl Avenue, but Discovery Avenue led to the gold mines. However, First Street swept broadly between the two rival districts, and along it too, rose hotels, restaurants, stores and offices.

The newspaper claimed enthusiastically that Atlin businesses were fully stocked and that it was possible to buy "everything from a needle to a fully equipped hydraulic plant".[171] This statement was borne out by the columns of advertisements in each edition of the newspaper. Clearly there was no shortage of entrepreneurs willing to accept the challenge of making a fortune in a new gold field.

Hotels were necessary and numerous. In the raw settlement teeming with strangers they provided more than basic food and shelter. Deals were made and partnerships formed in their barrooms, where nuggets fresh from the creeks were admired and passed from hand to hand. The first attempts at refined entertainment took place in hotel dining rooms, where furniture was pushed aside to make room for amateur musicals, serious recitations, debates or dances.

Larger more elabourate hotels followed the first canvas accommodations. The well appointed Grand Hotel commanded the corner of First and Discovery, and advertised an unsurpassed view. There were parlours, a reception room, and boudoirs en suite or single. The dining room was replete with the choicest viands and rarest delicacies. In the bar there were "appetizers, soothers and substantials."[172] Yet another amenity offered at the Grand was a telephone connection to the Gold House in Discovery.[173]

By contrast, there were many rough hewn but functional establishments like the Discovery Hotel at Discovery and Second Street. It was a plain log building, two stories high. The lower floor was a saloon, and upstairs a dozen bunks, arranged in rows along each wall, provided sleeping accommodations. When Popcorn Kate, who built the Discovery, was unable to obtain a liquor licence, she simply gave it to a friend and left Atlin.[174]

On the corner of First and Trainor, the Kootenay installed a billiard table, which may have drawn customers from the Vancouver Hotel down the street. A few doors away, guests at the Northern Hotel viewed activities along

First Street through huge windows that spread across the entire front of the building. The Northern Annex stayed open all night.

J. H. Russell built his hotel on Pearl Avenue near the town wharf. The amount of beer consumed in the area was not lost on Russell, who made plans to build a brewery. However, the North West Mounted Police stopped him and his brewing plant at the Customs Office at Tagish Post because he had no licence for the operation. Russell took his protests to Ottawa, and was eventually given his licence,[175] but there would be no brewery in Atlin in 1899.

There was no shortage of saloons, but many were primitive establishments tossed together to serve the first stampeders, and they offered just the basic amenities for thirsty miners. In July the police enforced building standards, and closed six saloons operating in tents.[176]

Atlin had two fully equipped drug stores, one of which opened a branch in Discovery. The Thomas Dunn Company of Vancouver, built a large hardware store and tin shop on First Street. Besides branch stores of other Vancouver businesses, such as McLennan & McFeely, there were bakeries, bath houses, two watchmakers, and hand laundries. Arthur Vogee was a professional photographer who recorded in crisp black and white the first chapters of life in Atlin and on Pine Creek. He had a studio on Second and Pearl Avenues, and another in Discovery. A sign painter set up shop as did a funeral director and embalmer. A carpenter built windows, doors, store fronts and staircases. There was a mattress factory on First Street which also manufactured wire woven springs.

Supplying food for the area was a formidable enterprise. Bringing in perishables from great distances while avoiding spoilage or freezing required perseverance, and the fastest attainable speed. Several companies thought the risk worthwhile, though. Parson's Produce, a Winnipeg company with branches in Vancouver and Dawson City, opened a store in Atlin. They brought fresh produce from the prairies,[177] but it came bearing high price tags. Kelly Douglas, a Vancouver supplier, established in Atlin, and remained in the north for many decades.

Men from the P. Burns Meat Company arrived in Skagway in February, 1899 with a huge shipment of goods, including 3,500 kg of freight and 20 steers. They were heading to Atlin to establish a meat market. Fourteen one-horse sleighs lined out with the cattle and started toward the White Pass trail. Fourteen days later they had progressed only about 16 km. The trail improved as far as Bennett, but from there the caravan leaders, Dominic Burns and Billy

AHS P-420

*Stampeders camped on the beach at the foot of Discovery Avenue as Atlin City took shape. The Grand Hotel, under construction, is visable above the head of the miner on the left. (1899)*

AHS Bourne Collection P-1020

*First Street developed as a broad avenue of businesses in tents and frame buildings that reached from Discovery Avenue to the saw mills at the south edge of Atlin City. (1899)*

AHS Henning Collection P-1207

*First Street, looking north, with the Bank of Halifax under construction at the corner of Pearl Avenue.*

AHS P-1244

*The three banks of Atlin were on Pearl Avenue. The Canadian Bank of Commerce is at upper right. The Bank of Halifax with the steep roofed tower at left, and the Bank of British North America is at lower center with a ladder on the roof. (1899)*

Henry, faced a month of travel over an unbroken route with temperatures that slid to -51 deg. C.

In March when they finally arrived in Atlin, they set up a butcher shop on First Street, slaughtered the herd, and quickly sold all of the beef. Even the poorest cuts brought nearly $2.50 a kilogram. Success in hand, they immediately organized another shipment. This time they were on the trail during the long pleasant days of April and May, but deteriorating ice, and disintegrating trails introduced new levels of difficulty.[178]

Patrick Burns and his brother continued bringing cattle from their ranch in Alberta[179] and from the Queen Charlotte Islands.[180] The Klondike and Atlin gold rushes provided the opportunity for them to establish a firm foothold in the north and from this lucrative beginning they laid the foundation for the corporation that became widely recognized in Western Canada.

The professional men who came to open offices in Atlin were mostly young, and looking for adventure and a springboard from which to launch a career. Edward Garden was a promising young architect who designed the Northern and Irving Hotels.[181] There were several lawyers including C. W. Sawers, many doctors[182] and a dentist. Mining engineers, civil engineers, and surveyors flocked to Atlin. There were at least 10 offices tendering the services of mining brokers, real estate agents, and notaries public.

Cautiously the banks allowed themselves to bend to the pervading optimism. If their arrival was more sedate than that of the stampeders, their first premises were in keeping with those of most other new enterprises. By April, 1899, Atlin had three banks: the Canadian Bank of Commerce, the Merchants Bank of Halifax, and the Bank of British North America.[183] One did business over a rustic plank counter with mesh fence wire enclosing the teller. All were little more than a safe surrounded by canvas or gaping boards, and security must have been more a hope than a reality.

Soon the banks settled into permanent quarters along Pearl Avenue. The Bank of British North America was a large building with lap siding and shingled gables. Huge main floor windows reflected passers by on the boardwalks. A second floor was circled with smaller windows. Pull-back curtains at some of them indicated that bank employees were probably quartered there. Bare of Victorian gingerbread, the bank was a dignified looking edifice anchored by two huge brick chimneys. In a small annex was a fully equipped assay room where there were five furnaces, gold rollers, retorts, crushers, and precision gold scales.[184]

Diagonally across the intersection of First and Pearl was the soaring tower of the Merchants Bank of Halifax, an "elegant structure (that was) large and commodious, and in every way an ornament to the city"[185]. The Canadian Bank of Commerce of Toronto, first opened on 1 March, 1899 in temporary quarters in the Opera House on Discovery Avenue.[186] Later it located across from the Post Office and Telegraph building on Second and Pearl. The new banks were an imposing presence in town and lent to its growing aura of permanence. As well they offered secure repositories for the accumulating wealth from Atlin creeks.

The business community bristled with activity and optimism, and its leaders, eager for the trend to continue, made several attempts to organize its forces. The Brokers' Association formed in May, was short lived, but late in the year the Atlin Mining and Investment Bureau was organized. Its directors numbered the most prominent businessmen in the community: J. H. Brownlee and R. C. Lowry, surveyors, W. A. Spencer, manager of the Merchants Bank of Halifax, J. Anderson of the Bank of British North America, Alex Godfrey of the Thomas Dunn Company, J. A. Fraser of Parson's Produce, C. Christopher, a mining engineer, E. N. M. Woods, Stipendiary Magistrate, and A. C. Hirschfeld, a newspaper man. Their purpose was to promote the camp, attract capital, and to act as mining brokers and a real estate agency.[187]

While Atlin had become the business centre, Discovery was the gold camp. It grew along two dirt streets, Discovery Road on the bench just above the Discovery claims, and Victoria Street which stretched behind it near the hill. Hulet Wells described his impressions:

> . . . substantial log saloons and hotels were going up, and soon gambling establishments lined both sides of the main street. Most of them furnished music as an accompaniment to the blackjack, poker and faro games, and all summer long we heard *She Was Bred in Old Kentucky* and *Just One Girl*.[188]

Gambling became an irresistible pastime for many with the source of gold just a stone's throw away, and the supply apparently endless. Nuggets and gold dust circulated freely, and games were easy to find in the saloons or on the streets. Blackjack was the favourite and crowds gathered whether the stakes were 25¢ or $1,000 a bet.[189]

The usual venue for business, church and social gatherings in Discovery was a huge tent owned by the British Miners' Protective Association. After a floor was installed weekly dances were held there throughout the summer.[190] Cultural aspirations in Discovery were fewer than in Atlin, and this may have reflected the predominantly male population.

With an eye to the lopsided gender distribution, and the opportunists' sense of a sure thing, prostitutes filtered into the communities to establish themselves on the fringes of the business sector. There were a number of "sporting houses" both in Atlin and Discovery. Some of these establishments considered themselves classier than others, and frequently offered patrons entertainment and refreshments. The girls were described by Hulet Wells as "something less than queens"[191] and scarcely met the day's standards for beauty.

The police tacitly acknowledged their existence, and simply endeavored to ensure that they conducted their business in an unofficially designated area. In Atlin the red light district was in Sawdust Alley, near the sawmills at the southern end of town.[192] In Discovery the "sporting houses" were "zoned" to the eastern outskirts. The *Atlin Claim* maintained that the Government wanted to keep them out of sight, but close at hand, so that they might be easily raided, and thus taxed.[193]

On April 15th, while the lakes were still ice bound, and snow lay thick on the ground, the first white baby was born in Atlin. Atlintoo Marie Lowry was delivered by Dr. A. S. Monro.[194] The occasion so moved one miner that he wrote a long and florid verse, published in the first issue of the *Atlin Claim*: "Hale and Welcome, Little Sunbeam."[195] The first white baby boy arrived in a tent on June 7th, 1899. He was Francis Henning, who spent much of his life in Atlin.

The boom of a gold strike resounds far and wide, and extends an irresistible invitation to audacious eccentrics looking for adventure and the limelight. Perhaps the most colourful character to arrive in Atlin that season was Count Emile Janne de Lamare of Paris, France. He was not without credentials. He edited a new publication the *Klondike Revue*, and was a correspondent to a number of other journals, including the prestigious Le Figaro.[196] De Lamare appeared also to represent extensive French financial interests and at one time reputedly carried a list of 140 claims in which he was willing to invest.[197] Known as the "Big Frenchman", he circulated through Atlin with a certain congeniality and a swagger. He lived in the very grandest manner that Atlin could afford, and he entertained with fine champagne to toast the future of the camp.

Mrs. Mary E. Hitchcock, was a wealthy matron of statuesque proportions, who enjoyed being the focus of attention. She gained this to a large degree from her unusual travels. Her first published book, Two Women in the

Klondike, related her adventures with a companion as the first female tourists to Dawson City.[198] In search of more writing material, she came to Atlin in May 1899, and settled herself in a cottage on Discovery Avenue. She immediately involved herself in "every movement with the district's welfare at heart".[199] Although an American, she did not hesitate to make forceful representations on Atlin's behalf, when she visited Victoria in the fall. The *Atlin Claim* reported that she would place the town's need for a school and a hospital before the appropriate Ministers in such a way as to "leave no room for argument in the matter".[200]

Aside from civic issues, Mrs Hitchcock was particularly excited about the mining potential of the Atlin gold fields, specifically hard rock mining. She claimed to hold a number of promising properties. For a time she employed eight men to prospect on her behalf.[201] She was prepared both to invest and to encourage others to do so. Socially too, she was the doyenne. Lord Ernest Hamilton, who was also in Atlin representing British mining interests, described her as the "only lady in Atlin".[202]

When the most prestigious personage in British Columbia spent two weeks in Atlin it was an obvious reflection of the government's optimism for the prospects of the new gold fields. Lieutenant-Governor McInnes was a bulky gentleman, with fashionable whiskers and starched collars, who was not entirely comfortable in a rough northern mining town. Yet, as the guest of T. R. Billett, Manager of the Bank of Commerce, he visited the creeks, and tried his hand at goldpanning. For the duration of his visit he was also the centre of attraction at teas and banquets.[203] He most certainly provided Mrs. Hitchcock with her finest hour, as she was frequently at his side during his visit. Before leaving, the Lieutenant-Governor said graciously that "the townsite of Atlin is an ideal one, and would make the foundation for a magnificent city."[204]

At first there was little finesse at community gatherings, and the settings were rustic. In March, a concert and dance were held to raise funds in support of a permanent hospital. Regardless of the fact that it was held on Good Friday, most of the community turned out to hear songs and recitations delivered with great ardour, and dedicated zeal for the cause. Later patrons danced away the evening in moccasins and working clothes. There were in the crowd only two "boiled shirts".[205] This and other events helped bind a social fellowship in Atlin that soon encompassed all the refinements of the late Victorian era. Some gatherings were remarkably sophisticated as the result of determined efforts to confirm the community's permanence. To celebrate the Queen's Birthday on May

*The gold rush presented an opportunity for businesses to prosper in a new territory. J. St. Clair Blackett, from Victoria, located his store on First Street in Atlin City. (1899)*

*The Atlin Post Office opened in the summer of 1899 on the corner of Pearl Avenue and Third Street.*

Yukon Archives #19, Vogee Collection

*Lumber and canvas were quickly turned into buildings for the multitude of businesses that appeared in the burgeoning young mining town. (1899)*

Yukon Archives #202, Vogee Collection

*The road to Discovery passed through Spruce Junction where the Half Way House and the Spruce Junction Hotel accommodated travellers or mining crews.*

24th the Australian and the Jolly Britishers' Social Clubs held a joint dinner in the most formal style.[206]

Her Majesty's Birthday was wholeheartedly celebrated throughout the district. Seven hundred dollars was offered in prizes for boxing, cockfighting, and pistol competitions. Hundreds of spectators lined First Street to watch foot races, pole vaulting, high jumping, and caber tossing.[207] Later a political oration was delivered by Rev. John Pringle, and a concert followed. Under the patronage of Mrs. Hitchcock and other leading ladies, a ball was held at the Russell Hotel where the floor was considered perfect for dancing.[208]

Meanwhile, at the Victoria Hotel in Surprise City, the celebrations were less restrained, and the crowds gathered to see, not only cockfighting, but the greasy pole climb, the ugly face competition, and a flapjack race.[209] The music at Surprise was a welcome addition to the celebrations - George Gabriel played a guitar made from a Sunlight soap box.[210]

There was no semblance of local government, so civic problems were left to the residents. As soon as the ground began to thaw in the spring, it became clear to alarmed citizens that there was a problem with sanitation. The waste created by the large influx of people and animals into the area endangered the purity of the lake which caught the spring run-off. Dr. A. S. Monro was the Medical Officer of Health, but he had no simple solutions.[211] Notices were posted warning citizens to remove all garbage and manure to the designated "nuisance grounds". Signs were also placed on Atlin Lake asking people to refrain from watering their horses within fifty feet of domestic water holes. The public was advised to boil all drinking water. When a horse died from glanders on the street in Discovery, the carcass was cremated to prevent further spread of the contagious disease.[212]

Concern was growing about the possibility of fire sweeping through the town where tents, and buildings made from lumber still oozing pitch stood in close proximity along streets edged with wooden boardwalks. Meetings were called and a committee was formed. A bucket brigade resulted,[213] but funds were needed for its support. The community's dawning social sophistication became evident when, in August, a ball was held to benefit the new Fire Brigade. Printed dance programs first appeared on that occasion to help the outnumbered ladies keep a record of promised dances.[214]

Clamorous construction in the towns, and the din from the creeks blended with the general bustle to belie the fact there was a pronounced shortage of jobs. Government interference and some serious bureaucratic blunders committed by local officials were acute impediments to many mining operations where work too often lagged or stopped, leaving men idle until settlements were reached. Unemployment was to become a grave issue. The imbalance was also felt by professional men.

Dr. Hicks had to discontinue his practice after two weeks, because he did not hold a British Columbia licence. Although he was willing to overlook the small matter of his licence, and to see patients at no charge, his services were not in demand since there were so many doctors. By his estimate, in the spring of 1899, there was one for every ten people in Atlin and Discovery. His recourse was to try logging. Throughout the rest of the spring and summer it was a constant struggle to find piecemeal work to stave off poverty. Even when he finally found a partner who operated a fairly profitable claim on Spruce Creek, their work came to a sudden stop when the mine became the subject of litigation and was awarded to another claimant.[215] The plight of some of the jobless became so serious that a Citizen's Relief Fund was set up to help those who could find neither work, nor a means to leave the district.[216]

Amid the frenzy and the euphoria of the boom were the underlying realities of everyday life. For most pioneers, living conditions were rough and cramped. Tents were cold and damp. Typical cabins were thrown together using unpeeled logs, and dirt or tin for roofs. Old socks, long underwear, and other castoffs were used for chinking when moss, the usual material, was buried under the snow. As the logs dried and shrank, bark and bugs dropped from the ceiling, and shreds of old clothing hung from the walls. Sometimes a muslin lining was tacked up to contain the debris and keep it out of the bean pot. Smoke from wood stoves and lanterns darkened interiors, and windows were expensive luxuries. Hauling water, cutting firewood, cooking and baking endless loaves of bread, were chores that stole time from a miner trying to work his claim. Little wonder men teamed up so chores could be divided. Lucky was the man whose wife agreed to come to the gold fields with him.

The road between Atlin and Discovery was busy with foot traffic as many people hiked for food and other essentials. All commodities were expensive and hard won. If food was not properly cached, marauding dogs or wild animals could play costly havoc with a man's supplies. The new Atlin Post Office opened on Second and Pearl in May, but there were no stamps. Long lines were made even longer by the residents from the creeks who had to do business there,[217] until another post office opened in Discovery on July 18th.[218]

Sundays were usually set aside for personal chores. Miners bent over scrub boards, mended clothing, cut hair, and wrote long wistful letters to anxious families and sweethearts. Money, however, could buy a reprieve from the

BCARS HP 76008

*Discovery City, first known as Pine Town or Pine City, was strung along the bank of Pine Creek where Miller and McLaren first found gold.*

daily drudgery and a number of businesses sprang up to fill the void. Mrs. Hayes' laundry in Atlin did a brisk business, and soon there were several others in both Atlin and Discovery. An hour at a bathhouse was far more satisfying than a wash in a gold pan. At the Pioneer Bakery a loaf of fresh bread cost 12 cents.[219] Bootmakers, barbers and restaurantiers all profited from men fed up with households of their own hasty design.

As the first year ended and winter set in again, sluice boxes froze, and the creeks grew still. There was little work to do, so those who remained in the district dug in to hibernate or set about to enjoy themselves. Dances were held weekly,[220] and these were often preceded by a concert, a lecture, or a whist party. Ice skating and snowshoe picnics were popular, and rambunctious teams held rousing soccer games on the lake in front of the town. The year closed with two well attended matches - "The Rooters" vied with "The Tooters",[221] and "The Atlinites" battled against "The Stikeen Trailers"[222], while spectators roared and shivered on the sidelines.

Just how many people lived in Atlin, Discovery, Surprise City and along the creeks that first hectic year is uncertain. The tremendous ebb and flow of people throughout the district made it impossible to make a close count. Estimates, however were made, and one pegged the area population as high as 22,000[223], while numerous others ranged down to 5,000. W. W. Bilsland[224] estimates the figure to be 5,000 based on the fact that about 4,000 Miners' Licences were issued; however, considering the amount of commercial activity in the area, and the probable numbers of men working for wages, this figure seems low. Figures published by the Atlin District Board of Trade for 1899 were 1,500 in Atlin, and 4,500 in Discovery and on the various creeks.[225] These estimates would appear to be realistic figures.

1899 was undoubtedly Atlin's "hour of glory", but even during that great year there were signs that all was not well. The "poor man's diggings," so widely heralded, did not exist throughout the district. However, the main source of the problems besetting the district was widely perceived to lie with the Government.

CHAPTER 5

# Bureaucrats and Bunglers: The Role of Government

Pine Creek belonged to British Columbia, but no one knew that until gold was found, claims registered, and a full-blown gold rush launched. Although it was quickly established that the creek was south of the 60th parallel, the inadvertent error made by Inspector D'Arcy Strickland affected all the initial claims, and stirred up a storm of trouble. Inept politicians and government officials failed to understand the problems which were compounding during the early months following the discovery as they malingered and created greater havoc instead of solving problems expeditiously. Postponements and uncertainty had miner pitted against miner, and citizens rallying against the government. Even the discoverers themselves were not exempt from the bureaucratic nightmare that gripped the young community.

Norman W. F. Rant began his work as the Acting Mining Recorder in Pine City on August 5, 1898. His office was in a tent, and his supplies consisted of one record book, and 200 miners' licences. Miners converged to record claims, and within two days Rant had run out of official forms, and he was issuing temporary receipts in lieu of claim certificates and free miners' licences.[226] During the rest of the summer and into the fall months, as gold seekers prospected throughout the area, and found other gold-bearing creeks, some 1700 claims were staked.[227] A chronic shortage of the proper forms, coupled with a marked lack of efficiency on Rant's part in handling the applications, severely crippled the process of recording claims.

Although the overworked Captain William Rant undoubtedly felt he had no alternative but to appoint his 19 year old son as Mining Recorder for the new district, it proved to be an ill-fated decision. Not only was Norman Rant young and inexperienced, but it was soon apparent he was motivated by self-interest and favoritism or, at best, capriciousness. His attempts to acquire claims, and his involvement in other ventures were clearly in conflict with his official duties.

As early as October 1, 1898, rumours of Atlin's problems were beginning to spread to the trails. One miner wrote in his diary that he had heard stories of ". . . government officials acting very badly . . . staking claims for their friends".[228]

Complaints of serious irregularities were sent to the Minister of Mines. A number of miners complained to the Minister that their claims had simply not been recorded for lack of forms.[229] I. E. Negus wrote to the Gold Commissioner in Victoria that, along with two others, he had located the Discovery Claim, on Willow Creek on August 9, 1898. He attempted then, and on several other occasions to record his claim, but was unable to do so. Finally ". . . Rant, God (sic) Commissioner, allowed (his) Discovery on August 22nd but in a few hours reversed his decision . . . saying that he would not call it a creek."[230] Willow Creek appeared to be little more than drainage from swampy ground above Pine City, although actually it was an old bed of Pine Creek, but its short length was rich with gold. While Negus was denied his ground, $10,000 to $20,000 was mined on a 154 metre stretch of Willow Creek.[231]

By far the most serious charges against Norman Rant were leveled by the British Miners' Protective Association, the only miners' organization in the camp at the time. They claimed that, not only did Rant refuse to record some properties, but that he was actively engaged in defrauding the rightful locators of certain valuable ground.[232] Later in June of 1899, when confusion still reigned, the editor of the *Atlin Claim* called for a special investigation of young Rant and his "questionable practices."[233]

Rant's attempts to acquire mining properties was a marked indiscretion, but he also became involved with a group trying to purchase land from the Provincial Government which was intended for the townsite of Atlin. Their scheme was to survey and develop the land for sale to the public. Since Rant was the only British subject in the company, his name appeared on the land application.[234] Clearly Rant's inexperience and poor judgment were getting him into trouble.

Rampant claim jumping was the worst problem arising from the initial mistake about the geographical location of Pine Creek. Captain William Rant had assured miners that the first claims registered erroneously under the regulations of the Northwest Territories, would be recognized by the British Columbia Government. Unfortunately, he did not make it clear, at the time, that the original claim owners would not be permitted to retain their entire claims.[235]

Rant should have instructed them to select, without delay, 30 metres of their original 77 metre claim, the amount of ground allowed by the British Columbia Mining Act. Opportunists immediately spotted the oversight, and staked over the original larger claims in order to gain the ground which would have to be relinquished by the first holders. Some properties were staked many times over, and a contemporary photograph shows fourteen claim posts securing the corner of just one claim. On another claim eight posts stood wrapped together with saplings. Someone with a sense of the ridiculous had written on the largest stake, "Together we stand. Divided we fall."[236] Over-staking constituted a challenge of the rights to a claim, and all work had to cease pending a judicial decision on ownership.

Norman Rant was severely criticized for accepting multiple applications for one piece of ground. This charge, however, was also leveled against subsequent Mining Recorders, and they, like Rant, consistently maintained that they were not empowered to refuse any miner who wanted to register a claim. Consequently the confusion was allowed to proliferate.

On 31 October 1898, Captain William Rant issued an order from Bennett that no mining was to be done during the closed season, in order to protect rightful owners from subsequent stakers who would mine illegally. All mining was to stop, not only on the disputed claims, but on all claims in the Atlin district. It appears, however, that the order was largely ignored and work was carried on by both owners and claim jumpers over the winter.[237]

On 15 November 1898, embattled Atlin miners held a meeting, and then issued a complaint to the Minister of Mines that Norman Rant had left the area, and all records had been taken to Bennett. They asked to be allowed to appoint a Mining Recorder from their own ranks. The government, however, was in the process of taking its first decisive step, and shortly afterwards it was announced that a new Gold Commissioner would be sent for the Bennett and Atlin Lake Mining Divisions.

Joseph D. Graham was selected for the job. He was a respected and qualified administrator with solid mining experience. He traveled north to Bennett with J. H. Brownlee, a surveyor, whose mission was to determine the 60th parallel, and plan the Atlin townsite. Also in the party were several members of the British Columbia Provincial Police force.[238]

Gold Commissioner Graham arrived at Bennett on 28 December 1898.[239] He did not proceed directly to Atlin, but stayed to take over from the former Constable, J. J. McKenna, who had been appointed as Bennett's Acting Mining Recorder several weeks earlier, when Captain Rant had been recalled to Victoria.[240] For three months government business in Atlin was in the hands of W. H. Vickers of the British Columbia Provincial Police who was appointed Acting Mining Recorder.[241]

The confusion Graham found at Bennett explained the many complaints already received in Victoria. He estimated that more than a thousand claim applications had been received by Norman Rant that were just "bundled up and put away for posting up later". These records were incomplete. Some applications had been paid for, while in other cases miners had tendered fees which Rant had declined to accept.[242] Throughout Graham's correspondence to Victoria there were many oblique references to Rant's inefficiency. He complained repeatedly of mistakes made in the recording of claims, and the fact that certificates were not prepared. There were several thinly veiled references to Rant's dealings with the miners which Graham apparently felt were less than honest.[243]

Before coping with the appalling tangle of paperwork, Graham set about insulating and heating the government offices in Bennett, which he described as "worse than a barn". He sympathized with McKenna who had been unable to obtain clerical assistance, and had endured conditions so cold that, ". . . the ink froze to the pen before it could be transferred to the paper".[244]

Graham's appointment to the district was received enthusiastically in Atlin. He had retired from a distinguished career in the British Army before entering the service of the Provincial Government in 1892. He was well respected in Revelstoke where he served as Government Agent and Gold Commissioner until 1897, when he became manager of the Waverly Mine.[245] With a wealth of experience in the administration of mining matters, the ability to deal with a firm hand, and a reputation for integrity, Graham seemed the ideal man for a difficult job. Coupled with his own merits was the fact that he had apparently been given "large discretionary powers" to deal with difficulties and disputes without delay or reference to Victoria.[246] Surely now the obstacles hindering gold miners in Atlin would be removed, and the industry would be allowed to flourish. Hopes were high when Graham arrived in Atlin on March 9.[247] Unfortunately, though, the government had not finished with Atlin and another blow had already been dealt.

This time the difficulty was the government's determination to capitalize on all the benefits of the Atlin gold strike. Bureaucrats had watched helplessly as the greatest share of the Klondike wealth slipped into the hands of Americans. More outfitting was done in Seattle and San Francisco than in Vancouver and Victoria. Americans greatly outnumbered British Columbians in the gold fields, so even the gold itself leaked away from the province. With all of this fresh in mind, politicians, encouraged by merchants in the coast cities, were determined that British Columbia should glean maximum profits from the Atlin gold rush. With that intent, the first heavy step resounded in January 1899, soon after the election of the Semlin government, when the Legislature passed an amendment to the Placer Mining Act. It restricted the issuing of free miners' licences to British subjects only. Known as the Alien Exclusion Act, it prevented all foreigners from mining anywhere in British Columbia. The one small concession was that an alien holding a free miner's licence could continue to work claims already in his possession. He could not, however, take up any new ground.[248]

Since Atlin had been an American camp from the beginning, the impact of the amendment was overwhelming. While the bill was under discussion, Americans rushed to Atlin to stake claims before it became law.[249] However it was more than two weeks after passage before Graham knew of it, and a number of miners, both in Atlin and the nearby Porcupine District to the west were disallowed claims registered after January 18th.[250]

The new amendment severely affected those whom it had been designed to benefit most. Early in 1899 the *Victoria Colonist* confidently announced that, "Victoria and Vancouver (business)men are predominant in Atlin . . ." [251], but it was not long before vehement protests poured into the legislature in Victoria. Businessmen moaned that double or perhaps three times the miners would have come to Atlin had it not been for the new law.[252] Miners, merchants and transportation companies in Atlin and on the coast all felt the impact, and deplored the loss of greater mining activity, and the expected influx of American capital.[253] Americans claimed the new law placed them on a ". . . level with the Chinamen," treatment they clearly did not expect from their Canadian neighbours.[254]

To further complicate matters the legislation was not well understood, and many did not realize that claims held previously by Americans were still valid. Consequently there was another spate of claim jumping, and threatened Americans claim holders warned that they would defend their properties with weapons if necessary.[255] They lodged a barrage of complaints with American authorities who made representations to the government in Ottawa.[256] The matter was also considered by the International Commission, a body whose function was to settle all the outstanding disputes between the United States and Canada.[257]

Within British Columbia there was great pressure from Richard McBride, a Member of the Provincial Parliament, and other influential men with Atlin connections for repeal of the Alien Exclusion Act,[258] but the Semlin Government steadfastly insisted that it was in the interest of British Columbians and common justice. In response to the petition sent by American residents of Atlin to the United States President, William McKinley, Semlin took the position that if indeed four-fifths of the claims in Atlin did belong to U.S. citizens, as the petition stated, then the act was justified.[259]

The Act was amended twice; once in January 1900, to allow aliens to hold claims purchased from Canadian locators,[260] and again in April 1900, to allow joint stock companies or corporations to hold free miners' licences under some circumstances.[261] Finally in April 1900, Ottawa intervened, and decreed that British Columbia had stepped outside its jurisdiction in attempting to legislate in an area of foreign policy, and so the Alien Exclusion Act was declared invalid and set aside.[262]

Over a period of 15 months much damage had been wrought, and many Americans who did not make it to Atlin in time to stake claims, headed immediately for the beaches of Nome, Alaska, where some exciting new finds were being reported. The initial boom could not be recaptured, and the injury sustained by the camp's reputation could not be undone.

Another matter which caused dissension between the government and the population of Atlin was the disposition of town lots. At first a reserve was placed on all land within 8 km of the Atlin Lake shoreline.[263] Later the reserve was extended to all land in the Atlin Mining Division.[264] These measures curbed the ambitious plans of men such as Norman Rant, and his partners. This group had gone to considerable expense to clear and level streets, and to lay out lots.[265] Some of the lots were sold, but shortly after these premature sales, both land and improvements were appropriated by the government and purchasers were reimbursed.[266] There were other would-be speculators; Sergeant W. Davis of the Mounted Police,[267] who applied for a land grant on the site of Discovery, and A. E. Ironmonger Sola who attempted to purchase 160 acres on the site of Taku City.[268]

There were further complications when Captain Rant issued an order that prevented anyone from building on an unsurveyed lot.[269] This prompted a businessman, L. D. McKinney to commission H. H. Eggerton to re-survey Norman Rant's lay out of the town which allowed for streets 18.5 metres wide, and lots 15 by 30 metres.

When Joseph Graham arrived in March, he ordered that all construction should stop. He gave further notice that when the government surveyor, Mr. Brownlee began plating the townsite, all streets would be increased to 28 metres, and lots would be 7.5 metres by 30 metres.[270] At the time there were about 200 buildings under construction, and Graham's order caused great consternation. It is not clear why he thought it necessary, since in January, the Commissioner of Works, C. A. Semlin, recommended to Graham that Brownlee's survey should follow the lines of Eggerton's previous work in order to protect existing buildings.[271]

The situation placed a burden of uncertainty on businessmen who had risked their investments, and started building in order to have their premises ready for the start of mining. Now it appeared their money and effort could all be lost.

In May, 1899, Brownlee's plot plan was complete, and all townsite land was put on the auction block.[272] Values were placed on existing improvements, for which the original owner would be reimbursed, should he lose his property to a higher bidder. However, there were plenty of available lots, and except for the most desirable commercial properties, competition for them was not keen.

The selling prices ranged from $125 to $500, and the government made money, a total somewhere between $40,000 and $50,000.[273] That in itself was enough to cause resentment. There were loud complaints that the upset prices of lots were too high and that inflated values were placed on existing improvements. There was also considerable indignation that as lots with log buildings sold, the government added the royalty for the logs to the price of the lot.[274] Opinions were reinforced that the government's principal interest in Atlin was to milk its residents for taxes. Nevertheless, despite the dissatisfaction, business and home owners now had secure tenure.

It was late in February, 1899, before the government took a decisive step to unravel the tangle of confusion over claim ownership, that had grown steadily worse since the previous July. The Bennett-Atlin Commission Act, 1899, made provisions for the appointment of a Special Commissioner to settle summarily all matters referred to him by either the Lieutenant-Governor, or the Government Agent. His mandate was to settle the disputes that had arisen as a result of the incorrect registration of the initial claims on Pine Creek. He was also instructed to solve the problems resulting from the conduct of government officials before the arrival of J. D. Graham. He was to be bound only by the principles of equity and the spirit of the Mining Act, rather than by the letter of the law.[275]

On June 7, a commissioner was actually appointed, but by this time the open season date of June 1 was past, and mining was still prohibited pending the arrival of a judge.[276] After further postponements, opening day was finally set for August 1 - halfway into the short mining season. There is little wonder that miners were angry, hungry and frustrated.

The Government's stated reason for the delay was that they wanted to give all miners a chance to get to Atlin in time for the hearings, and the long delay was unavoidable.[277] Embittered miners believed that there was some ulterior motive causing the hold up. They charged that by delaying, the government was protecting the interests of certain groups. Many were convinced that the extended ban on mining was designed to freeze out the small claim holders so that their holdings could be amassed by the large companies which were making their powerful presence felt in the new gold fields.[278] The whole question engendered deep seated resentment.

During this time many who had the means gave up all idea of staying in Atlin. Typical of these was Joe Clark, a real estate broker who left for Dawson City, calling Atlin, "a government-cursed town".[279] Potential investors hesitated to put money into an area where the tenure of claims and property was insecure, and government administrators were actually preventing mining men from doing their work.

Judge Aemilius P. Irving arrived in Atlin on June 16. He was accompanied by E. M. N. Woods, who had been appointed Stipendiary Magistrate for Atlin, and who initially, would act as Commission Secretary. Also in attendance were a number of prominent members of the British Columbia bar, including Richard McBride, M. P. P., Gordon Hunter, F. McBride Young, and A. C. Belyea.[280] The hearings convened on June 19 in a large tent borrowed from the Presbyterian Church.[281]

All the recriminations about the delay in opening the season, cast the Commission in a poor light from the beginning. To cover expenses a $25 fee was imposed on each plaintiff. This demand was not appreciated given the difficult financial circumstances of many miners, and the fact that the government was the cause of the mess. Yet once Judge Irving began to deal with individual cases, he quickly won public confidence, and was esteemed by Canadians, Americans, and Englishmen as a wise and fair man.

With respect to the miners who staked the original 77 metre claims, he awarded each man 30 metres, chosen with consideration for any work done on the ground.[282] Most of these claims had already been re-staked under British Columbia law.[283] Miller and McLaren were awarded a joint Discovery Claim of about 180 metres long and 300 metres wide, but denied certain bench claims that would otherwise have been theirs.[284] Irving also dealt with the 647 applications which had never been processed by Norman Rant. He settled cases of over-lapping and conflicting boundaries, and handled a multitude of disputes created by claim-jumpers.

Irving's judgments were generally considered fair, and there were few protests. Nonetheless, Judge Irving had hardly left the district, when claim-jumpers began to use the merest technicalities to claim properties which were already registered. Gold Commissioner Graham reported that such people would continue to raise havoc since he did not have the power to take action against them.[285] Obviously his range of influence was considerably less than at first indicated.

This lack of power in the hands of the Government Agent was a contentious issue prior to the Commission hearings, and so it remained for some time. An ineffectual Gold Commissioner was caught in the crossfire of discontent. Miners and businessmen alike saw J. D. Graham as being primarily an efficient tax collector. The miners did not always see the problem as being a lack of power, but rather a misuse of it.

After miners held a mass meeting on 21 June 1899, their representatives reported to Victoria saying of Graham,

> . . . that he has shown what appears to us to be much indecision of character; that he has postponed cases repeatedly, which . . . he ought to have settled on a first hearing; and further, that he, as Gold Commissioner, has allowed and sanctioned the nefarious practice . . . of taking successive fees from many miners for recording one and the same claim.[286]

Certainly some of the ire and mistrust generated during Norman Rant's time as Mining Recorder still smouldered. The British Miners' Protective Association accused Graham of making unfair decisions.[287] Graham had been at odds with the Association for some time.

One of the main problems and sources of dissension was the rights of the individual miner versus those of the hydraulic companies. It seemed impossible to please everyone. Both laid blame on the government for protecting the other. There were charges that the government was steadfastly turning down all applications for hydraulic leases so that they would not fall into the hands of aliens.[288] In fact the law allowed for a hydraulic lease to be staked over a placer claim, so that if a claim expired, or was abandoned, it automatically fell into the hands of the hydraulic leaseholder. In this way good paying claims were being lost to the individual miner forever. Free miners, who thought the government favoured large companies, wanted no hydraulic leases issued until they had truly abandoned the ground.

There were other officials under the gun. E. M. N. Woods, Stipendiary Magistrate was frequently accused of partiality. William Queen lodged a complaint, involving assault, against J. T. Brackett, the manager of a hydraulic company. The charge was dismissed in Woods' court. When a disgruntled Queen left the Court House he told anyone who would listen that he had been treated unjustly. Woods had him arrested and held for the remainder of the day. Queen complained to the Attorney General in July 1900 that Wood's conduct was "most outrageous during the trial, and before and after the same." He further alleged that he and other miners believed that Woods routinely ruled in favour of the richer hydraulic companies, who "were able to bestow more favours." Queen was calling for an official investigation, and there were others who made similar demands.[289]

Mary Hitchcock complained not only of a lack of justice in the court, but also that a clerk in the Mining Recorder's Office had jumped her claim on Munro Mountain.[290] There were also charges of highhandedness and unfair dealings by the Mining Recorder, E. W. Bickle,[291] who was eventually dismissed by Order-in-Council.[292]

Graham, representing the Provincial Government in Atlin, bore the brunt of local frustration with bureaucracy and the law. While the behaviour of some other Atlin officials was questionable, Graham was probably the victim of imperfect and deeply resented legislation and policies, to which he strictly adhered.

On 12 August 1901, a Commission of Inquiry into the conduct of public officials in the Bennett and Atlin Mining Divisions was convened in Atlin before Chief Justice A. J. McColl, of the British Columbia Supreme Court.[293] The hearing lasted two days[294] but the outcome of the investigation was never known, since Chief Justice McColl died in January 1902, before he had reported his findings. There was some talk of Graham resigning, but the question of impropriety on the part of government officials remained unresolved.[295]

Shortly afterward, Atlin lawyer, C. W. Sawers, charged that Graham had used public monies improperly on the purchase of two small buildings. Clearly a thorough investigation was necessary this time, so a Select Committee of the

House was appointed to look into the charges.[296] The hearings which were held in Victoria commenced on 16 April 1901, and lasted 18 days.

In March 1901, there had been an outbreak of scarlet fever in Atlin, and Graham had purchased, for $750.00, two small buildings, for use as an isolation hospital. Sawers claimed that the buildings had in fact belonged to Graham, that they were not worth more than $200.00, and that they were not used for fever patients, but for a nurses' kitchen and living room.[297]

Among those appearing in Court were the Honorable J. Dunsmuir, various carpenters and residents of Atlin, and James Stables, M. P. P. for Atlin. Stables had been president of the British Miners' Protective Association, and in that capacity had had a number of unpleasant encounters with Graham. Though he placed a higher value on the buildings than Sawers' other witnesses, his figure of $450 was still well below the figure paid by Graham.

Sawers was determined to use this forum to prove Graham's wrongdoing, and in the process he sorely tried the patience of the Committee with delays, drawn out cross-examinations, and irrelevant evidence.

The Select Committee's report was presented to the House on May 30.[298] They found that Graham's ownership of the buildings could not be established, they could not determine if there was any market value for the buildings at the time, and the buildings were used for the fever patients before being used by the nurses. Although the buildings were acquired at the request of Dr. Lewis, the Medical Officer of Health, the Committee decided that they were not a necessity, since patients could have been housed in tents, as they had been in previous years. In summary, however, the Committee found that:

> . . . the buildings in question were acquired at a price much greater than the real worth of the buildings, and the Committee cannot but feel that . . . Mr. Graham was not as diligent in the public interest as he might have been."[299]

By this time Graham had resigned.[300] Despite his apparent indiscretions his departure was lamented by many, as he had been a firm and able administrator when Atlin needed one. He had just never been able to persuade the government to give him sufficient powers to administer his district as he saw fit. A dedicated man, he had been brought down by those he served.

Immediately there was a clamour in Atlin, from both miners and businessmen, who once again demanded the appointment of a local man as Gold Commissioner and Government Agent to fill the vacancy left by Graham.[301]

The appointment of J. A. Fraser was welcomed as a triumph for the local community.[302] He had been President of the Atlin District Board of Trade since its inception in 1900, he was a school trustee, and a well-known businessman. All sides welcomed him to an important position in which he served with justice and honour for twenty years. The choice of Fraser in 1902 seemed to be a masterful move on the part of Victoria, but more likely it was a happy accident.

Except for the occasional fracas, Atlin's legal nightmare was over, but the wreckage could be tallied in lost population, business growth, investment and gold production. We can only speculate about what Atlin might have become without such serious interference and bungling. Still, despite the hindrances, deep roots took hold. The gold camp image was disappearing, and no committed citizen in 1902 considered Atlin anything less than a permanent community.

CHAPTER 6

# The Placer Creeks

*"Atlin has only two good years - 1899, and next year!"*

- A sometime miner.

Atlin miners sold their gold for $17 - $19 an ounce, a price that ranged from $2 to $5 higher than gold from the Klondike or from other mining camps in British Columbia.[303] It was coarse and full of nuggets of notable size. Gold from Pine Creek was pale yellow, and prized by jewellers; from Birch it was shot through with quartz. Spruce Creek gold was darker in colour which told of its copper content. Miners who knew gold learned to identify its source by its appearance, and some could tell where on a particular creek it had been mined.

Pine Creek gold lay just below the gravel. McKee gold was hidden under 24 metres of overburden. Some paystreaks followed streams through narrow ravines, where tailings could scarcely be stacked without interfering with neighbouring claims. Rarely was there adequate water for sluicing or hydraulicking. This condition was attendant to a semi-arid climate where annual rainfall measured only about 30 cm, and snow depths added up to little more than 1.5 metres each winter.

By the end of the mining season in 1899, dozens of creeks had been staked from mouth to source. Even streams that would never produce gold had been claimed by eager Cheechako stampeders. Some claims would yield gold beyond the craziest dreams of the miners who worked them. Other creeks gave up little to the pick and shovel miners, but later poured the yellow metal into the coffers of large companies who dug for it with powerful machines that moved great volumes of earth.

It has been estimated that in the season of 1899, there were from 1500 to 2000 men working individual 30 metre claims on Atlin creeks.[304] The most successful miners were on the section of Pine Creek that stretched from about

AHS P-1246

*A washing plant separated gold from gravel.*

AHS Baker Collection P-2783

*On Spruce Creek, a steam shovel worked its way through the gravel and boulders overlaying the pay streak which was deeply buried on bedrock. (1932)*

AHS P-2361

*Nate Murphy, hunting guide, trapper and fur farmer, followed the stampede to the O'Donnel River in 1912. He made the O'Donnel his home and mined gold from its banks for nearly forty years.*

a kilometre and a half above Discovery Claim to a few claims down stream from it.[305] Here were the "poor man's diggings" that had been so widely publicized in the fall of 1898. In this golden mile, a man with little more than a wheel barrow, a sluice box, and a strong back could wash out $30-$40 a day. On the Discovery Claims, Fritz Miller recovered approximately 55 oz in one "clean up" after only 40 hours of work by 5 men.[306] One miner, working a bench claim, recovered $15 per day, and he had to haul every wheelbarrow load of gravel down to the creek to wash it out.[307] It was slow hard work, but certainly he could not have complained, because at a time, when $2 a day was considered a fair wage, he undoubtedly earned more than ever before in his life.

In 1899, 40,000 oz of gold was recorded from the Atlin District, making it a record year. Yet, it was generally believed that more than half of all recovered gold was never declared by miners, so they could avoid paying the 1% royalty tax demanded for each ounce.[308] That first year was also the year of the individual miner in which he had the run of the whole camp, without serious competition from the large companies.

It soon became clear that most of Atlin's creeks were not "poor man's diggings". Even when several claims were amalgamated and worked together, ordinary placer methods were not adequate to release the wealth from the ground. Bedrock and the pay streak were too deep, and the boulders were too large to be handled with pick and shovel methods.

By the end of 1899 the truly disillusioned had left for Nome and elsewhere, but many others remained to fight the mounting pressure from the large companies. Most were inexperienced, and few made fortunes, but the majority entertained enough faith in their holdings to hang on for a while. The Provincial Mineralogist, in 1900, noted the absence of the "broke miner" about the camp.[309] Indeed, until 1904, the individual miners collectively produced more than the companies. By 1904,[310] however, the hydraulic companies were firmly in control, and ground that had been previously considered worked out, or uneconomic, once again had good prospects.

Every creek was as different as the methods used to mine it. Each had its stories, its heyday, its characters and its miners.

## Pine Creek

Pine Creek flows 26 km from Surprise Lake to Atlin Lake. After meandering from its source, past the site of Discovery and Spruce (or Halfway)

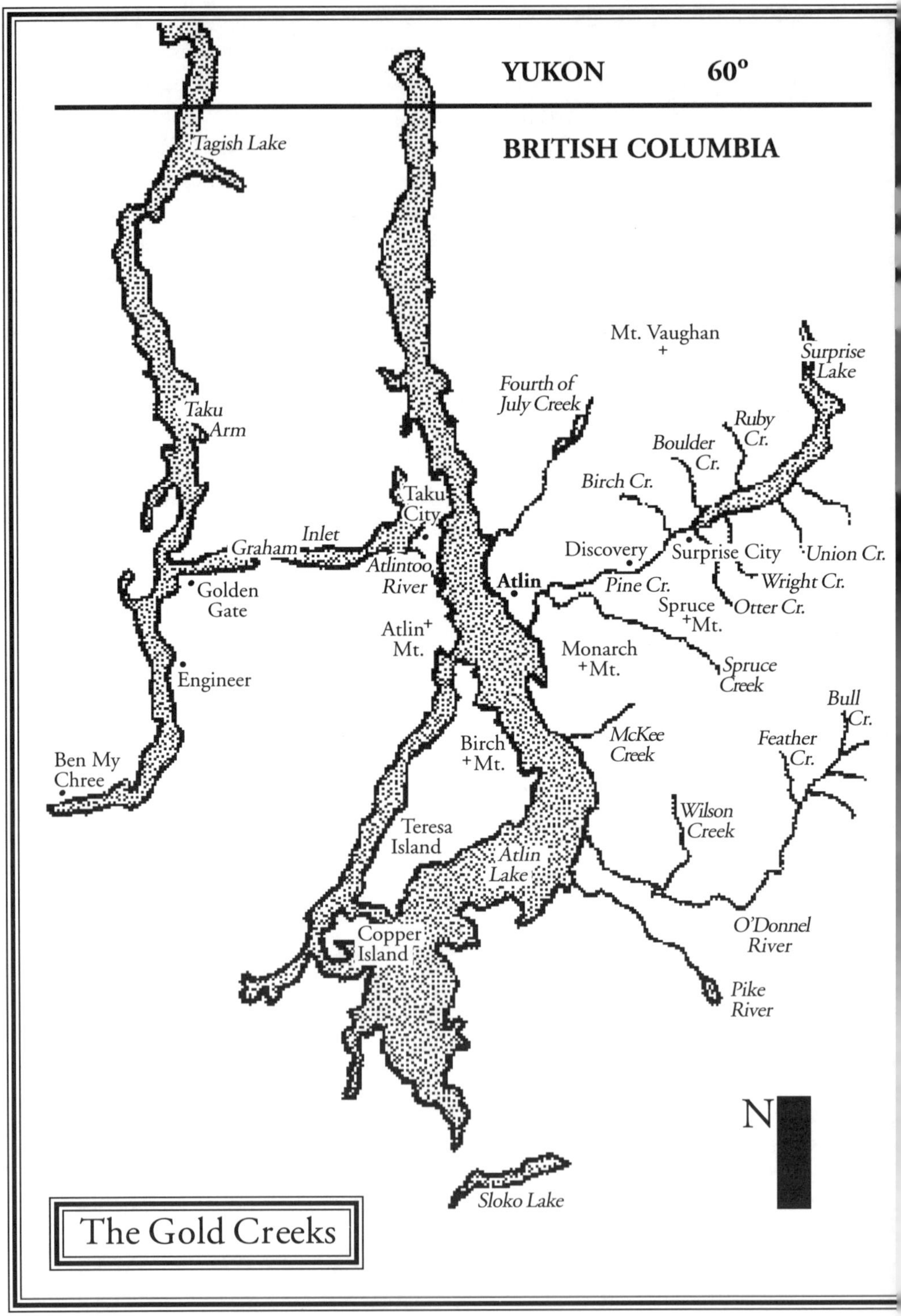
YUKON
60°
BRITISH COLUMBIA
Tagish Lake
Taku Arm
Graham
Inlet
Golden Gate
Engineer
Ben My Chree
Taku City
Atlintoo River
Atlin Mt.
Atlin
Birch Mt.
Teresa Island
Copper Island
Atlin Lake
Sloko Lake
Mt. Vaughan
Surprise Lake
Fourth of July Creek
Ruby Cr.
Boulder Cr.
Birch Cr.
Discovery
Surprise City
Union Cr.
Pine Cr.
Wright Cr.
Spruce Mt.
Otter Cr.
Monarch Mt.
Spruce Creek
McKee Creek
Bull Cr.
Feather Cr.
Wilson Creek
O'Donnel River
Pike River
N
The Gold Creeks

Junction with little drop in elevation, it plunges through a steep and narrow canyon creating the spectacular Pine Falls. From there it flows through small gorges and flats until it fans out on alluvial gravel at its mouth. It descends 275 metres over its length.

Both the riches and the problems of Pine Creek are closely associated with the yellow gravels which cover large portions of the Atlin District.[311] The shallow gravels around the Discovery Claims made for easy mining, and pioneer miners, including Miller and McLaren, found impressive amounts of gold there with relative ease. In most areas, however, the grinding glaciers of the Pleistocene left deep blankets of drift over the ancient creek beds where the gold lay hidden for millennia.

Pine Creek produced well for the pioneers. In 1899 J. D. Harrigan recorded a yield of $21,000 from numbers 8 and 9 below the Discovery Claims.[312] Another group of successful miners found rich ground on the high bench near Willow Creek at a place appropriately named Nugget Point. Sam McCaulley worked his ground with a wooden rocker. It was an ideal contrivance for separating gold from gravel where water was severely limited, as it usually was on bench claims, because the same water could be used over and over. Using his rustic hand operated separator, McCaulley took out 8 - 10 oz per day.

Captain Sperry and his son-in-law were working on claims called the Caledonia group. They built a 450 metre flume and the largest water wheel on Pine Creek to raise sufficient water to wash the gravel.[313] By this method they recovered 25 to 35 oz per day per man.[314]

Despite these successes there were, as well, the ongoing problems of tailing disposal, and water shortage,[315] and these difficulties often resulted in lengthy litigation, and lost working time. These were setbacks the average miner could seldom afford, and after just a few years most of the ground on Pine Creek was in the hands of the hydraulic companies.

Sunrise Hydraulic Company, organized by J. M. Ruffner and J. H. Rose was one of the first hydraulic operations on Pine Creek.[316] Ruffner bought up claims until he controlled two and a half kilometres of the creek including its north side benches between Discovery City and Spruce Junction.[317] Twenty-one mules harnessed to scrapers[318] gouged out a flume 6½ km long, 4 metres wide, and 4 metres deep to bring water to the bench workings.[319] For a time Sunrise Hydraulic was the largest single operation in the area, but by 1901 the operation had to be abandoned. Their tailings and those of the Willow Creek Hydraulic Company, which were being run into the creek, were seriously interfering with the water flow to other creek claims which had not yet been worked out.[320] The Placer Mining Act expressly prohibited hydraulic leases from interfering with workable placer claims.

Gold Run was a dry gulch situated about a kilometre above Discovery City, and for most purposes was considered part of Pine Creek. Since gold was never found in the area of Pine Creek which paralleled the gulch, it was supposed that Gold Run was an ancient creek bed, and testing supported the theory.[321] Summer mining in the gulch was beset by water shortages, but during the winter of 1902, about 70 men were at work with steam pumps and hoists, dropping shafts and stock piling gravel for spring clean up.

A small operation on four claims owned by Thomas and Brown caught the attention of the Provincial Mineralogist, who in 1904, considered it the model drifting mine of the camp. A shaft had been sunk to bedrock at about 12 metres, where the ground was mined to a height of one to two metres. Timbering inside the mine was so carefully done that it prevented the intrusion of most surface water, a problem that had already forced the abandonment of several nearby mines. A small steam pump drained what little water did accumulate. Wooden buckets for bringing gravel to the surface were operated by a double acting hand windlass. This sensible method, seldom seen in British Columbia at the time, simply raised a full bucket while it lowered an empty one. Water was brought to standard 30cm wide sluice boxes from a neighbouring ditch. Thomas and Brown's recovery system used a gold saving table that was another first in the camp. One man could run the efficient little operation, and by taking out as little as 3 oz a week, he would make a profit.[322]

While the upper part of Gold Run was worked by individual miners,[323] the British American Dredging Company acquired the lower end of the area toward Discovery City. In 1902 the entire camp watched with increasing interest as the company, managed by O. T. Switzer, prepared to launch the first dredging operation in the district.[324] At first the company had focused its exploration on the area stretching from the confluence of Spruce and Pine, and west to the shore of Atlin Lake. There the search for the old Pine Creek channel had occupied miners since the first gold strike. The Racehorse group, along with other claims, had been worked extensively at great cost, and produced little gold. The expanse was dubbed "the poor farm." British American had, at first, planned to dredge in "the poor farm," but, probably because the route of the ancient creek continued to elude them, they instead located the dredge at Gold Run.[325]

In 1903 they acquired more property, and at the same time invested in a Bucyrus Dredge, a Keystone Drill and a power plant. The plant was located

AHS P-2360

*This forest of claim posts at Nugget Point indicates the extent of claim jumping, a rampant problem on Pine Creek. Tents and buildings of Discovery City are visable in the upper left background.*

*Gold miners worked their claims on Pine Creek behind the main street buildings of Discovery City. (1899)*

AHS P-1026

*The powerful monitors of the hydraulic plants were extremely efficient at washing away deep overburden from the bedrock where gold had settled. Pine Creek.*

*The British American Dredging Company installed this dredge in the Gold Run area of Pine Creek. (1903)*

AHS P-1021

*A profusion of China wheels, or pumps, raised water for the sluices. Pine Creek. (1899)*

*Atlin Consolodated Mining Company, owned by the Guggenheim family of New York, operated on Pine Creek for several years. (1906)*

AHS P-1103

below Pine Creek Falls, and ditches, flume, power line, and dams were constructed at a cost of $300,000.[326] Electric power was generated and transmitted at high voltage to the dredge. The company had plans for a number of dredges on Pine, and the power plant was designed to supply power to all of them, from its two 60 cm turbines.[327] The dredge which had 96 buckets, each with a capacity of a cubic metre, could handle 2500 cubic metres daily. In tandem with the dredge was a large gold saving scow.[328] It soon became clear that the dredge was unable to handle the gravels which were solidly cemented in clay, and so the Keystone drill was used in advance of the dredge to drill, blast and loosen the ground.

Since dredges of this type, which had been pioneered in the Otago goldfields of New Zealand, were at the time proving very successful in the Klondike, there were great expectations for this project. Nonetheless the dredge design proved to be unsuitable for the terrain, and the whole scheme was an expensive failure. Breakdowns were frequent, and by the end of the first season, the lips on the buckets were damaged beyond repair. In addition to operational problems, the gold saving system was shown to be less than 50% efficient.[329] By the end of 1905 the whole dredging project had been abandoned before the bed rock, which lay as much as 9 metres deep, was reached.

The Pine Creek Power Company commenced operations in 1902 by acquiring the interests of Sunrise Mining Company. Working that year with a crew of 40, the company held more ground, and accomplished more work, than any other on Pine.[330] By setting off dynamite charges of enormous force, and then hydraulicking the broken earth with three 17.5 cm monitors, they were able to process great amounts of gravel. In 1906, with a crew of 25, the company reported a recovery of approximately 4,400 ounces of gold worth $75,000, the largest output in the camp.[331] Eventually Pine Creek Power changed hands, and amalgamated with others, but carried on to be an important player on the creek for many years.

The new North Columbia Gold Mining Company, managed by Julius M. Ruffner, had its office in Discovery in 1903.[332] At first their hydraulic operations were not particularly successful, and in some years results did not justify expenses. In 1907, Ruffner acquired the interests of the Pine Creek Power Company[333] and in some years as many as 70 men were employed.[334] Under Ruffner's enterprising management, a plan was made to relieve the chronic shortage of water on Pine Creek, by bringing in water from Surprise Lake. They began by building an 8km ditch, using a small steam shovel aboard a scow which floated in its own excavation as it dug forward. The duct was 8 metres wide and about 2 metres deep, and it would carry 15,000 miner's inches of water.[335] This ditch, it was claimed, would have the largest capacity of any mining ditch in the world. By the end of the 1908 season, about 5 km of ditch was complete. In 1909, however, a spring flood did extensive damage, and carried away some of the ditch, causing a great deal of expense and lost time. Despite the setbacks, Pine Creek Power and North Columbia Gold Mining, both in Ruffner's capable hands, had a very successful season. Their combined incomes exceeded $70,000.[336]

Even though water was scarce in the summer months, spring floods such as the one of 1909 were not uncommon on Pine Creek. Year after year, the Minister of Mines Reports chronicle extensive damage to flumes and ditches, which cut short the time left to mine between spring thaw and freeze up. Time lost to flooding was especially frustrating at the beginning of the season, when the supply of water, so essential for hydraulic work, was most plentiful.

The benches on the south side of Pine, from Discovery to Spruce Junction, had proved early to be rich diggings. At a place known as Stevendyke Gulch, it was determined the ancient channels of Pine and Spruce converged, then ran north toward Thrond (or Trond) Gulch and emptied into Atlin Lake when its shoreline meandered along the flats around Spruce Junction. Paydirt at Stevendyke was found in two runs, one in the old yellow gravel, and another in the blue.[337] The Stephendyke Hydraulic Company, organized by D. Todd-Lees, and Frank Weir began operations in 1901. Their company controlled about a kilometre and a half of Pine Creek, and 227 hectares of bench land.[338] In 1902 profits were good, and the next year the claims were bonded to Ruffner's North Columbia Mining Company.[339]

During 1906 and 1907, the largest operator in the camp was the Atlin Consolidated Mining Company, owned by the Guggenheim family of New York. This powerful company had widespread holdings in Alaska, and mined with great success in the Klondike. On the north side of Pine Creek they held the property between Discovery City and Gold Run, an area called Tar Flats. Here they installed a giant 70 ton Bucyrus steam shovel with a 1.6 metre bucket, capable of handling 3,000 cubic metres every 24 hours.[340] Each day the huge machine devoured 5 - 6 cords of wood to keep the boilers fired.[341] Remarkably every log was cut by hand. Power was drawn from the plant below Pine Falls to run a 5 ton locomotive and 40 dump cars, that carried gravel to the screening and washing plant. The big shovel began operation in mid-August, and rumbled 24 hours a day until October 25th. Thirty-six men were employed and during their first short season the company cleaned up $25,000 in gold. Results for the following two years were excellent, and in 1907 they recovered the largest output in the camp.[342] In 1908, W. F. Copeland, a new manager, got mining off to an early start, then began considering improved methods to cut production costs. Since the company was experimenting similarly somewhere else, all operations were closed down awaiting those results.[343] The closure of such a large operation, together with the failure of the dredge, signalled a drastic slump for the entire camp.

Perhaps the Guggenheim's huge success in gold mining resulted from knowing when to quit, yet from 1910 on, and for a number of years, North Columbia Gold Mining Company worked the Guggenheim property with good overall results, the best year being 1913, when profits exceeded all other seasons.[344] It was obvious that hydraulic mining was the best method for Pine Creek, and Ruffner's company used from 6 to 15 big monitors at various times.[345] In 1914 the companies were reorganized into Columbia Mines. They remained the largest interest on the creek, employed as many as 50 men, and consistently produced good values. In 1918 Ruffner's company, Atlin Gold Mines, acquired the Guggenheim properties, and the remainder of the group was sold to F. H. Mobley and Louis Schulz, who named their company Discovery Mining and Power.[346]

In the entire district gold production slumped dramatically during World War I, when miners went overseas and investment capital disappeared. After the war, largely because of the high cost of materials, freight and labour, mining did not rebound to the levels of former years. By 1924 there was but one small hydraulic operator on Pine Creek, and Discovery City was a ghost camp.

In 1925, a quartz-streaked nugget weighing 49 oz was found. Its value was $600.[347] For hard-bitten miners who refused to abandon their dreams, it must have provided incentive to continue the search.

Discovery Mining and Power, owned by Charles V. Bob of New York, continued to be the major operator on Pine well into the 1930's. Bob had acquired all the major properties, including those of Ruffner's Atlin Gold Mines. In 1925, the company under the direction of Andrew Sostad, began major development work to uncover the old channel of Pine in the area above Pine Falls. Walter Sweet, the foreman, had over 5 km of ditch built, and installed 215 metres of flume made from used hydraulic pipe reclaimed from all over the area.[348] With water tapped from the lower end of the Ruffner ditch, two monitors were used to cut a huge trench, 1150 metres long by 60 to 90 metres wide. It became known locally as the Panama Canal. The project continued for five years, but no preliminary drilling or geological work had been done, and the end results were insignificant amounts of gold. The company was finally sold to Golden Centre Mines.[349]

At the onset of the great Depression, Atlin's fortunes improved as gold prices remained stable, and investment interest revived. When prices and labour costs sagged with the nation's economy, gold mining became financially attractive once more. In 1934, the situation further improved when President Franklin Roosevelt set the gold price at $35 per ounce.

Throughout the period small numbers of individual miners, including Kenneth McLaren, continued to work summer and winter on Pine. Many did well by turning to old workings, where areas of rich ground had been overlooked by greenhorn miners, but Pine Creek's glory days seemed to be over.

## Willow Creek

Willow Creek was not far from the east end of Discovery City. I. E. Negus staked the first claims in 1898, only to have his application denied by Mining Recorder, Norman Rant.[350] Willow Creek was little more than a dry gulch which enters Pine Creek near Nugget Point. One miner described Willow as a creek his father liked to mine, because it was so small it was not necessary to get his feet wet. He also claimed it was so rich in coarse gold at the surface that miners could dig with a trowel and their bare hands.[351] While Willow was officially designated a bench of Pine, it was generally considered a separate entity by the early miners. It may have been an old bed of Pine, and bedrock did lay just centimetres beneath the surface in some places.

Until late in June, 1899, the ownership of many Willow claims was in dispute, and most mining was stalled. Settlements could not be made until Judge Irving arrived to rule on complaints. The ground was so extremely rich that claim jumpers had wasted no time in moving in amid the confusion which reigned in 1898. Once the properties were finally free of litigation, many inexperienced pick and shovel miners prospered on Willow. Fred Cotterell a 14 year old from Juneau, mined with great success, along with his uncle and a partner.[352]

With work on both Pine and Willow in full swing, water proved to be something of a problem, and in August of 1899 the Willow Creek Miners' Cooperative Water Association was formed by claim holders, under the leadership of Thomas Bateman and Alfred Carmichael. The Gold Commissioner gave them a permit to build a ditch from Pine Creek which would assure a steady flow to the workings.[353] A work force was gathered, but not all of those working were claim holders. Some were unemployed miners who volunteered on the strength of jobs promised to them as soon as the flume was delivering water.[354] All the workers were to earn coupons representing 50¢ per hour of work. The intention was that the coupons would be redeemed from ditch revenues.[355] Just a week after the co-operative was organized, 100 men began digging, and less than 10 days later nearly 2.5 km of ditch was completed.

It was a splendid piece of co-operative effort, and the ditch was opened with great fanfare and celebration on 22 August 1899. In the spirit of camaraderie, and with a sense of real accomplishment, most of the camp participated in a football match and a picnic, and danced into the night.[356] High spirits began to plummet later, though, when some of the expected jobs did not materialize.[357] Part of the problem was that the ditch failed to provide reasonable profits, and hence the coupons were difficult to redeem.[358]

Within a couple of years the ground had been thoroughly worked, and there was a hydraulic company waiting to move in, in 1901. George Brackett owned the Atlin and Willow Creek Mining Company, and it was managed by his son, Frank. The company acquired the rights to the ditch by purchasing a large number of the co-operative's 50¢ coupons, and the claims of many of the shareholders. Brackett's right to the ditch was hotly disputed by some of the original shareholders, however, and a legal contest over the question lasted for two years. In 1902, a decision ruled in favour of Brackett,[359] but other legal battles continued to sap the resources of the company.

Brackett, a former mayor of Minneapolis, Minnesota, had received a severe blow during the financial collapse of 1893. A few years later he resumed the role of city leader, but this time as the second mayor to serve notorious Skagway. He was also the construction engineer who built and financed the toll road from Skagway to the summit of the White Pass, only to have it made obsolete by the railroad. In Atlin, his mining operation provided significant employment, and produced good returns until 1906, when he sold out to the Guggenheim interests.[360] The Guggenheims worked the Willow Creek property in conjunction with their other properties on Pine Creek.

## Spruce Creek

Gold was discovered on Spruce Creek in 1898 by Fred Markus,[361] who was also one of the earliest stakers on Pine Creek. Later he would become a prominent building contractor in Atlin and Discovery. Spruce is about 25 km long, and joins Pine Creek from the south, approximately 4 km above Atlin

Lake. As it flows its north westerly course, it crosses deep moraine deposits, which are the source of its characteristic blue gravels. Its main tributaries entering from the south are Little Spruce and Dominion. Rose Creek joins it from the east.

The Minister of Mines report on Spruce Creek in 1899 was decidedly pessimistic. More work had been done on Spruce during the season than on any other creek, but bedrock was deeply buried, and the 30 metre claims allowed little room for tailings. Poor drainage presented additional problems, even when two or more claims were worked together.

It was the mining inspector's opinion that Spruce could only be successfully mined by drifting or ground sluicing, which, because of the expense, would have to be left to the large companies.[362] Spruce did not look like a good prospect for the individual miner. By 1900 the Minister of Mines reported that the creek claims were already worked out.[363] With such gloomy forecasts development on Spruce was slow, but gradually the individual miner gained a belated foothold, and in fact Spruce was the last of the major creeks to be dominated by the companies.

Even during the winter months of those early years many miners were drifting with success. In the winter of 1904-5 there were 200 working on Spruce, and amid so much activity a small self-reliant community developed. A boarding house run by Mrs. Clay catered to 16 miners.[364] The Spruce Creek orchestra, led by John Cartmel, was often hired to play at dances in both Atlin, and on the home creek.[365] For all the bleak predictions, Spruce Creek was destined to become the number one gold producer in Atlin, and it would rank second, only to Lightening Creek in the Cariboo for Province wide gold output.

Winter drifting was always hazardous as miners dug deep into the towering banks of the creek. Knowing too well the danger of a tunnel caving, or a bluff collapsing they were ever watchful, but even the most carefully engineered mines were potentials for disaster, and on Spruce two of the area's tragic accidents occurred.

In January 1907, the temperature dipped to minus 40 degrees, and smoke from cabins in the Spruce Creek community wafted upward in thin columns. Andy Broe looked for some sign of activity at his neighbours' cabin, but saw no light in the window, and no smoke. The two partners who worked the Texas Bench claims should have been at home preparing supper. Matt McEwan and Daniel Burnyeat were from England. The pair had been working hard, and their labours had brought success. The 33 year old McEwan had been in Atlin since 1899, and finally he had saved enough to realize his dream of buying land in Alberta. This would be his last season of digging for gold, and in the fall he would leave the rustic cabin and begin his life as a rancher.

Broe must have felt a deep stab of fear for his fellow miners as he bundled up and dashed out to investigate the Texas workings. Inside the drift, he found a tangle of collapsed timbers and dirt. His alarm spread from cabin to cabin, and anxious miners came to help with the rescue work. More help arrived from town, Pine, and the other creeks. Time was crucial, and the most knowledgeable men issued orders as the dangerous recovery operation continued through the bitterly cold night. By noon the next day, two crushed bodies, one on top of the other, were finally uncovered.[366]

In March the next year three men were at work in another Spruce Creek drift when timbering near the portal began to crack and sag under the immense weight of the overburden. Dirt started falling into the tunnel, and one of the miners, a man named Anderson, yelled a warning to Ed Bowley and Billy Alliott, who were further back in the tunnel. Anderson escaped, but Alliott, who was close behind him, never reached daylight. Perhaps he turned back, unsure if Bowley had heard the alarm, or to see if he had been hurt. Both men were trapped behind the collapsed entrance. For several anxious hours Bowley and Alliott could be heard calling as rescuers worked feverishly to reach them. The tunnel continued to slough in, dashing any hope that the miners might escape on their own from inside the drift. Finally a rescue attempt was launched from above, but it took several days, and for Bowley and Alliott it was too late.[367]

Memories of accidents like these, and the cries of trapped men, would haunt the close knit community on Spruce Creek, but they served to reinforce the tight bond miners had for one another.

Spruce's West Nugget has been accepted as the largest nugget ever found in Atlin's history, although it is not generally considered the most valuable. Its golden bulk was interlaced with a significant percentage of quartz, and resembled a huge ostrich egg. The spectacular West weighed in at 83 oz. 5dwt. 15gr., and was unearthed on 126 below Discovery by William West and G. Hoffenden in July, 1899.[368]

Great sums of money were repeatedly ploughed into mining ventures which showed no returns. Although the local workforce and businesses usually gained from these short lived operations, shareholders often watched their investments melt away. The Columbia Hydraulic Mining Company whose claims were located just opposite the mouth of Dominion Creek, operated the first monitors on Spruce Creek. The company invested $100,000 in 1901 to lay

AHS P-1100

*At a typical placer mine, such as this one on Spruce Creek, miners lived on their claims where their work was just outside their cabin doors.*

*A large gold mining operation on Spruce Creek was a jumble of logs, lumber, water wheels, and flumes used in the ongoing pursuit of coralling enough water to sluice the gold bearing gravel.*

AHS P-2551

13,000 metres of flume, 30 metres of sluices, and to install a water powered generating plant.[369] They also erected a 20 metre derrick.[370] W. F. Gore, the manager, employed 10 to 13 men over each of the next four summers, and using two giants, or large monitors, they moved some 40,000 cubic yards of gravel. This work, it turned out, was little more than expensive prospecting since they had no idea where the pay streak was, prior to starting the operation, and they failed to find it. Unfortunately they had squandered a large capital investment without carefully testing the ground. Good pay was mined from properties both above and below their claims,[371] but their aimless search served only as a financial drain. In 1906 the assets of Columbia Hydraulic Mining were seized and sold by the Sheriff.[372]

Although poor drainage, due to relatively level grade was a problem for hydraulic miners on upper Spruce, some companies, such as the Blue Canyon Partnership, Consolidated Spruce Creek Placers, and Spruce Creek Power Company were good producers, and employed several hundred men each year.

Despite the absolute failure of the Gold Run dredge, in 1904 preparatory work was being done to launch another dredging operation near Blue Canyon on upper Spruce Creek, and once again the community's hopes soared. The company was the British American Dredging Company, a Philadelphia syndicate allied with the British Columbia Dredging Company.[373] Horse teams hauled the components of a Bucyrus type dredge from Atlin to the Company's property. It was a more powerful machine, expected to handle the tenacious overburden which had spelled defeat for the first dredge. An 11.5 km pole line was installed from the power house on Pine. Very little, if any, preliminary exploration was done on the ground, and once again in 1906, after only a few weeks' work, an expensive experiment was abandoned. Another dredge had failed, but in a different way from the one on Pine. It could handle the gravel, but it could not, for some reason, save gold.[374] This was the final test for the syndicate.

The dredging companies were not the only ones found lacking in the matter of planning. Northern Mines, managed by Richard. D. Fetherstonhaugh, brought a steam shovel to Spruce Creek a year before the Guggenheims installed theirs at Tar Flats. It was an ordinary railway shovel, however, and totally inadequate for the cemented clays and gravels that ran deep on Spruce.[375] After the shovel operation was abandoned in 1907, men working with picks and shovels proved the ground to be rich.[376] Despite failed plans and lack of experience, many Spruce Creek mines operations made profits, and by 1909,

*Flumes snaked through the creek beds to bring vital water to washing plants or sluices.*

Spruce had earned the reputation of "banner creek" of the area, a reputation which has been sustained.[377]

A long time operator on Spruce was Isaac Matthews, who worked a number of leases from 1911 until the late 1930's. Productive bench claims such as Poker, Joker and Croker were drifted by Matthews himself, or by lay miners. He frequently worked summer and winter employing as many as 60 men, and producing consistently high values.

Another fixture on Spruce was H. G. Marshall, a model of patience and perseverance, who mined his claims with pick and shovel for 39 years. He seldom produced less than 30 cents for every cubic metre, but that made him no better than a living wage. The work was a back breaking, solitary, grind which he continued until 1939 when he retired at age 77.

During the thirties individual miners continued to drift among the old workings that riddled the benches along Spruce. It was dangerous work since no records or maps had been kept in the early days, and miners had to rely on the memories of oldtimers, or hearsay accounts to track early workings. Many original tunnels had collapsed and there was always the possibility of flooding if old drainage adits had sloughed in. Nevertheless, Carl Lykergard had particular success in 1933-36, and in his last year produced a 44oz. nugget.[378]

The Depression brought large companies back to Spruce - Colpe Mining Company of Vancouver, Bratt, Morse and Company, and Columbia Development, controlled by A. R. Kaufman of Kitchener, Ontario. Atlin men such as Jack Tintinger, John Noland, sometime big game outfitter, and James Eastman, inventor and aviator, all played roles in the history of Spruce Creek.

## McKee Creek

McKee Creek rises near the headwaters of Spruce, and flows swiftly west for 13 km to empty into Atlin Lake, about 17 km south of the mouth of Pine Creek. Its banks are well over 30 metres high in places, and made up of solidly bonded materials.[379] Gold was first discovered there by Robert McKee and four others in October of 1898.[380] McKee had served with the Union Army during the American Civil War, before making his way north. He was a prospector and a Cassiar miner with a good background in the business. By the time he made his discoveries on McKee Creek, most miners had left the area for the winter, but the newly opened creek was fully staked by proxy before the year ended.[381]

In 1899, McKee worked his discovery claims, and with four men, washed out from 25 to 40 oz per week.[382] That winter he tunnelled into the bench where he again found good pay.[383] The next summer he installed a "baby" hydraulic plant, and by September, when he sold out, he claimed to have taken out almost 400 ozs with the small nozzle.[384]

During the first few years, most claims were acquired by companies who intended to set up hydraulicking operations. As a consequence there was little conflict between the companies and the individual miners on this creek, although the division of water caused some problems. Almost 100 claims, including Robert McKee's, were purchased in the fall of 1900 by the Nimrod Syndicate in the name of the Atlin Mining Company.[385] The company was managed by Richard D. Fetherstonhaugh, and the next year he acquired another large section on McKee, placing them in a dominant position on the creek.[386] Another sizable block of claims was owned by McKee Consolidated Hydraulic Limited, whose manager was F. T. Hamshaw.[387]

A third group of claims was purchased in 1900 by the Christopher Syndicate.[388] Charles Christopher was a well-known mining man who had worked all over North and South America. His opinion was highly respected, and he had little trouble attracting capital to finance his ventures.[389]

It was widely believed that the terrain on McKee, with its steep 11% grade,[390] and apparently good water supply, was well suited to hydraulicking. It was difficult mining, however, because the ground was hard, and when the very large boulders could not be removed by water pressure, dynamite had to be used. Mounting expenses, due in part to blasting, had operators searching for ways to cut costs. Cheap Oriental labour was one solution, but when Fetherstonhaugh hired Japanese labourers, he evoked an angry demonstration from the rest of the mining community. Still the yields were worth the investment, and sometimes the recovery ran from 2 - 3 ozs per man per day.[391] The first clean-up of 1902 produced about 180 ounces of gold, worth $3,000 after sluicing for only 50 hours.[392]

As early as 1900, it was apparent that McKee's abundant supply of water was a myth, and Christopher's group, restructured as McKee Consolidated, and the Atlin Mining Company became locked in a dispute over water rights. While the Atlin Mining Company held more ground, McKee Consolidated had rights to the best water supply.[393] For a few years Atlin Mining carried on its operation, while McKee Consolidated seemed more occupied with preparatory work. In 1904, the companies merged to become Amalgamated McKee Hydraulic Mining Company, managed by Hamshaw.[394] That year the company produced $52,000.[395]

Immediately work began on a bench which was apparently an old channel. Severe water shortages persisted and this meant season closures in mid August, and sometimes a maximum of two hour's hydraulicking per day. Still, profits were handsome, and the new company was successful. Indeed their results were sufficiently rewarding that in March, 1905, the *Atlin Claim* announced that the company was the first hydraulic company in the district to pay a dividend.[396]

Meanwhile McKee Consolidated Hydraulic Mining Company and Ginaca and Company were working in close quarters on the upper reaches of the creek, near a tributary named Eldorado. McKee Consolidated had no dumping space for their small hydraulic plant because a number of creek claims just below were not yet worked out.[397] Finally McKee Consolidated closed down in 1905 to allow Ginaca to work past the point where the operations of the two companies were in conflict. Ginaca used the other company's plant and water to move about 27,000 cubic metres of gravel, and posted good returns for the season.[398] The next summer McKee Consolidated pursued an older, deeper channel of the creek. The elusive channel remained hidden, and water supply was again a factor, but enough gold was recovered to pay expenses. In 1908 compromise was reached on the matter of water supply when McKee Consolidated and Amalgamated McKee came under joint management. The new management proved to be most satisfactory, and despite the necessity for a great deal of dead work that year, 25 men were able to produce $28,000.[399]

In 1909 the Pittsburgh British Gold Company acquired both companies and all other private holdings, McKee became the first creek in the area to be fully controlled by one interest. For the first time there was an excellent supply of water, and the company enjoyed an increased output.[400] During 1910 one clean-up yielded $25,000.[401]

In 1912, George Adams, who had previously been foreman, became company manager.[402] The following year the company was reorganized as the Delta Gold Mining Company, and with Adams still as manager, profits increased by 50%.[403] During this era Delta Gold Mining was enterprising, innovative and successful. They built a large reservoir about 7 km upstream, an improvement to the water supply which allowed for a substantially longer season. Delta Gold was also the first in the area to use manganese steel plates and riffles in their sluice boxes. These replaced wooden riffle-blocks and allowed for a more efficient rate of gold recovery.[404]

The impact of World War I was felt on McKee as elsewhere. Explosives were almost unobtainable, and the work force depleted. In 1919 the plant fell idle.[405] In 1921 some small scale prospecting was done, but the shareholders in Pittsburgh were reluctant to commit any large investment until another pay streak was found.

George Adams, in the meantime, had returned to his own claims in the Cassiar country, but in 1924 he came back to Atlin to work for Delta Mining, which had taken over the McKee properties.[406] For several years results varied as Adams continued to search for the lost channel, convinced it would reveal rich pay dirt. He drifted and hydraulicked with crews of 14 and 17 men, working two shifts, most summers, but reported only fair results.

In 1936 Atlin Gold Mines Limited was formed with George Adams as the senior of several partners. He managed the operation until 1938 when he was killed by accidentally stepping into the lethal jet stream of one of his own monitors.

## Boulder Creek

Boulder Creek was known as Musket Creek when Discovery claims were staked there by Pete Clark and C. C. Sutherland in 1898.[407] Immediately it took its place among Atlin's major creeks. Boulder is about 9.6 km long, and flows from the north into Surprise Lake, near the outlet to Pine Creek. Any miner who worked on Boulder Creek came to understand all too well the origins of its accepted name. The boulders which line the valley floor were even larger and more difficult to move than those on neighbouring creeks, and pay dirt was deeply buried on almost all areas of the creek, often up to 18.5 metres. This made for expensive and difficult mining but still the work was worthwhile, because once reached, the recoveries were rich.

For a time during 1901-02 Boulder was the creek where individual miners were most productive and successful. During those years 150 independent miners were at work during the height of the season.[408] In 1902, collectively they paid royalties on $70,000 worth of gold.[409] Claims 32, 33 and 34 below, owned by Black and Rayle had a nine day clean up worth $750, all of it coarse gold with many nuggets worth $5 and $10. No attempt was made to save fine gold.[410]

Hydraulic companies made an early appearance on Boulder. In 1899 the Pendugwig Syndicate,[411] acquired all the properties from the mouth up to 30 below, and numerous bench claims. Charles Christopher, well known on McKee Creek, was company manager.

The *Atlin Claim* announced formation of the Atlin Consolidated Hydraulic Mining Company on August 2, 1899. Notable principals of this group included Lieutenant-Governor McInnes, Richard McBride, as well as Alice Houghton, Charles Christopher, J. H. Brownlee, Capt. W. A. Spencer of the *S. S. Scotia*, and hotelkeeper J. H. Russell, all prominent Atlin business people. Almost immediately a half interest in the group was acquired by Count E. J. de Lamare, who, without delay, returned to Paris to float a company. To attract capital, and perhaps because he was a confirmed showman, he purchased a 48 oz. nugget from the Discovery claim to exhibit at the Paris Exposition of 1900.[412]

The result of de Lamare's energetic promotion was the company most closely associated with Boulder during the first decade of mining, the Societe Miniere de la Colombie Britannique. It commenced operations on rich ground in 1900. Originally the company was managed by the ubiquitous Charles Christopher,[413] but in 1901 de Lamare, and his French associates bought out Christopher and his partners.[414] Henri Maluin arrived from France to take charge of the operation, and he stayed with the French company, as it became known locally, until the end of the 1904 mining season. The French company faced all the attendant difficulties of mining Boulder, from annual spring flooding to the awesome barriers presented by enormous rocks, yet they scored some profitable years. In one week during 1902, they recovered 400 oz of gold, and over the entire year produced $100,000. For the next several years the company mined its ground systematically, running two shifts per day during the season, and reported good profits to its French shareholders.[415]

Although the company held hydraulic leases it spent several years working out the rich placer claims.[416] The grade on the lower parts of the creek is slight, and so there were problems of dumping space for a hydraulic operation. Beginning in 1903 this ground was leased to J. H. Black and W. Pollard Grant for a number of years. At first they struck extremely rich ground at bedrock where 450 metre drifts were made in an area 30 metres wide. Careful timbering was an expensive necessity under the heavy overburden. The steam hoist used to raise gravel 18 metres to the surface, consumed 700 cords of wood during the season.[417] Gold was recovered in substantial amounts, though, and left a good profit, even after the high costs of operating.[418]

In 1911, after several successful years, events began to turn against the French company. The pay streak was lost, and so were the labour and expendi-

*Count Emile de Lamare, known around Atlin as the Big Frenchman, had a hydraulic operation on Boulder Creek, the Syndicate de Lamare. (1900)*

AHS P-1248

tures of a whole season spent looking for it.[419] About this time, Thadie Obalski, who had managed the operation for a number of years, died suddenly.[420] The company seemed unable to recover from these disasters, and eventually it was sold to other French interests.[421] The buyers, however, encountered shortages of labour and capital created by World War I. In 1915 the company was sold again, and so ended the French association with Boulder Creek.

Over the next several years, Boulder was always worked successfully during the winter by individual miners, who were usually drifting on lays taken on company property. Year after year these operations paid winter wages and better.

Boulder Hydraulic Mining Company, under the managership of C. D. Newton, began working the upper reaches of the creek during the early years of the century, and their profits were generally good.[422] The French property was worked again in 1917 by Charles Miller who carried on both hydraulic and drift mining on the old French company claims on the lower creek area.[423] His crew worked underground over the winter and recovered $16,000 in gold.

Meanwhile on the upper portion of the creek Gus Anderson and J. H. Black worked with some 12 men.[424] In 1920 J. H. Black sold out to W. F. Gore who represented Chicago interests.[425] Determined to overcome the water shortage, this company built a 46 metre dam across the creek, which allowed them to operate two giant monitors.[426] Their success was limited, and except for the work Gus Anderson did alone, Boulder saw little work in the next couple of years.

In 1927 the outlook for Boulder began to change when Consolidated Mining and Smelting took a lay on a large number of claims, and commenced hydraulic operations with the benefit of good financial backing and expert advice.[427] In charge was Herbert Peplow Pearse, a distinguished and very experienced Australian miner, who had been a pioneer on Birch Creek.[428] That year there was a flurry of work, as the company built a flume, installed a pipe line, and built new sluice boxes.[429]

This was the beginning of many seasons of work on Boulder by Consolidated Mining and Smelting. In 1930 MacLeod White, a mining engineer who would have a long association with Atlin, replaced Pearse, and continued the company's ambitious program.[430] Bedrock was finally exposed in 1934, and the company turned to working underground on the lower creek.[431] Employing about 12 men, the company had several good years. In 1936 the underground operation was abandoned, the camp was moved further up the creek, and another hydraulicking operation became the company's focus.[432] Time and a persistent search proved that Boulder had by no means surrendered all her gold to the early miners.

## BIRCH CREEK

Birch Creek joins Pine from the north about 3.2 km below Surprise Lake. It originates in the ravines of Mount Leonard, and skirts the eastern toe of Munro Mountain. It drops over 600 metres in a short descent through a narrow and rock strewn channel, before spreading into flat swampy ground at its mouth. A small tributary, Plum Creek, enters high up on Birch from the east.

Initially, because of the great depth of its bedrock, very little work was done in the early years by individual miners.[433] In 1899, however, a large number of claims, including Discovery, were taken up by an English group, the Atlin Lake Mining Company, which was managed by Mostyn Williams. Because of the nature of the ground, this company made immediate plans to begin hydraulic work,[434] and established the first hydraulic operation in the area.[435] The experiment proved quite worthwhile, and in 1900 Williams was successful in his bid to buy up most of the remaining claims on Birch. A small dam was built to raise sufficient water to operate two monitors.[436] There was an average work force of 16 men.[437]

The company changed hands in 1902 and again in 1903, and there followed a great deal of dead work before the rich deposits previously indicated could be reached.[438] Once again persistent water shortages hampered the operations of both the hydraulic company, and the individual miners who had located on the upper creek.[439] In 1905 Atlin Lake Mining Company had to close down on July 1 when the water simply dwindled away.[440] Ironically, the following year flooding caused by spring freshets did considerable damage.

That year the mine was sold to the Dominion Trust Company who gave H. Peplow Pearse, a lay on the property. He became a long standing figure on Birch Creek. For the next two decades he mined through good years and bad, with the unswerving faith that rich pay was waiting only for him to find it. The first season, although water was in short supply, Pearse's miners moved about 16,000 cubic metres of gravel, and recovered gold amounting to $5,000.[441] The gold recovery from Birch improved as the company moved into the upper areas of the channel. In 1908 they reported $10,000, the best showing from Birch to date.[442] For a while Pearse's fortunes increased annually as his monitors tore at the gravels on higher claims.[443]

A very large nugget of almost pure gold, weighing about 73 oz was found on Birch in 1913.[444] Given its purity, it was perhaps the most valuable ever found in the district, and it definitely gave credence to the belief that Birch still held vast riches, which awaited only the solution of the water problem.

In 1914 extreme water shortages, and a shift of the old channel away from the existing creek slowed mining and returns dropped.[445] Pearse was about to face some lean years as the situation on Birch took a turn for the worse.

There was a tragedy on Birch in 1916, when an old timer, Frank Charlebois, ignored warnings from the Gold Commissioner, and persisted in mining under an overhanging bank. Inevitably it collapsed and killed him.[446]

Throughout the war years Pearse stubbornly continued his annual battles with Birch Creek's unpredictable water supply and unyielding gravels. Some years he worked with a crew of only two or three men. Returns were generally poor or nonexistent until, at last, in 1922 bedrock was exposed, and this was Pearse's most successful year yet.

Pearse finally left Birch Creek in 1927 and turned his attention to Boulder Creek. Only a few small one and two man operations continued on Birch Creek. During the thirties Billy McDonald, George Gash, and George Bouquist worked separately and successfully on the creek.

Sometimes a miner's life became entwined with one creek, and such was the case with Joe Yonaites and Birch Creek. For a while he raised pigs behind a piney hill, and lived in a one room cabin sunk into a slope, just east of Birch. He sometimes travelled in winter on huge homemade skis. Eventually he left the skis at the little cabin and moved to a solid two story log house which he built on a rise not far from the swampy mouth of the creek. Windows overlooked the barn and fields of a comfortable farm. Farming and Birch Creek gold provided him with the means to acquire an impressive number of Atlin buildings during the economic slump of the 1940's. He bought at rock bottom prices, and became a real estate tycoon in a town whose future appeared bleak, except perhaps, to him. Gum Boots Joe, as the locals nicknamed him for his habit of wearing rubber boots in all seasons, was a man of many ventures, not the least of which was successfully finding the distinctive coarse gold of Birch Creek.

## Otter Creek

Otter Creek was almost fully staked in 1898, but placer miners who tested their claims in 1899 were disappointed. In fact, Otter became something of a laughing stock to other miners who were reaping rich returns on Pine and elsewhere, and the expression "light as a clean-up on Otter" was coined.[447]

Otter Creek originates near the upper reaches of Spruce Creek, and flows in a northerly direction around the flanks of Spruce Mountain to enter Surprise Lake opposite the mouth of Boulder Creek. Along its course it goes through two canyons. Downstream from the upper canyon bedrock is buried deeply beneath glacial gravels, and individual miners quickly abandoned work in that area.

A 25 year old Irish cheechako miner found out the upper part of Otter Creek was not easy diggings either. Alfred Carmichael arrived in Atlin in the spring of 1899, only to learn that almost every creek had been staked from source to mouth. To his relief he finally located some open ground on Otter above the upper canyon, which he promptly staked without even prospecting the ground. Sometimes home ties, longings, humour or the loyalties of the miners were revealed in the names they gave their claims, and Carmichael commemorated the name of his old boarding house in Victoria when he recorded the Roccabella Group on behalf of former fellow residents. Carmichael set to work on his new ground, but was soon forced to face the reality that the bedrock and the pay streak were too deep to mine without capital. Muscle power alone would not be sufficient to persuade Otter to surrender its riches.[448]

That first year's toil on Otter's upper claims though, was not a complete loss. The gold was tantalizingly coarse, and there were some large nuggets. One worth about $250 was purchased by the British Columbia government for display at the Paris Exposition of 1900.[449]

Carmichael met John Eaton Moran during the summer of 1899, and they formed a loose partnership that lasted for seven years. Moran had already consolidated claims, and together they set out to secure a number of additional hydraulic leases, and to raise the funds necessary to develop the property. Carmichael spent part of the summers of 1899 and 1900 working on a survey gang. Meanwhile their property was worked by George Alverson on a lay. Alverson was a knowlegeable miner, and later the young pair were happy to give him charge of their hydraulic operation in the spring of 1901. During the first year of operation, 430 metres of ditch was constructed, and the 5.7 cubic metres of gravel that was washed, paid well. The crew of four worked in two shifts, day and night, amid hordes of mosquitoes.

During the winter of 1901-02 the partners, working alone, drifted into the gravels above bedrock and below glacial clays. They systematically prospected in every direction while inexperience and daring put them at the edge of several near disasters. Moran was a good natured and easy going fellow, who never backed away from a challenge. He solved the pesky problem of thawing frozen dynamite by either hanging it behind the stove pipe, or warming it in a bread pan of water, a process that could have given them dynamite headaches or worse, had they spent much time indoors.

Otter Creek began to gain respectability when their prospecting results of the 1902 season were good enough that J. H. Brownlee and some partners formed a syndicate to work a 5 km stretch of creek and bench claims, leased from Carmichael and Moran.[450] Over the next two years, the group, known as the Otter Creek Hydraulic Mining Company, did a great deal of work. They built endless flumes and ditches, and installed 600 metres of pipe in descending sizes to provide pressure for three large monitors. In preparation for the 1903 season, they built wagon roads and a camp with a large dining hall, a bunk house, a cook house, a blacksmith shop, tool and powder houses, a cabin for the foreman and a stable. Mining began in July, but just as they reached the old channel previously located by Carmichael and Moran, the water supply dried up.[451] Unfortunately these problems were not overcome, and in 1904 the company dropped its lease.[452]

In 1905-06 Carmichael decided to spend the winter on the Otter Creek property. On October 31, he heard the far off lonesome whistle of the *Scotia* as it churned away from the Pearl Avenue wharf for the last run of the season. Carmichael settled in to do some drifting, to read Shakespeare, and to indulge in dreams of Christmas in London the following winter. His companion was a cat who went with him to the mine every day, and followed him in and out of the tunnel with every trip of the wheel barrow. He sometimes walked down to Surprise City, where he received his mail at the Surprise Lake Hotel, run by John and Catherine Roxborough.

Rev. F. L Stephenson was a good friend of Carmichael's, and in early December they travelled to Atlin together with a cutter and horse hired from the Surprise City hotel, and there enjoyed some early Christmas cheer. January was so extremely cold that the thermometers on Otter and other creeks remained frozen, so the lowest temperature during the cold snap was never known.[453]

In the summer of 1906, Carmichael and Moran once again worked their group of claims, known as the Otter Creek Development Company. In 4½ months with 5 men, they removed a large quantity of overburden, and washed out $4000.[454] These results showed promise, but Carmichael was restless, perhaps as a result of his solitary winter. In the fall, he left for greener pastures on the Queen Charlotte Islands. In his possessions was one 19 oz nugget of gold, worth about $300. He had picked it out of the sluice box at the close of the season, and it represented all he had to show for seven years on Otter Creek.[455]

Moran, more dedicated and persistent, stayed to manage the Otter Creek Development Company, but even with storage dams to combat the water problems, results were disappointing. Finally in 1914, when reworking old ground, the gold recovery doubled the two years previous.[456] Gold was apparently present, but the water supply, despite improvements, was insufficient, and even the most optimistic miner was taxed to carry on in the face of war time shortages, and in 1917 operations were suspended.

Meanwhile nothing had been done on the lower creek until 1905 when Henri Maluin and M. R. Jamieson began prospecting close to the mouth. They reasoned that mining might be more feasible if bedrock could be located at lake level. Their search continued for two more years,[457] and in 1907 Maluin formed and managed the Otter Creek Hydraulic Gold Mining Company with W. H. Brethour as superintendent. The next year they began hydraulicking with two 15 cm nozzles,[458] and discovered there was a considerable amount of gold in the surface gravels, but bedrock and the old creek channels were very deep. In 1910, two upstream reservoir dams broke within a month, destroying ditches, about 185 metres of sluice boxes, and the water supply for the season.[459] Not discouraged, Brethour brought in a keystone drill in 1911, and at last located what he thought was the elusive bedrock at about 13.5 metres. As he had hoped each drill hole proved the existence of gold at this level.[460] On the strength of this discovery, a considerable amount of development work was done in 1911-12, employing as many as 30 men.[461]

In 1914, the company was reorganized as the Mines d'Otter, with Henri Maluin as General Superintendent and W. H. Brethour continuing as Manager. Just as they were on the point of exposing the proven gold deposits, war broke out in Europe, French financing dried up, and mining supplies became unobtainable.[462] Maluin was a French reservist, and had to return to his country, leaving the mining operation to Brethour. At this point the company had 2900 metres of supply ditch, three reservoir dams, and a sawmill representing

*Bags, bottles and a bowl full of gold from the Noland Mine on Spruce Creek.*

an enormous capital outlay,[463] and it was imperative that work continue. Despite the difficulties, the final clean-up of 1916 was most rewarding,[464] and in 1917, for the first time, operating expenses were met.[465] By then a labourer was paid $7 per day,[466] and with all other costs inflated by war, it would be several more years before the mine paid its way and profited.

J. E. Moran closed down his operations on upper Otter Creek after 17 years, and went to work for Maluin. Throughout the twenties the company experienced several good years, but true bedrock was never reached.

Except for the war years, Henri Maluin travelled each year from France to remote Otter Creek to oversee mining operations. He and his wife brought their infant daughter, Rene, over the Fantail Trail in a hand sled in 1905. She would spend nearly all of her girlhood summers until 1925, in the midst of the gold mining on Otter. The family had a two story frame house near Surprise Lake. Rene lived an outdoor life, a tomboy existence, and she became a good shot with a .22 rifle. She often hunted grouse for her mother. A half century later she was drawn back to Atlin, and found the weathered shell that had been her family home. She purchased a small cabin in Atlin, and spent a few last summers in the land of her unusual childhood.

In 1927, Maluin remained in Europe because of ill-health.[467] The next year the company was taken over by the Compagnie Des Mines d'Or du Canada,[468] which became known locally as "Coffee Can."[469]

With Moran in charge of operations, the new company attempted to do what Maluin had failed to do in 1907 - locate bedrock at lake level.[470] Moran was also determined to solve the water shortage once and for all. Drawing on his 29 years of experience on Otter, he developed an ambitious scheme which would become the Union Creek Ditch. To execute his plan he hired Walter Sweet, who had stampeded to Atlin along with most of the White Pass railway builders in 1898. Sweet was a self-taught miner with considerable ingenuity.[471] He had gained experience building a ditch on Pine two years previously, and he knew how to build Moran's ditch.

For almost 16 km his crew carved the Union Ditch around the rugged flanks of Idaho Peak to carry the waters of Union, Wright, Idaho, and Casino Creeks to a dammed reservoir high up on Otter.[472] Wooden flumes lined with galvanized iron spanned areas of rocky terrain where trenching was impossible. Horses delivered all supplies to the site over steep and broken ground.[473] Lumber came from the company's mill, and flat sheets of iron were bent to shape at the site.[474]

By 1932, with ample water, the "Coffee Can" monitors had torn into three separate layers of gold bearing gravel,[475] and profits were substantial, but bedrock remained mysteriously buried. The company maintained one of the largest camps in the district, and employed 25 men during the summer season.[476]

Late in the fall of 1933, just after the monitors had been shut down, J. E. Moran was tragically killed when the drift he was working, caved in. Once again the loss of a key man seemed to stagger a company. Although Walter Sweet took over as foreman, the French group never again operated on the same scale. By 1936, Otter Creek was in the hands of a few laymen who worked underground.[477]

## Ruby Creek

Ruby Creek flows around the eastern side of Ruby Mountain and enters Surprise Lake about 5 km from its mouth. An unusual feature of the Ruby Valley is the presense of volcanic debris, in places 43 meters deep, over glacial morain.[478] Gold was discovered on the creek in 1898 by John Kirkland and Alec Brown[479] who also discovered small rubies buried deep beneath the lava flow.[480]

The early individual miners were quickly discouraged because of the great depth of bedrock. Count de Lamare's syndicate attempted to mine a number of claims at the mouth, but found the venture unprofitable.[481] The first hydraulic leases were taken up in 1900 by P. W. Law who organized the Atlin-Ruby Creek Hydraulic Mining Company with $48,000 capital. Most of the capital was spent unwisely, little work was done, and the leases were cancelled.[482]

There was little activity until 1903 when the Ruby Creek Partnership was formed by the discoverers Brown and Kirkland, joined by two others, a Mr. Coutts and James Lumsden, to mine 900 metres of creek bed. In 1904, under the direction of foreman, Robert McKee, bedrock was struck at 14 metres, but gold was found in the 7 to 9 metres of the gravel above it, making the mining proposition a good one. However, a good deal of water was found flowing on bedrock, and this made mining difficult with the methods the Partnership was employing.[483]

For a number of years the partners looked for the financing that would allow them to install a hydraulic plant to mine their ground profitably. Finally in 1908 the property was taken over by a group of Seattle investors represented in Atlin by T. M. Daulton.[484] The company, known as Placer Gold Mines wasted no time in installing modern hydraulic equipment and the necessary ditches and flume. An average of 20 men were employed throughout the season, and hopes were very high that this well capitalized company, working on proven ground, with a good water supply and a steep grade, would be a profitable venture.[485]

Unfortunately though, Placer Gold faced a good deal of dead work and the water supply was not always reliable.[486] The company did not doubt that gold in paying quantities was present, and so work continued with crews averaging 18 men. In 1914, they installed manganese-steel plates in the sluice boxes improving the efficiency of the operation and production was increased significantly.[487] In 1915, their operation was moved onto richer ground, and there profits were most satisfactory.[488] They managed to maintain their crews, even during the war years, when mining on other creeks was in a state of limbo, and in 1917 Placer Gold Mines reported a recovery worth $80,000.[489]

The company's successes during the war and immediately afterwards resulted in a new focus of interest in Ruby, and during the winters of 1915-17, a number of men began drift mining on the upper creek.[490] In 1919, Placer Gold Mines lost the paystreak under basalt lava too deep to be removed hydraulically, and Daulton moved the plant higher up the creek above the lava.[491]

T. M. Daulton died in 1920. He was an amiable gentleman and a highly respected miner, who had installed the first hydraulic plant on Ruby, and worked on that creek for 12 years. Management was assumed by C. E. Leonard and C. H. Titus. During 1920-24, Placer Gold Mines enjoyed notable profits. In 1922, Ruby Creek was the camp's top producer, with $58,000.[492] By 1924, it appeared the upper creek had been worked out, and the company moved its equipment once again to lower ground.[493] Two year's work resulted in only the promise of good paying results, and the company reorganized as the Lake Surprise Mining Company in 1927.[494] In 1928 the company had ceased operations.[495]

In the spring of 1929, Peter Matson and Louis Schulz had taken a lay on the same lower creek ground. They reckoned that the ground had never been worked thoroughly, and they set out to hydraulic on the creek bed, and drift beneath the lava flow.[496] By 1932, they were running one shift of five men, with only moderate profits.[497]

During the war years, and the early twenties, when Ruby Creek ranked high among Atlin producers, a number of smaller operations began to mine above the ground held by Placer Gold Mines. Emil Turnquist, John Roxborough, Charles Miller, and Arno Krumbeigel all worked the upper reaches of Ruby Creek at various times during the twenties and thirties.

## Wright Creek

Discovery Claims on Wright Creek were granted in 1898 to H. O. Blankenship, J. & C. Welsh, and C. Wright.[498] The creek originates on the slopes of Idaho Peak and flows into Surprise lake from the south about 3 km east of Otter. It had a good fall, and great hopes were held for its hydraulic potential in the early days.

Early stakers on Wright fell victim to disputes over claim sizes and the claim jumping that plagued the district until ownerships could be resolved by the Special Commission of 1899. Nevertheless, a significant amount of work was done that season on about 50 claims. The results were so good that in proportion to the amount of ground worked, Wright produced more than any other creek in the district.[499] In absolute terms, the miners on Wright in 1899 paid royalty on 3,729 oz, which was second only to Pine.[500] The gold won from Wright was very coarse, and highly prized. In September 1899, a 44 oz nugget, the largest found to that date, was picked up on Wright Creek by A. Moorhouse.[501]

Some consolidation of claims had taken place, by 1900, but most of the mining was still being done most successfully by individuals. Production fell off sharply in 1901, and the heyday of the small scale miner was over.[502] The upper creek produced only patchy returns until about 1908.

The Pendugwig Syndicate made an appearance on Wright in 1900 with plans to commence hydraulicking near the mouth of the creek.[503] Sinclair and Brown's sawmill received an enormous order from the company for materials to build a camp,[504] a two kilometre long flume, and sluice boxes.[505] A wagon road was built from Surprise City to Wright.[506] The familiar overburden problems and water shortages spelled failure for the Syndicate, though, and by 1902 the operation had been abandoned.

That same year, the English Counties Hydraulic Mining Syndicate spent $7,000 in preparation for work which they planned to begin the following year.[507] This company's manager was a Victoria lawyer, C. Dubois Mason, who was just one of many professional men lured north by the promise of gold. No doubt he was disappointed by the meagre returns shown for his efforts, and, in 1904 management was handed over to Messrs. Gierke and Company, a change which did not improve the company's luck. For five seasons they were singularly unsuccessful. With a crew of 12 men, and an outlay of $2 200 in 1905, they were able to show the shareholders just $100 worth of gold.[508] Just the same, in 1906, perseverance paid off, and the company struck rich pay dirt.[509] The following year was also extremely profitable, but from then profits dwindled. In 1915, the mine operation was shut down in midsummer when most of the owners and employees enlisted for military service in Europe.[510]

Wright Creek lay virtually dormant until 1920 when another good pay streak was located by a small group of individuals.[511] It was not until after 1929 and the generally renewed interest in mining, that Wright ranked again as one of Atlin's important creeks. By this time though, attention centred on the upper reaches of the creek. Lyman Hodges, Fred McLennan, Frank Brown, John Hyland, J.E. Moran Jr., and Vic Lindahl all worked on small hydraulic operations throughout the thirties.

## The O'Donnel River and its Tributaries

The O'Donnel River is more than 80 km long, and flows from a large plateau covered deeply with a gravel wash. Its headwaters consist mainly of Bull, McKinley and Feather Creeks, but it gathers water from many small tributaries as it flows in a fairly level southwesterly course to enter Atlin Lake about 35 km south of the Atlin townsite. Known also in the early years as Moose or Caribou Creek, the O'Donnel River is by far the largest of the gold bearing streams in the area, with an average width of 18 metres, and it certainly rates the designation "river."

Claims were staked on the O'Donnel and its tributaries in 1898. Little work was done on the O'Donnel itself, but fair values were found on Dixie, Burdette, Feather, Bull, Wilson (also known as Slate), Fox, Caribou and Cavell Creeks.

From 1902 to 1905, there was a flurry of interest in the O'Donnel, Dixie and Bull, but supplying mines so far from Atlin presented problems with no easy answers. About 1904, the Guggenheim Company decided to mine on the O'Donnel, but were faced with the job of getting supplies to their camp, when no road existed. It was possible to travel with a steady team and a supply wagon above timber line, along the upper slopes of the Johnson Range, if it was done before erosion from spring runoff cut too deeply into the hillside. It was a tricky route, however, and A. M. Morris, who represented the Guggenheims, hired Alfred Carmichael to locate a wagon road, and to survey the company's property.[512]

In 1905, about 60 leases on the O'Donnel and Bull Creek were bonded to O. B. Perry of San Francisco, who had plans to install a dredge. After a season's prospecting, the idea was apparently abandoned.[513]

*The miners addressed problems by holding mass meetings. First Street, Atlin. (1899)*

AHS P-681

By 1906, the major company working the O'Donnel system was the O'Donnel Syndicate, owned and managed by Robert McKee, who was working his Gold Hill leases.[514] In the spring of the year, McKee experienced a serious set back when his winter dump (ore which had been stockpiled at the surface, but not yet sluiced) was swept away by an unexpected flood. This loss was offset later when sluicing provided substantial returns amounting to about 4 oz. a day. Since McKee's prospects looked good, and because there was great enthusiasm at the time for dredging, he looked for a way to install either a dredge or a steam shovel,[515] but the idea was never a practical one, and it was not carried out.

In 1911, McKee's new company, the Canadian-Alaska Exploration Company, began work with 15 men. They built a stable and two bunk houses, installed a steam-powered sawmill, and steam pumping and hoisting equipment. A 900 metre supply ditch was dug, and a shaft 3 by 4 metres wide, and 7 metres deep was sunk.[516] The steam pump could not control the ground water flowing into the shaft, and the problem persisted when larger pumps were installed. The next year McKee changed tactics, and, using a Keystone drill, struck bedrock at 28 metres, a depth much greater than he had anticipated.[517]

Working at such depth was clearly an expensive proposition, but the ground was proven, and so, in 1913, it was taken over by J. M. Ruffner, who employed 30 men to do some very thorough prospecting and development work.[518] Known as O'Donnel Placers, Ruffner's company poured $62,000 into further work in 1914, and although returns were excellent, they did not cover expenses for the season.[519]

When Ruffner left the river in 1914 to attend to other interests, the management of O'Donnel Placers was taken over by Frank Breeze, but the fortunes of the company began a steady decline over the next few years. The bed of the O'Donnel was particularly flat, and the shortage of water, despite a large supply ditch system, was a recurring problem. In years when returns were not good, the problem was always a lack of water. 1916 was not a good year, and in 1917 the company did not operate.[520]

Wilson Creek, or Slate, depending on the whims of typographers, was staked entirely in the first rush, but abandoned soon after. In 1906, the discoverers, Andrew Grier and Jack Tintinger, began freighting supplies to the creek via Blue Canyon and the high route along the Johnson Range. Their activity was enough to spark a small stampede of eager miners.[521] Before long 90 miners, including Kenneth McLaren, had claimed ground that had laid open for two years.[522] Most work done that summer was in the nature of prospecting, but on the Discovery claim, and a few others nearby, it was believed that the operators were taking out from $25-30 each per day.[523] The following years returns were not as good, and even though the creek was worked by about 30 men in 1908, few made good profits. The pay streak was hard to locate in a valley which ranged from 300 to 600 metres in width.[524]

1912 was the year of the big stampede to the O'Donnel area. Late in the season rich ground was struck on a bench 9 to 15 metres above the river bed.[525]

There were also new discoveries allowed on Burdette, Silver, Trout, Johnson, and Moosehorn Creeks, which emptied into the White Swan River (now called the Teslin River). Many of these were staked by Indians who said there was gold from the "grass roots" down. They made a good business of locating claims which they did not pretend to prospect, before they transferred them immediately to willing white miners.[526]

In 1913, about 100 hydraulic leases were granted in the O'Donnel area,[527] and another 70 were granted in 1914.[528] Results from hydraulic operations over the next few years were patchy, and water problems prevalent. A few individuals, however, persisted and made good livings.

Nate Murphy worked his O'Donnel claims for some 30 years. Often his stepson, Melvin Beckman, helped him drift in the winter months, and wash the "dump" in the summer.[529] Murphy's work was systematic and steady, and in 1932 he averaged $5 for every cubic metre he brought out of the ground.[530] During the years following World War I, there were others who worked and lived on the O'Donnel, all neighbours of Nate, and his wife Marie, who regularly treated them to her home cooking. Tom Prpich, Marco Pini, Marco Viola and Louis Piccolo were part of the O'Donnel community. Although it had run undisturbed for many years, by 1934 the O'Donnel was one of the busiest mining sites in the area.[531]

Lost to history are the names and locations of dozens of creeks that were found, tested and staked by the first prospectors in the district. Local names too, were often changed by map makers, and other creek names simply faded from memory after the early frenzy of discovery had past, and the reality of disappointment set in. Interests focused on easier to reach locales as trails grew in and cabins mouldered. No doubt there is gold in many of the "lost" creeks, but finding them with modern maps might be difficult. There are no old timers to question now, and if today's seasoned miner were asked, "Where can I find gold?", his wry answer might be, "Gold is where you find it," and the evidence would not prove him wrong.

CHAPTER 7

# The Japanese Incident

In the 1860's, double rows of steel rails were laid with amazing speed across the United States to join two seaboards with swift freight service and comfortable passenger coaches. The boom to transportation made distances diminish, and gave a nation something to celebrate, but hidden in the euphoria was a disturbing fact. A brief era of frenzied track laying had spawned a century of racial tensions and animosities that were destined to ride the rails into Canada.

American railroad builders had employed Chinese from the California gold fields, where the diminutive men had proven themselves the physical equals of brawny Europeans. At first, a few were brought in to replace striking workers. More were hired as their value to the project was recognized. The Orientals never struck, never refused a job, did not drink or carouse, and proved to be daringly capable with explosives. Soon they were brought by the thousands from China.[532]

The first Chinese came to British Columbia in 1858. They were miners who left the placer tailings of California to sift the gold bearing sands of the Fraser River. When gold was scarce they settled in the cities. By 1878, when Chinese were banned from employment in public works, there were about 3,000 of them in British Columbia.[533] A decade later Victoria's Chinese community had a range of tradesmen and professionals, which included merchants, importers, contractors, druggists, doctors, bankers, manufacturers, craftsmen, and of course launderers and restauranteurs.[534] Clearly they were capably self supporting, but British Columbia's population of 35,000 whites wanted to remain "all white."

Despite the fact that the Canadian Pacific Railway, which brought prosperity to the Province, could not have been completed without Chinese labour, British Columbia passed Exclusion Acts against all Orientals in 1884, and again in 1885. These Acts were specifically aimed at preventing them from working in the mines and the forests, but on both occasions Ottawa vetoed the legislation.

The antipathies of the white population encompassed the Japanese, who had begun to settle in British Columbia in significant numbers during the 1890's. By the turn of the century, most North Americans drew no line between Chinese or Japanese; the Oriental issue was all inclusive. It crept north like a hungry dog, following those who moved from one gold strike to the next, and along strings of construction and mining sites.

Anti-Oriental sentiments were a hallmark of turn-of-the century British Columbia. Chinese were called the "beardless children of China," or "celestials."[535] Newspapers of the day referred to Japanese immigrants as "Japs" or "Nippons" and Trade Unions openly campaigned for a halt to their immigration. They were accused of being unclean, immoral, unwilling to integrate, and of sending most of their money home. The real outrage centred on their willingness to work for low wages. In some cases, it cost less to hire Japanese than to keep horses. Inevitably though, even in the face of heated protests, companies continued to hire the cheap and steady Oriental workmen.

One such company held a large number of promising claims on McKee Creek. The Atlin Mining Company was run by Superintendent Richard D. Fetherstonhaugh, and the owners were a group of Englishmen who did business as the Nimrod Syndicate. After two years of promising results on McKee Creek, they decided in 1902 to invest $30,000 in a hydraulicking operation to wash down the creek's massive overburden. With due caution, the English directors instructed Fetherstonhaugh to hire Japanese labour for their expensive new venture.

Fetherstonhaugh could not have doubted that the decision to employ orientals would incite an uproar. Just a few months earlier George Brackett had considered hiring Japanese to work at Willow Creek, and sparked an immediate outcry. The *Atlin Claim* voiced the hope that the ship carrying these controversial passengers might sink.[536]

As soon as the news of the Atlin Mining Company's plans reached the creeks, the miners formed an association, and planned a confrontation. A delegation headed by Jack Kirkland, the hotelier, met with Fetherstonhaugh who was in no position to reverse the decision, even if he so wished. The Japanese were already on their way to Atlin, but he was prepared to make a concession. He said his company would import no more cheap labour, and the Japanese would be sent back by the end of the year if the ground proved profitable.

Kirkland was not satisfied with the offer, and spoke for the entire community when he warned of further resistance to any Oriental labour in Atlin.

The Board of Trade endorsed the anti-Japanese stand, while Dr. Henry E. Young tried to persuade the angry miners to contribute funds to reimburse the company for the losses incurred if the Japanese were returned to Vancouver.[537]

On the evening of March 20, 1902, 17 bewildered Japanese, each wearing identical winter boots and parkas, and carrying identical .44 rifles, arrived in Atlin after a cold trip across the ice from Carcross, on two horse drawn sleighs. Early the next morning they were rushed to McKee Creek and settled in bunkhouses. The deed appeared to have been done. However, the day before when the citizens of Atlin saw that the newcomers were armed, they were furious. The startling news spread to Discovery, and up and down every creek. The next day a miners' meeting was held in Discovery, and dozens of irate miners decided to march to Atlin.[538] A somewhat larger, but orderly meeting reconvened on the corner of First Street and Discovery Avenue.

To forestall the trouble that was clearly brewing, J. A. Fraser, then still President of the Board of Trade, met with Fetherstonhaugh. He returned to the gathered miners with the news that this time Fetherstonhaugh had promised that the Japanese would be gone by June 1. He had also guaranteed that should another employer attempt to bring in Japanese labour, he would immediately dismiss his employees, and assist in turning back the newcomers.[539]

Fetherstonhaugh's lame promises were not acceptable to the miners, who resolved to march to McKee en masse, but unarmed, to rout the Japanese. Marching in single file, 150 men set out immediately along the 19km trail. Ned Thain, who was Acting Gold Commissioner, in the absence of Joseph Graham, refused to take sides in the dispute, but he sensed the ugly mood in the camp was likely to become dangerous.

Determined to keep the peace he swore in twelve special constables.[540] Young Guy Lawrence was one of them who nervously wondered what might happen if the situation flared. Later he described the bizarre scene. Thain positioned himself beside the trail, and began reading the Riot Act. He repeated it over and over again as the single file of men trudged down the lake in resolute silence. No one stopped to listen, so each must have caught only a few words of Thain's peculiar monologue.[541] Meanwhile Constables Walter Owen and Billy Vickers, overtook the marchers with their dog teams and mushed to McKee Creek, where they waited for the grim procession.[542]

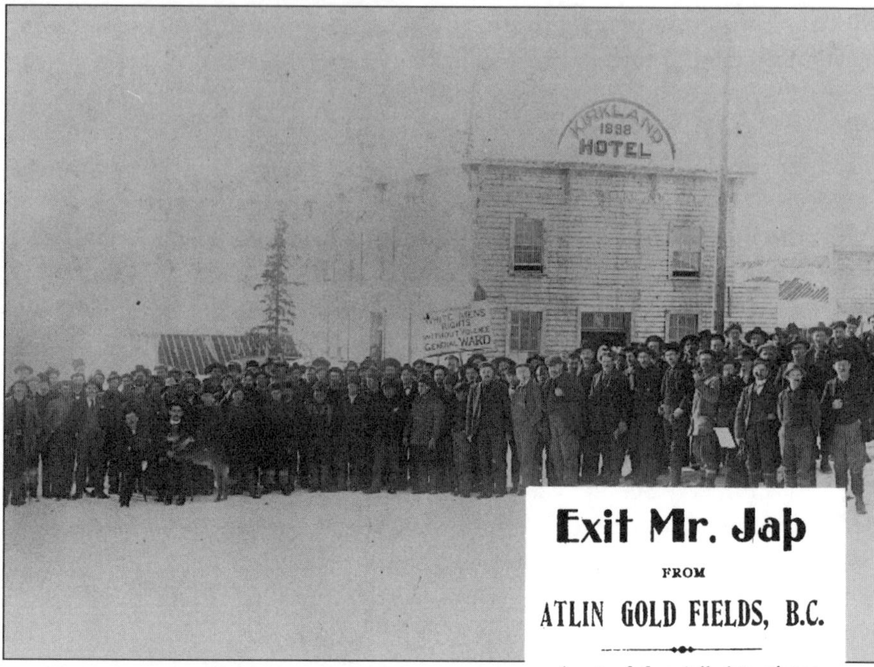


Vancouver Public Library #30669

# Exit Mr. Jap

FROM

## ATLIN GOLD FIELDS, B.C.

A set of five full-plate photos, marked consecutively in order of procedure of above happenings, can be obtained upon receipt of Two ($2.00) Dollars from

MUIRHEAD BROS.
SPRUCE JUNCTION,
ATLIN, B.C.

NO AGENTS.

Atlin Claim, April 5, 1902

AHS-P-544

The unsuspecting Japanese were sleeping in their bunkhouse when the miners arrived at 6 p.m. They listened fearfully as an interpreter passed on the miners' message, and explained that they were expected to leave early the next morning.[543] For some reason Fetherstonhaugh was not at the creek to intervene.

With the ultimatum delivered, the Japanese were allowed to go back to bed, although it is doubtful that many of them slept. Outside the miners built fires and gathered around for a long night's vigil in the snow.[544] Owen and Vickers stood by. In the gray morning light, at 5 a.m., the miners were loading the immigrants' luggage onto a sleigh they had brought for that purpose, when Fetherstonhaugh arrived. He was furious at the presumptuous action. By then the terrified Japanese most assuredly wanted only to escape the wrath of the irate men, and the strange, cold, creek valley, but Fetherstonhaugh hurried them back inside the bunkhouse, where he told them their safety would be guaranteed if they would stay.

Outside, a heated discussion took place, and the mob flatly refused to leave without the Japanese. Eventually a small committee, headed by H. E. Ward from Discovery, met with Fetherstonhaugh. The watchful constables were also present. It was clear the miners had the upper hand, and finally Fetherstonhaugh backed down. He agreed to dismiss 15 of the Japanese but asked to keep the two cooks. In return, the newly formed Miners' Association agreed to raise no objection if he brought in white labour at $2.00 per day. In fact, Ward promised to use the influence of the Association to find men who could complete the Atlin Mining Company's planned work within the cost estimate allowed for the Japanese workers.[545]

The entourage of confused Japanese and tired miners made their way back to Atlin. The little group of 15 Orientals were a sorry sight as later they set out on foot across Atlin Lake to Carcross, farewelled by a jeering and unfriendly crowd waving banners from the shore.

Considering the intensity of the feelings aroused, the unpleasant incident was settled with remarkable restraint. Ward effectively controlled the miners so neither the Provincial Police, nor the Special Constables needed to intervene. Undoubtedly the miners realized that they had the unofficial support of the government,[546] and the wisest course of action was to stay on the right side of the law. They managed to force their will by the sheer weight of numbers.

There were repercussions for some time about the events of that spring. The British failed to understand the intensity of Canadian workers' resentment, and an English newspaper warned that without cheap imported labour,

Canada could never be properly developed.[547] Fetherstonhaugh was later dismissed by the Nimrod Syndicate, and he fought a lengthy and involved court battle against the company on a number of charges, including wrongful dismissal.[548] The reasons for the dismissal are not clear, but obviously the English owners would not have been pleased at his capitulation in the face of public pressure.

The miners won the first round, and for the next few years there were no challenges to their firm position. For a while the Gold Commissioner, J. A. Fraser, was able to ensure that every hydraulic lease granted included a clause prohibiting the employment of Japanese labour.[549] There were occasional references in newspaper editorials to the possibility of a Japanese work force coming to the mines, but it was not until April, 1907 that Julius Ruffner hired 21 Japanese to work for the Pine Creek Power and Northern Columbia Gold Companies.[550]

Predictably, loud protests rose from the creeks. The editor of the *Atlin Claim* referred to the newcomers as "objectionable people",[551] and a large crowd, representing every creek met to discuss the matter. A Miners' Protective Association was formed, and their stated aims were to fight the invasion of Japanese by persuading white miners to not work with them, and convincing merchants not to do business with the offending companies. They also planned to help those who were caused hardship by the presence of the Japanese.[552]

There was a good deal of talk and a call for the formation of a branch of the Asiatic Exclusion League, but even though this was the year of the Oriental Riots in Vancouver, Ruffner's companies were able to hold firm for the mining season. The miners' fear for their jobs was unfounded. In Atlin, labourers were in short supply, and wages remained high.[553] The Japanese deprived no one of work.

CHAPTER 8

# Quest for a Mother Lode

The two indomitable prospectors who staked the first placer claims in the Atlin Mining District, also located the first quartz claims. Even as their Pine Creek ground was producing handsome profits, Fritz Miller and Kenneth McLaren set 4 x 4 claim posts on Surprise Mountain, at the west end of Surprise Lake, and recorded the property on 24 August, 1898.[554] Assays showed the presence of silver, gold and lead in promising amounts on their ground,[555] but the site was never developed by the discoverers.

A certain aura surrounds almost all placer creeks, and the men who work them, even as they recover fortunes from the gravels, cannot resist pondering. Where did the gold come from? Where is the mother lode? Atlin miners knew the grinding glaciers of the Pleistocene had gouged the precious metal from somewhere, transported it unknown distances, and left it in the trail of their passing. Eons of erosion then washed and tumbled it in the creeks. Regrettably all clues to its source were washed away, but that did not deter prospectors from searching, and countless fortunes were squandered in futile quests for the elusive mother lode.

By late September 1898, 25 quartz claims had been recorded, and some were given promising reviews.[556] By the next summer, the interest in quartz properties increased as available placer ground grew scarce and it became obvious the easy diggings found near Pine City were not a widespread condition. The frenzy associated with the first placer rush turned into a dash to stake mountains and high valleys. Atlin was surrounded by quartz locations.[557] There were even claim posts on the corner of Discovery Avenue and Second Street, in the heart of Atlin's business district.[558] Storekeepers, speculators of every stripe, and even the usually reserved women of Atlin fell victim to the fever. They were tantalized by the possibility of gold beneath their feet.

Despite the widely held conviction that quartz mining was the real future of Atlin, it was an extremely expensive proposition, and could not be done without large scale investment, even when good returns were assured. From all the excitement of 1899, just one mine developed, and it was on the outskirts of Atlin.

The property, known as the Anaconda Group, was located by Robert Florman[559] at the edge of the lake, a half kilometre south of Atlin. Florman's partner, W. J. "Sailor Bill" Partridge, known in Dawson City as the man who changed his clothes several times a day and only wore a suit once. A wily promoter, he lost no time seeking a buyer, and he soon caught the attention of the Nimrod Syndicate.[560]

The sale of 20 claims, constituting the Anaconda property was concluded after young Lord Ernest Hamilton, considered the moving spirit of the syndicate,[561] arrived in Atlin in September, 1899, to represent the Nimrod interests.[562] Initially, the British group had been prepared to pay $40,000.[563] Locally, the asking price was believed to have been $10,000.[564] Lord Hamilton maintained that after a good deal of haggling the Nimrod Syndicate paid $4,000.[565] Some gold was visible, but he doubted the quality and quantities claimed by Partridge. In spite of these misgivings, once the deal was made, preparations commenced and mining would begin in the spring of 1900.

Richard Featherstonhaugh, who had run the Syndicate's Blue Bell mine on Texada Island, was put in charge.[566] Confident of the mine's future, he started construction of bunk houses, a dining room, a blacksmith shop and battery shed.[567] Unrealistically the Anaconda property was compared favourably with the successful Treadwell Mine at Douglas, Alaska. Such optimism flew in the face of the generally acknowledged fact that Anaconda ore was low grade.[568]

Plans called for the installation of a steam-operated five stamp mill, and 20 tons of mill machinery was brought in across the ice. A 30 metre tunnel was dug. By 17 May 1900, before a crowd of curious and hopeful spectators, the operation commenced. As some suspected all along, the time, effort and money spent to set the mine in motion was sadly misdirected. Anaconda ore was low grade, and could not be mined profitably. There had been no basis for the claims made on its behalf. Nimrod abandoned the property, and it was never worked again. Although the first quartz mine failed, it inspired a certain confidence in Atlin's future as an important mining centre.[569] Mining was, after all, a gambler's game of trial and error.

The Nimrod Syndicate had invested a good deal in Atlin, and they were not ready to leave. They took an option on the Imperial Group of claims on the south slope of Munro Mountain, during the 1900 season. Ore from the Imperial was processed at the Anaconda stamp mill which kept it in operation.[570] At the end of the season, the mill was moved to Munro Mountain[571] where it also processed ore for several other mine owners. Nimrod shareholders were becoming disillusioned about quartz mining, though. The Imperial Group was not showing good returns, and the company was embroiled in litigation surrounding its Yellow Jacket claims on Pine Creek. Nimrod did not renew the option on the Imperial property. Instead, they concentrated efforts on their placer holdings on McKee Creek.[572]

The Imperial Mine was not abandoned. It was well situated near Atlin with a good road and plenty of accessible timber and water. Until 1904, the property was worked by another English syndicate, and managed by H. Peplow Pearse. Production figures were encouraging,[573] and serious consideration was given to the installation of a 10 stamp mill, powered by the water from a nearby lake and electricity from the power plant below Pine Creek Falls.[574] However, the cost of freighting the sacks of ore to the smelter in Tacoma was prohibitive. Although little work was done, interest in the Imperial Mine property lingered for several years.[575]

The stories of the Anaconda and the Imperial were repeated with variations many times over. The Lakeview Mine, west of Boulder Creek, the Big Canyon on Fourth of July Creek, and the White Moose, Kirtland, Black Bear, and Big Horn Groups on Taku Arm petered out in similar fashion. At first the mining men fed local optimism and small investors eagerly volunteered capital, but then the costs of development soared beyond expectations and the anticipated profits remained ever elusive. The amount of gold mined was never enough to meet the expense of concentration machinery, construction of wagon roads or aerial tramways, camp buildings and the mining itself. The obstacles of labour, distance, freight, power and water were insuperable in every case.

Of all the early prospects for development of a lode mine, the most promising was the Yellow Jacket, which was considered by many, and for a long time, to be the mother lode for Pine and Spruce Creeks.[576] The property, located east of Pine City, was first located as placer claims by miners who were working in the stream bed. In 1900, four overlapping claims on the Yellow Jacket[577] caused a long period of litigation until the court decided in favour of Charles Christopher in 1901.[578] The following year Nimrod once again assumed control of the Yellow Jacket property, and sank a 14 metre shaft on the ledge, but their work was stopped by other legal complications as Julius Ruffner ran a 30 metre cut, revealing low values in free gold.[579] By 1905 the property belonged to the North Columbia Mining Company which operated a powerful mill on the site. Production figures indicated some very rich ground, and interest in the property remained high. In the end, though, the profits were not enough to sustain a mine.

The Beavis Mine, located less than a kilometre north of Atlin on the lake edge, was discovered by William Beavis. He worked it alone for a short time[580] until 1902, when it was bonded to Henri Maluin, the French engineer, for $10,000.[581] Three shafts were sunk on the nine claims, and gold was visible to the naked eye.[582] Sample shipments of ore were sent to the smelter the next year, and results were good.[583] However, by 1904 the property was under bond to C. E. Wynn-Johnson of Skagway, and samples tested by the government assayer were reported to indicate the presence of gold, but not in sufficient quantity "to admit of shipment."[584] After that the Beavis property lay idle, except for assessment work, until the seasons of 1907 and 1908, when there was a flurry of development work, during which three shifts of miners were employed. It ended when it was determined that, although there was good ore, diamond drilling was necessary. It was never done. By 1910, two shafts had filled with water, and the mine was inactive.[585]

Coupled with the search for gold, was the ongoing search for other marketable minerals. The discovery of native copper by the Lavadiere brothers, Noel, Frank and Thomas, was made on Copper Island in 1899.[586] The apparently extensive deposit caused quite a stir when it was found. A fine specimen, a copper nugget weighing 50kg, was displayed at the Paris Exhibition of 1900.[587]

That a mine never developed on the Lavadiere's claims, was not for want of time and effort spent on their part. At various times the property was bonded to southern capitalists, who undertook extensive development work. In 1908, the brothers, always strong in their belief there was good ore in quantities worth shipping, began building a canal to connect Torres Channel with their mine on Copper Island.[588] The same year talk circulated about building a 50 km road along the western lakeshore.[589] The Lavadieres' optimism and persistence was typical of mining men and their work went on. In 1917 there was a scheme to barge out sacks of ore, but the expense would have been too great.[590] The next year they bonded their entire holdings, those on Copper Island, and copper-gold claims on Hobo Creek, to a representative of American investors who expected to start operations the next season.[591] Certainly older, and perhaps wiser, the Lavadiere brothers finally turned away from their claims. Transportation quandaries were never solved; the copper still lies embedded in the rock.

One of the most original bids to dig a fortune from the earth was the attempt to sell the hydro-magnesite which lay in large surface beds adjacent to Atlin townsite. Locally the deposit was referred to as the lime beds, and later as the alkali flats. It was very pure, and extremely accessible, however the market was small. In 1904, A. C. Hirschfeld, publisher of the *Atlin Claim*, staked it, and shipped 200 tons to San Francisco, at a cost of $8 per ton.[592] It was intended for use in the manufacture of magnesia brick for furnace linings.[593] The purity of the hydro magnesite impressed the purchasers, but its moisture content was high, as were the freight charges. Hirschfeld could not compete with European suppliers.[594] In 1908, British interests acquired the property. They planned to dry the magnesite in a rotary furnace to reduce weight and save on shipping costs.[595] The idea was never carried out because the owners and their American buyer could not reach an agreement on price.[596] During World War I, when European suppliers could not ship, the Atlin deposits were re-examined.[597] Freight costs remained the deciding factor, and no mining was done.[598]

Since the market for magnesite always remained small, the easiest to mine mineral in the district was never exploited. Atlinites, however, were practiced at making use of materials at hand, and they found several good uses for magnesite. In a wet stage it resembles plaster of paris, and sets as hard, so it was used for chinking log cabins. It was also suitable for glazing windows. A less obvious use was as a wood preservative. Early fence posts set in holes containing a couple of shovels full of magnesite mixed with the backfill, never seemed to rot.

Throughout the mining history of Atlin the most controversial quartz property, worthy of large scale development, has been the Engineer Mine. Plagued by misfortunes and lengthy litigations, it has fascinated the intrepid and intimidated the superstitious.

In July 1899, a group of engineers were working on the White Pass railroad near Caribou Crossing. Nearby two Swedish prospectors set up camp. They told the engineers about a pale yellow metal they had found on the eastern shore of Taku Arm, which in their estimation was at once too yellow to be silver and too pale for gold. Several of the White Pass men, however, felt that the discovery was probably gold, and at the very least, it bore investigation. To a man they were ready for a diversion, particularly after weeks of battling swamps, mosquitoes and stubborn pack animals on the railroad job. An adventure with the possibility of making a rich gold strike was all the more interesting. A deal was struck with the Swedes, who, in exchange for a grubstake and half of any profits, agreed to take on the engineers as partners.

Hank Diers left with the prospectors to investigate. Three anxious weeks passed. The engineers pressed on with their work, and speculated. When Diers returned he had news that stirred the gold lust, and triggered an intemperate celebration around the campfire.

The discovery was about 15 km south of Golden Gate, at the entrance to Graham Inlet, and about 42 km from Atlin. It was on a small stringer of quartz in which gold was visible to the naked eye, and it led to a larger body 5 to 15 metres wide. There were many other promising stringers. The prospecting party had recorded the discovery, consisting of 13 claims, in Atlin. Ore samples assayed by the Bank of British North America showed incredible results; gold yields as high as $17,000 to the ton.[599]

Later assay reports gave results up to $13,000 per ton.[600] Although lower than first reported, the values signaled a very promising prospect and optimism was widespread. Captain John Irving offered to carry the first shipment of ore to Bennett, free of charge, aboard the *S. S. Gleaner.*[601] A public company was formed, the Engineer Mining Company of Skagway, with the town's Mayor, John Hislop, as President. When 30,000 company shares were offered at 10¢ each, there was no shortage of investors.[602] It was commonly assumed that since so many White Pass men were involved, it was a virtual certainty that a rail spur would be built connecting the mine with Log Cabin.[603]

Work started with tunneling and crosscutting, and continued over the next several winters.[604] In 1902, a Hendy stamp mill was installed at a cost of $15,000.[605] It appears the money began to run out at this point. No work was done for two seasons, and although the company was reorganizing in 1904 to gain financing, the move was not a success.[606] While the richness of the property was never doubted, the large amount of capital required to bring it into profitable production was never found by the railroad engineers who started the mine.

In 1906, due to a complex set of errors, oversights, and a shortage of money, some of the claims were allowed to lapse. They were immediately staked by a man named Brown, but somehow a small group of Atlin men, known as the Northern Partnership, acquired them. One account tells of trickery and schemes associated with the quick take over of Brown's property, and of a curse of death and disaster placed on the mine, and on the head of Captain James Alexander, a principal in the Northern Partnership.[607]

Development work began, and rich ore was found in small stringers as usual. Hopes were ever high. In 1910, it was reported that 800 pounds (364 kg) of select ore yielded 240 oz of gold.[608] The word "select" is key to the way that high values were obtained. High yields from systematic mining was not typical of the Engineer property. The ground was a network of faults. Rich seams pinched off, disappeared abruptly, and finding them again was an extremely expensive guessing game. The Gold Commissioner's report to the Minister of Mines in 1910 explained that, although shipments of ore to the smelter brought returns of about $6,000 to the ton, and the 800 pound lot yielded $3,000, "the season's run of all classes of rock, about 140 tons, yielded about $8,000," breaking down to a more realistic $57 to the ton. In 1910, the company had a two stamp mill operating on the property.[609]

For the next couple of years a portion of the property was under litigation, and little work was done.[610] By 1912, the mine was in the hands of one member of the Northern Partnership, Captain Alexander, who had about 30 men working during the summer on surface prospecting. They revealed a larger ore body than at first suspected.[611] Alexander's carefully planned development moved at a busy pace for the next few years. The richer ore was shipped for smelting, and the stamp mill at the site crushed low grade and development samplings.[612] Profits were substantial enough to persuade both the Provincial Government[613] and potential investors[614] that the mine was worthy of full scale development.

Over the next years, Alexander poured work into the mine, while a procession of potential buyers visited the site. By 1917, the war was affecting mining, and available labour dwindled as it progressed. In 1918, Alexander was apparently ill, and a smaller working force appeared to be "marking time." By fall, though, he seemed to have made a sale. He and his wife started south with representatives of the company planning to buy Engineer. With the prospect of a pleasant winter outside, and the sale of the mine he had promoted for so long, Alexander left his companion of many years, a parrot named Polly, at the Caribou Hotel in Carcross, and caught the train to Skagway. From there he sailed on the *S.S. Princess Sophia*. North of Juneau the *Sophia* ran hard aground on Vanderbilt Reef. While rescue boats hovered nearby, the skipper, thinking it would be too dangerous, refused to transfer his passengers. He was certain the high tide would free the ship. Instead a vicious storm blew up, sending rescuers scurrying for the nearest shelter. The *Sophia's* great hull slammed the reef for terrifying hours, until it split. She disappeared into the seething water of Lynn Canal, and all aboard were lost, 343 passengers and crew.

Captain Alexander's death had a profound effect on the immediate future of the Engineer Mine, although ironically the Minister of Mines' Report for 1918 made a glowing statement in its behalf:

> Altogether it is a wonderful showing of gold, and there is every reason to believe that it can be developed into one of the greatest gold producers on the continent.[615]

Yet the Engineer, fraught with trouble, ceased production again. Five years of litigation began during which the ownership, and even the title to the property, were challenged. Only a small amount of development work was done. In 1924, a group of New York investors, the Engineer Gold Mines Limited, whose president was Charles V. Bob, acquired the mine, and work began again on a grand scale. A power house was built on the Wann River, 5 km south of the mine, and transmission lines strung to the mine site. A mess hall, bunkhouses and several residences were built. Extensive underground work was accomplished with a crew which peaked at 60 men.[616] The activity at Engineer was a boon to the whole area, and with the cloud apparently lifted from the mine, the prospects for investment on other properties in the district improved.

In 1925, the labour force averaged 100 men, with a maximum of 140. To meet the increased demand for electricity a new dam was built at Wann River. The company's shares shot up to $100 each in New York,[617] but however exciting the profits sounded, they were misleading. Mostly high grade ore was being reported. Management was apparently more interested in the property's showing on the stock market than they were in systematic mining. For a few years profits continued to be good, and about 20 men were employed each season until 1930. By that time, accessible rich stringers were fairly well exhausted, and in 1931 the mine closed for a year and a half.[618]

In 1933, the mine was again the subject of litigation. Subsequently it changed hands a number of times. What little mining occurred, was done selectively with a view to tapping the richest ground. During almost 40 years of operations, 18,000 oz of gold was taken from just 17,000 tons of ore.[619] It was an exciting record, except for the fact the mine was virtually high graded. Experts then and now maintain that the property has never been fully explored, and that the promise of 1900 still awaits realization.

As for the curse, it seems that fate did frown darkly on many men associated with mine. One undeniable case was the untimely demise of Captain

*Engineer mine on Taku Arm.*

Alexander who had such a long and close association with the mine. And there were others later connected with the mine who died under circumstances that, while not indicative of foul play, were somewhat strange. According to one source, Charles V. Bob was eventually jailed for shady market dealings.[620] Polly the Parrot lived at the Caribou Hotel for more than half a century after Captain Alexander left her there that fateful fall. She died naturally in 1972, at an age believed to be well over 100. The alleged curse of the Engineer Mine obviously did not include her.

The Engineer and other mines on Taku Arm had good connections to rail transport to the coast, and to Atlin. Freighting ore was a costly consideration, but it would not be a deterrent in opening any really promising property in that area. The *S. S. Gleaner*, and later the *Tutshi*, made regular visits, and delivered mail to a wooden box nailed to a tree at Golden Gate. That tree was the only "Post Office" for residents of Taku Arm. For nearly half of each year, the frozen lakes were the "highways" connecting the mines with both Carcross and Atlin.[621]

In the early years, profitable work was done on the 13 claims at Baker Lake,[622] which were a continuation of the Engineer claims.[623] P. F. Scharschmidt and Harry Price,[624] operated the Gleaner group, and work was done on the Kirtland and Black Bear Groups.[625] The owners of the White Moose, a group of prominent men from Butte, Montana, were optimistic enough to consider building a smelter on Taku Arm.[626]

West of the White Moose, where Bighorn Creek flows from pocketed glaciers into the Fantail River, Fred Lawson and his partners, one of them Thomas Kirtland, acquired claims including the Bighorn in 1909. The property, first staked in 1898, passed through several hands before Lawson stepped in. In 1910, the Provincial Government built a wagon road from Kirtland, a stopover on Taku Arm, up the Fantail River to Bighorn Creek, and along the creek valley to the lower terminal of an aerial tramway located on the Lawson claims, a distance of 16 km.[627] Fred Lawson took advantage of the road to haul in machinery for a one-stamp mill.[628] Lawson, a colourful veteran of the Battle of Little Bighorn,[629] made an adequate living for himself and his wife, for more than 20 years, on the high barren slopes of the mountain that bears his name. He eventually sold out in 1934.[630]

A property of anticipated potential was located at the extreme south end of Taku Arm on the lower flanks of White Moose Mountain. The claims, eventually known as Ben-My-Chree Mine were first staked by Stanley McLennan, who was unable to develop them without financial help. He approached Otto Partridge, who was at Bennett City managing the Bennett Lake and Klondyke Navigation Company. Partridge was interested and, from then on, the magnificent glacial valley where McLennan found the gold would become the focus of his life. Partridge found other investors including the Honourable Maurice Egerton, an English aristocrat who was in the area on a big game hunt, and Mrs. Partridge's companion, Emily Dalton.[631]

By 1911, development unearthed ore with high assay values, and plans were made to build an aerial tramway to convey it from the mine entrance to the foot of the mountain from where it could be hauled to the lakeshore for shipment. On 5 October tragedy struck. Heavy early snow created an avalanche which thundered down on the mine, burying Stanley and Annie McLennan in their cabin at the foot of the mountain. All work stopped and would not resume until the McLennan estate could be settled.[632]

Over the next two years crews varying from 10 to 61 men worked at Ben-My-Chree. More claims were added to the property, new buildings were erected, and the tramway rebuilt. However maintaining the tram on a mountain slope subject to snow slides seemed impossible, and Partridge and Egerton were advised to move it to another of their properties. All the time they were leveling complaints at White Pass about the high cost of incoming freight. They claimed the cost on the all-water section from Carcross to Ben-My-Chree was three times more than the tariff from Skagway to Liverpool, England, a distance of 40,000 km. The "intolerable" transportation costs, they stated, were prohibiting real development of mineral reserves in the entire district.[633] On marginally good properties, this was undoubtedly a factor. By 1915, the Ben-My-Chree Mine had been abandoned.[634]

Joseph T. Mandy, in his summary for the Minister of Mines Report of 1936, considered the year 1917 was the beginning of the decline of placer mining in Atlin. At that time the large mining companies began leaving, and it was generally thought the creeks were worked out. War time conditions also paralyzed the industry, and momentum was lost. Prosperity would not return to the creeks until the mid-1930's, when it was established there was much gold left to mine. In the meantime, though, mining interests turned to lode mining. During this phase of mining history, Julius Ruffner, a success in Atlin placer operations, became interested in a silver-lead property in the vicinity of Crater Creek, a tributary to Fourth of July Creek.[635]

Ruffner came to Atlin about 1903 to manage the North Columbia Gold Mining Company, which had an office in Discovery, a number of claims on Gold Run, and a bond on property held by the Stevendyke Hydraulic Mining Partnership. He was described as a bright young fellow, managing a large operation.[636] He became a colourful local character who usually traveled the camp on horseback, wearing black bearskin chaps and a pair of six guns. He was a man of energy and enthusiasm, who attracted more outside capital to Atlin than any other mine owner or promoter.

The Crater Creek property was on Vaughan Mountain, 23 km from Atlin. In 1920, Ruffner organized Atlin Silver Lead Mines with a group of Seattle investors. Although assays indicated silver-lead values of $400 a ton, capital was hard to find.[637] Ruffner, however, was an experienced promoter and his persistence over the next few years paid off. In 1927, he had a crew of about 20 men working the property.[638]

Ore was hauled to the lake over a wagon road built in 1926 by the Department of Mines.[639] To avoid White Pass freight costs, Ruffner used the *M. V. Gladys* to barge the ore to the railroad portage at Scotia Bay. From that point on, though, he had no choice but to give White Pass his business.

Atlin Silver Lead Mine was doing well in 1929 when Julius Ruffner died. At first the company was taken over by C. V. Bob of New York.[640] After a series of financial problems, bad luck, and another change of ownership, the mine was finally abandoned by Bobjo Mines of Ontario in 1934.[641]

During the 1920's the Tulsequah River area began to attract attention. The Tulsequah flows into the Taku River from the north at a point about 10 km east of the international boundary. The area lies about 250 km south of Atlin. Interest increased steadily, until by 1930 there were 100 prospectors working in the region.[642] The Tulsequah Chief Group, was first located by W. Kirkham of Juneau, and bonded in 1923 to the Alaska Juneau Gold Mining Company. A series of companies prospected and developed it, but isolation and difficult access, coupled with a shortage of capital following the stock market collapse of 1929, eventually killed all hope for a viable mine. Similarly the Whitewater Group, situated 10 km up the Tulsequah River from its mouth on the Taku River, was located by a group of Juneau men in 1929. It too passed from hand to hand until it was acquired by the Polaris-Taku Mining Company in November, 1936.[643]

Early in 1934, when President Franklin D. Roosevelt, with a stroke of the pen, raised the price of gold by almost 100% to $35 per oz, he became an instant hero in gold mining communities, including Atlin. Suddenly many quartz properties which had held good promise, but were not financially feasible to work, became potential mines. Exploration work began on the Whitewater claims and development proceeded, until by 1939, the Polaris-Taku Mine was in full production, A full work force demanded a town for families, and Tulsequah, British Columbia came into existence. There was no road connection between the mining town, and either Juneau or Atlin, but reliable air transportation served the people there. In summer there was regular launch service to Juneau. In terms of facilities, however, the community lacked nothing. Residents even enjoyed a splendid downhill skiing facility complete with night lighting and a challenging ski jump.[644]

The success of the Polaris-Taku Mine, meant the Atlin Mining Division had finally realized the potential pursued by miners since 1899. The isolation of Tulsequah, the high cost of freight, the lack of easy access were easily absorbed by the profits earned. The promise of "next year" was finally being realized.

CHAPTER 9

# The Political Scene - Unions, Campaigns and Alliances

Never aloof from affairs of state, the isolation and strong independence of Atlin's first citizens served to intensify their interest in political matters. The impact of government decisions on their everyday lives was of acute concern to them. Further, Atlin itself was of considerable interest to the Provincial Government in Victoria.

At the centre of the community's earliest political activity was the British Miners' Protective Association. On April 22, 1899, 75 miners met at Charles Oubridge's Pine Tree Hotel in Discovery.[645] The chairman was James Stables, a highly respected miner. The intention of the group was to address Victoria directly, and demand action on a long list of issues. Among their many grievances, the most notable were centred on the confusion caused when the Discovery claims were thought to be in the Northwest Territories, and staked according to its regulations, the Alien Exclusion Act, and the irregular practices and partisanship of local government officials.

Membership in the British Miners' Protective Association was apparently restricted to British subjects, but this was not an attempt to exclude Americans and others. The idea was that British subjects would have stronger influence in Victoria. Present at their first meeting was a young American, who was destined to become a best selling author, Rex Beach. He called upon fellow Americans to contribute to the cost of sending delegates to the provincial Legislature.[646]

A couple of weeks passed, nothing changed, and there was even greater unrest in the community. Because of the confusion over the legal size of claims on Atlin creeks, claim jumping was endemic. There was still no judge appointed to untangle the bureaucratic mess which had been growing steadily worse for months. The mining season remained closed. Finally there was a growing conviction that every government decision was designed to obstruct the miners and businessmen of Atlin.

In early May several hundred citizens assembled around a hastily built platform on First Street in Atlin. Notably absent from this angry gathering was Gold Commissioner Graham. Probably more important was the presence of the Member of the Provincial Parliament for Atlin, C. W. D. Clifford. A miner himself, Clifford, sympathized with the plight of the Atlin men and appreciated the difficulties they faced in dealing with the government. After a lengthy debate, Clifford, along with Gordon Hunter, and C. H. Caton were elected as a delegation and dispatched on the next boat to Victoria.[647]

Premier C. A. Semlin, not a particularly strong leader, was struggling with a divided cabinet and a slim majority. He was preoccupied with greater problems than those of a few cantankerous miners. The delegation returned to Atlin with no promise of redress. With every passing day the list of grievances grew, and in June a group of miners marched in from Discovery to present their demands to the Gold Commissioner.[648] Graham was unable to offer any solutions or appeasement, but a flood of letters from the British Miners' Protective Association finally had some impact in Victoria. The Special Commissioner, provided for in the Bennett-Atlin Commission Act, passed in February, was at last appointed, and the long delayed mining season was opened.

The British Miners' Protective Association remained active and throughout 1899 was a particular thorn in the flesh of local officials. Certainly it played a major part in discrediting Gold Commissioner Graham, and, along with the Board of Trade, was successful in lobbying to have him replaced with a local man. Claiming 90% of the district's miners, as members (which probably means they admitted Americans on some level), they frequently took a stand in local controversies and mining disputes, and stood firmly against the importation of oriental labour in 1902.[649]

Later, when the major mining issues had been settled, the Association slipped into a more social role. Members erected a large tent, 9 x 20 metres. It served as Discovery's Presbyterian Church, concert hall, dance hall and sports arena.[650]

When Miller and McLaren found gold on Pine Creek, the uninhabited Atlin region was part of the huge Cassiar riding. It was represented in Victoria by C. W. D. Clifford, as well as the ambitious Captain John Irving. The gold strike, the rush, and the founding of Atlin immediately drew Irving to the area. He began establishing extensive business interests in the town. Through the John Irving Navigation Company he attempted to gain a monopoly of the transportation systems in and out of Atlin. His aggressive heavy-handed tactics made him extremely unpopular. The *Atlin Claim* laid the accusation that, "he never missed an opportunity of bleeding miners and storekeepers on fares and freights."[651]

John Irving was defeated in his bid for a seat in the Legislature in June, 1900. It was principally the voters of the Atlin district, who in their first election as a new community, toppled his plans.

Added to the strong feelings about local issues, was the volatile situation surrounding Joseph Martin in Victoria. Martin, a member of Semlin's Cabinet, was primarily responsible for the Alien Exclusion Act, and had for three stormy months served as Premier with literally no support in the House. Although Martin attempted to appeal to the voters of Atlin by promising to have the Alien Exclusion Act repealed,[652] candidates such as A. L. Belyea, Capt. Langley and W. C. McCraney, who were known to be Martinites,[653] received relatively few votes.

Clifford, who had championed Atlin miners, was re-elected. The immensely popular young James Stables won narrowly over his friend Alex Godfrey, a former Vancouver alderman, who was supported by many Atlin businessmen.[654] Irving received some support in Bennett and in the fishing communities of the coast, but he was soundly rejected in Atlin, Discovery and on the creeks.[655]

Before the next election in 1903, the Cassiar riding was split into two, Atlin and Skeena, each to be represented by one member.[656] This change was universally welcomed in Atlin. The strong feeling had always been that the coast with its dependence on fisheries and the Atlin district with its interest in mining, had absolutely nothing in common.

Another important change had taken place in the interim. For several chaotic and unstable years the British Columbia legislature had seen a series of leaders emerge and decline as alliances shifted and dissolved. After several constitutional crises Richard McBride emerged as Premier. One of his first initiatives was to organize British Columbia politics on party lines.

Political parties had not been particularly visible at the first election in the Atlin area in 1900. However, by 1903, the community, like the rest of the province, split into Liberal-Conservative and Liberal camps. The Liberal-Conservative nomination was won by Dr. Henry Esson Young over John Irving. James Stables offered his services to the Liberals, but for some reason his earlier support had evaporated. The party selected prominent businessman and community leader, John Kirkland.[657] The race between Young and Kirkland was a tight one.

The Liberal-Conservative leader, Richard McBride, had practised law in Atlin in partnership with W. Pollard Grant during 1899-1900.[658] In August he brought his campaign to Atlin to support his friend, Henry Esson Young.[659] For several days the popular politician was royally entertained. Large crowds heard him speak at several meetings. He offered Atlin voters radical changes to the Placer Mining Act, and assured them he was most certainly opposed to the importation of oriental labour.[660]

Kirkland was well supported, but his political leader, Joseph Martin, was a liability. The election was decided in favour of Young by a narrow margin of 34 votes out of 450 cast.[661] After delaying his departure to attend to a dying patient, Young set out for Victoria in November, 1903.[662]

The respect and admiration earned by Henry Esson Young as a physician followed him in his political life. Elected in three subsequent elections, he worked actively for Atlin and served the province well. His constituents were generally pleased when, in 1907, McBride appointed him a Cabinet Minister. At various times he held the portfolios of Provincial Secretary and Minister of Education.

Young was an innovator and a clever legislator. He believed firmly in the future of British Columbia and the potential of the north. His time in office was a period of great prosperity for the province, providing him with opportunities to launch important projects. He founded the sanatorium, later known as Essondale. He is remembered for "fathering"[663] the University Act, for reorganizing the school system, for providing free text books, and for establishing Victoria's Normal School and a number of night schools.

Young was especially successful in improving Atlin's relationship with the provincial government. He was instrumental in bringing about a number of welcome changes to mining legislation. At the local level he saw to it that Atlin received her share of grants for roads, schools and the hospital.[664] Soon after McBride stepped down, Young resigned from office in 1916.

In 1933 Atlin again elected a member who became prominent in the government of his day. William J. Asseltine, a mining man, served as Minister of Mines, Trade and Industry in the Patullo Government from 1937 to 1941. When Atlin began suffering from severe economic difficulties and a declining population, Asseltine made forceful representations to his government for relief. He later became a member of the Social Credit party.

There has never been a Member of Parliament based in Atlin, and Atlin itself has always been a small component of a very large riding. Nevertheless federal politics have always attracted considerable interest, especially at election times. In 1900 the district was part of the Burrard riding. The sitting member for that riding, Liberal, George R. Maxwell was re-elected. He had spent some time in the community the previous year. When the community's frustrations with government were at a peak, in September 1899, he had attended a Miners' Meeting at Discovery. He listened to loud complaints about the mail service, and the need for a County Court Judge, among other matters.[665] He did not take the Atlin miners for granted.

Maxwell died in 1902, and he was not replaced until the Federal election of 1904. By that time Atlin was part of the new Comox-Atlin riding. The new member was another Liberal, William Sloan, who had struck it rich in the Klondike in 1896.[666] Sloan worked hard for his Atlin constituents. He won improvements to the postal service, an all-important communication link for the people of Atlin. He wrestled with the difficulties which hampered navigation in the area, such as the obstacles in the Atlintoo River, and the largely uncharted waters of Atlin Lake. He was re-elected in 1908 but, as a loyal party member, stepped aside to leave the seat open for Hon. William Templeman, a member of Laurier's cabinet who had been defeated in Victoria.

The elections themselves were demanding for candidates and returning officers alike, and on all accounts they demonstrated a dedication to democracy for every corner of Canada. Candidates who traveled to Atlin during an election campaign committed themselves to many days en route when their time was most valuable. In winter the task of carrying ballot boxes to the polling stations was no simple matter. In 1907, two men snowshoed with the ballot boxes on their backs for over 375 km from Atlin to Telegraph Creek. Temperatures hovered around -40° C and Deputy Returning Officer, W. Pollard Grant, was not accustomed to snowshoes. By the time he and a companion arrived at Nahlin Telegraph Station, he was lame and near exhaustion. The linesmen pitched in to help and Grant and the ballot boxes were carried the rest of the way by dog team.[667]

Ballots for the federal election of 1903 were cast two weeks after the results of the election were known in the rest of the country. In all probability this anomaly was because of the difficulty of travelling in the winter.[668]

Atlin never had an official local government, but its politics were always intricate. It has been noted that Atlin is remarkable for having developed, over a long period, an unofficial but complex political structure at the local level.

Atlin's most enduring organization was undoubtedly the Board of Trade. It was as much a quasi town council as it was a promoter of trade and commerce. Merchants and businessmen, were the backbone of the group, but the public at large were frequently urged to attend meetings.[669] A wide range of normally municipal responsibilities were assumed by the organization - fire protection, road maintenance, and licensing of merchants. The Board of Trade advocated for the schools, lobbied to change mining legislation, solved the problems of garbage disposal, and later developed power and lighting for Atlin.

At other times fraternal orders, friendly societies and unions were prominent and influential. First among these was the Arctic Brotherhood. This was a group founded aboard the *S. S. City of Seattle* as she steamed up Lynn Canal early in 1899 with a load of stampeders bound for the Klondike. The passengers, who had enjoyed the jovial fellowship of their companions for several days, resolved to form a brotherhood. It would consist of northern miners and prospectors from both sides of the international border. "No boundary line here" became their motto. They pledged to leave cabin doors open, to care for sick and destitute comrades, and to prevent cruelty to animals.[670] A uniquely northern order had been born.

The first lodge, Camp #1, was opened in Skagway. The initiation rituals were elabourate, if somewhat tongue-in-cheek. The badge was a beer bottle cork, and the dues, the price of a bottle of beer.[671] In line with its ideal of camaraderie, the main purpose and function of the order was undoubtedly to host first rate entertainment, an obligation taken most seriously.

Camp #3 was founded in Atlin in April, 1899.[672] A hall was built on Third Street, where, for a short time, the organization flourished. In less than a year, though, many of the founding members had left the area, and the Atlin "Camp" was defunct. The Atlin Arctic Brotherhood building and assets were sold,[673] but elsewhere in the north, the organization remained active. In 1905 Camp #19 was established in Discovery,[674] and a hall was built next to the Nugget Hotel. The membership rose to 182 and included all of Discovery and Atlin's most prominent lights.[675] Once again there were banquets, dances, concerts, and picnic excursions organized in lavish style.

In Atlin the Arctic Brotherhood had been replaced to some extent by an organization more serious in intent. The Ancient Order of United Workmen, Lodge #15, was formed in May 1901 largely through the efforts of Richard Fetherstonhaugh.[676] The United Workmen was a fraternal order based on equality, which offered mutual support and a form of life insurance to its members. The brotherhood built a large building on Third Street, which became the focus for many of the community's social events.

Dating from the summer of 1899, there are regular reports of meetings of the Masonic Brotherhood in Atlin and Discovery. In July, 1899, seventy-five "sojourning brothers" gathered in Atlin.[677] The Grand Lodge was not convinced of the permanence of Atlin, and Atlinto Lodge #42 was not formed until December, 1904. A permanent charter was issued in June, 1906.[678] Many charter members had been prominent on the rolls of the Arctic Brotherhood and the Ancient Order of United Workmen. The first home of the Masonic Lodge was the A. O. U. W. hall.[679] Later a more permanent home was established opposite the Kooteney Hotel on First Street at Trainor.

The Moose Lodge formed in 1914, along with the Masonic Lodge were both prestigious organizations which underpinned the social, and political hierarchy of the community. Such organizations allowed Atlin to function as a well-regulated community without an official local government. It was through these organizations that community leaders were identified and given credibility.

A different but equally significant power base lay within the unions and mining associations. These organizations were not permanent entities, but were prominent at various times, and for the most part issue-oriented. The Atlin District Miners' Union championed the cause of better wages, and adopted a standard wage of $5 for a 10-hour day in 1901.[680] Another union was formed in 1903 by men working for a contractor to the Pine Creek Power Company. They demanded an increase of 50 cents over the $2.50 per day they were being paid.[681] The Atlin District Miners' Protective Association sprang to life as a result of the 1907 attempt to bring Japanese labour into the camp. Especially in the early years, Atlin miners were an independent assortment, and none of these unions endured. The miners could, however, apply strong leverage when motivated by a common cause.

Employers had their forum too for a short time. The British Columbia Mining Association, formed in January 1903, was a provincial organization, with forceful representation from Atlin. Fetherstonhaugh, Brownlee, Hirschfeld, Pearse, Dockrill and Ruffner played leading roles as representatives of the large mining companies and businesses in Atlin.[682] The group was controversial. Small operators mistrusted the motives of the hydraulic operators, and there was a sufficient number of them to carry a vote to disband the provincial Association in April 1903.[683] The organization reformed, however, and in Atlin it played a significant role in promoting the interests of the large companies as they gradually took control of mining in the camp.

Alliances of various types have always played a significant role in Atlin's history. Only by understanding the dynamics of its political, economic and social forces can we catch a glimpse of the unique entity that was, and is, Atlin.

CHAPTER 10

# The Board of Trade, Commerce, and Industry

By mid September 1898, barely ten weeks after Fritz Miller recorded the Discovery Claims, there were two hotels, five restaurants, six general stores, and two barber shops doing business in tents at the edge of Atlin Lake.[684] There were still others with firm plans to open in the spring. By the time the ice thawed, allowing the creeks to flow, and as soon as the miners began gearing up for the all-important first season, a small but functioning town was already rooted and eager to serve them.

Mining was the raison d'être for Atlin. It was the industry on which the town expanded and gained permanence. A steady stream of mining men arrived with grand schemes and vast capital. Too often, though, they raised expectations amid great hullabaloo and optimism, only to vanish eventually, leaving disappointment in their wake. While the "high rollers" came and went, it was usually the businessmen who endured. The vital role played by businessmen and merchants in Atlin's history must not be overlooked.

The government's stultifying grip on mining development during the first year or two was confronted more effectively by the business community, than by the miners themselves. Both commercial and mining interests benefited as the Atlin District Board of Trade identified and addressed their concerns. From its founding on 27 February 1900, the Board of Trade proved to be an enduring body which represented the town on every front.[685]

J. A. Fraser, a practical and competent Scot, became the Board of Trade's first President. He was assisted by J. St. Clair Blackett as Vice President. Both were merchants, and Blackett was well known in the business community of Victoria. They were supported by members who worked with a common will to resolve community issues. Within a month they succeeded in securing cheaper freight rates from the White Pass Company for the shipment of mining machinery.[686] They published a 24 page pamphlet, outlining the advantages of

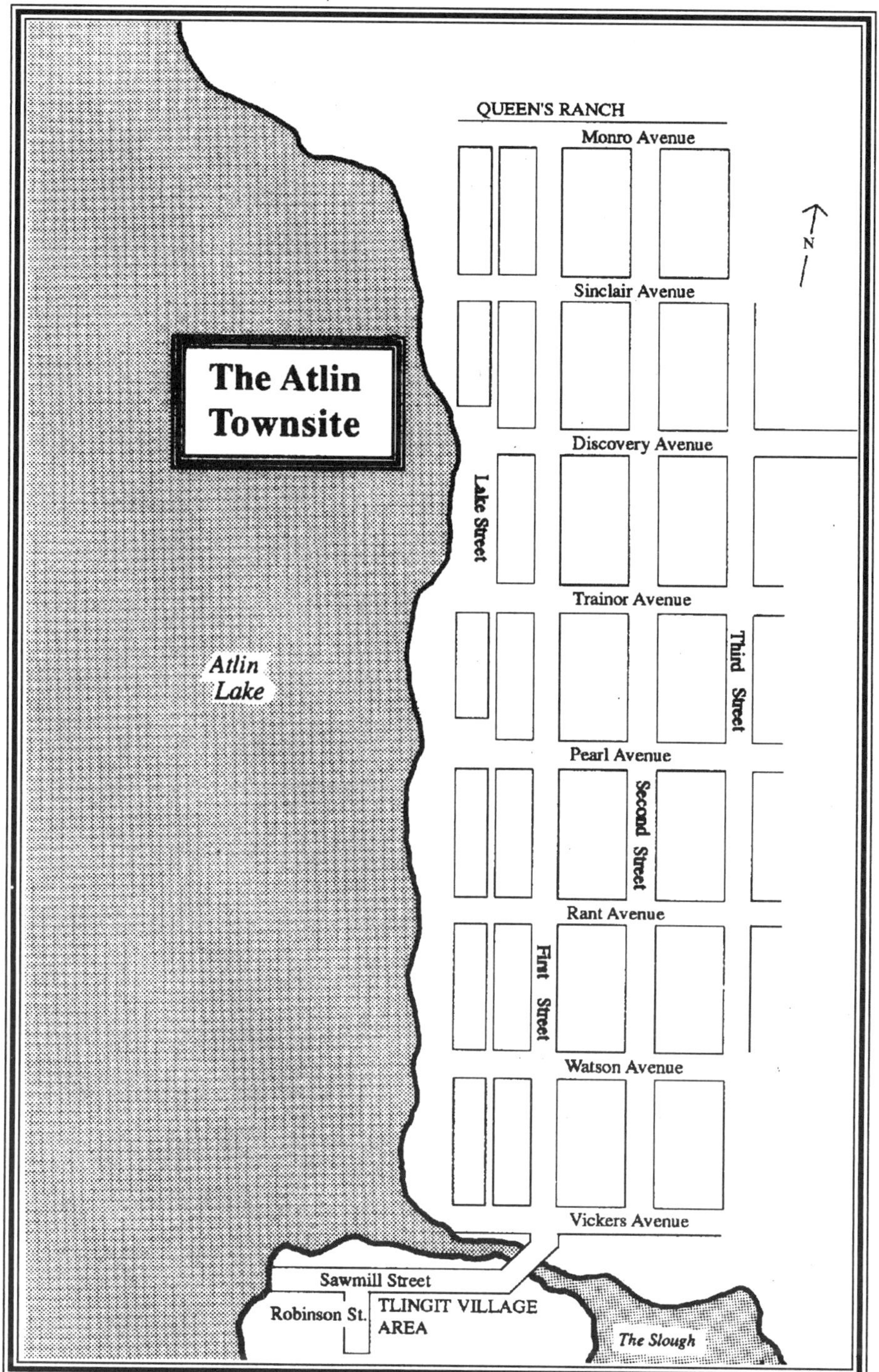
The Atlin Townsite
QUEEN'S RANCH
Monro Avenue
N
Sinclair Avenue
Discovery Avenue
Lake Street
Trainor Avenue
Atlin Lake
Third Street
Pearl Avenue
Second Street
Rant Avenue
First Street
Watson Avenue
Vickers Avenue
Sawmill Street
Robinson St.
TLINGIT VILLAGE AREA
The Slough

investment in Atlin.[687] By midsummer 1900, the influence of the 40 member Board was being felt in Victoria where their suggestions for changes to the Placer Mining Act were being heard.[688]

Over the next year the Board tackled the urgent issues of schooling, medical services, the need for a county court judge, improvement of winter trails, and the mail service. They attended to matters which were municipal in nature, due to the absence of an official town government. They dealt with fire protection and sanitation. They aimed a successful campaign at property owners in the business district to erect suitable buildings and to upgrade and maintain the business centre appropriately.[689]

When the Premier, Hon. James Dunsmuir, visited Atlin in August 1901, it was the Board of Trade which welcomed him on behalf of the community. The following year they pursued a wide variety of issues: a jail for Discovery, the demarcation of winter trails, unlicensed peddlers, the persistent problem of claim jumpers, increased wages for the teacher and the postmaster, and fire protection.[690] They were heard again later in 1902, when Gold Commissioner Graham resigned amid bitter recriminations, and charges of government bungling. The Board of Trade seized the opportunity to demand he be replaced by a local citizen. The Provincial Government responded quickly with the announcement that the new Gold Commissioner would be J. A. Fraser, president of the Board of Trade.[691]

A. S. Cross replaced Fraser as president.[692] A. C. Hirschfeld,[693] and James Stables, a former M.L.A. followed.[694] After 1907, and to a greater extent after 1914, when gold production slumped drastically, the Board of Trade was less active, but it continued to serve as a form of local government. Their concerns with freight rates and road access were ongoing. Major C. W. A. Nevile, a charter member was elected president after World War I, and held the office for 25 years.[695] Due largely to his efforts the Board of Trade was sustained, and easily resumed its prominent role in later years.

A clear sign of a community's prosperity and commercial success is the establishment of a bank. The Merchant's Bank of Halifax moved into temporary quarters in Atlin in 1898.[696] The move was consistent with its desire to associate with the gold rush and its vigorous push westward in that year. In their haste to be the first bank in Atlin, they encountered the problem of hauling a four and one half ton safe over the White Pass well before the railroad was finished. It was done at a cost of 13 cents a kilogram , using horses and block and tackle.[697]

By April 1899, two other banks had followed the rush to Atlin. The Canadian Bank of Commerce, and the Bank of British North America had crude but functional offices. In one case, the interior decor was rough sawn lumber with chicken wire enclosures for the tellers. One by one they moved into impressive permanent buildings, complete with assay rooms.[698]

Banking in the north had its own special problems, not the least of which was security. To transport currency in, and gold out of Atlin, meant shipments of great value had to pass through Skagway, where the criminal element was well known. Usually two men would be assigned to escort the shipment to the next bank at Skagway, a trip of two to three days in fair weather, but much longer if the conditions were bad. It was a grueling trip since the pair could never eat or sleep at the same time. One would have to be with the shipment at all times.[699]

Banks managers were high profile, important personages in the life of the community. W. A. Spencer, of the Bank of Halifax, and T. R. Billett and his accountant Leonard De Gex, of the Bank of Commerce played influential roles in social, church and business groups.[700] Billett acted as host and guide to the Lieutenant Governor when he visited Atlin.

The three banks competing vigorously for business, saw Discovery as a potential branch location, and attempted to established there early in 1899. In each case the move proved impractical as the start of mining was delayed due to government intervention. None of the Discovery branches lasted more than a week.[701] By the close of mining in 1900 season, it was apparent that there was not enough business to sustain three banks. In January 1901, the Bank of Halifax closed.[702] The Bank of British North America ceased business in 1905.

The Canadian Bank of Commerce stayed until the spring of 1907,[703] when low levels of investment signaled an end to the boom and a slump in Atlin's fortunes. J. A. Smith, the manager, and all the bank's assets were transferred to the Skagway Branch. The loss of the bank was a blow to the business community, which undermined its confidence and changed its status in the eyes of the outside business world.

The lumber business, vital to the building of Atlin, always played a role in the economy, although to a lesser degree in later years. The scream of a saw and the crash of falling timber were the first industrial sounds to break the wilderness silence. The short limbed spruce and scruffy pines of latitude sixty were poor cousins of their majestic coastal relatives. Their spindly trunks could be turned into 6 x 6, but milling 8 x 8 timbers was almost impossible. Still the sub-arctic forests were an essential resource for building, mining and fuel. Lum-

AHSP-1077

*The Canadian Bank of Commerce had rustic quarters in a tent and the opera house before building on Pearl Avenue and Second Street.*

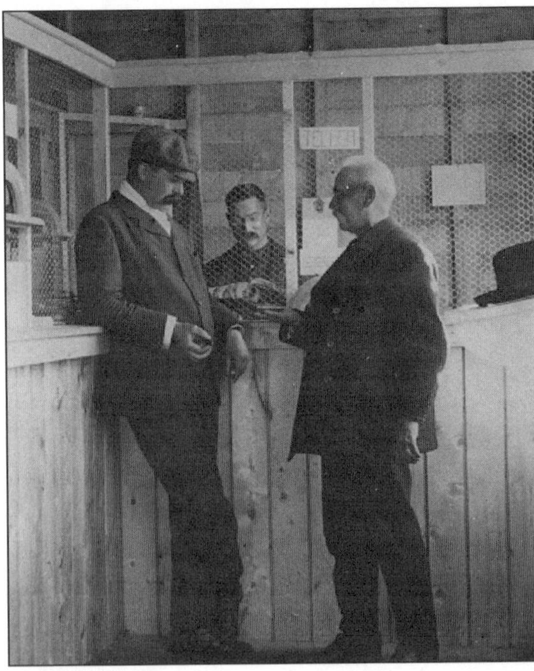

AHSP-1086

BCARS #91365

*The Bank of British North America at the corner of Pearl Avenue and First Street. (1901)*

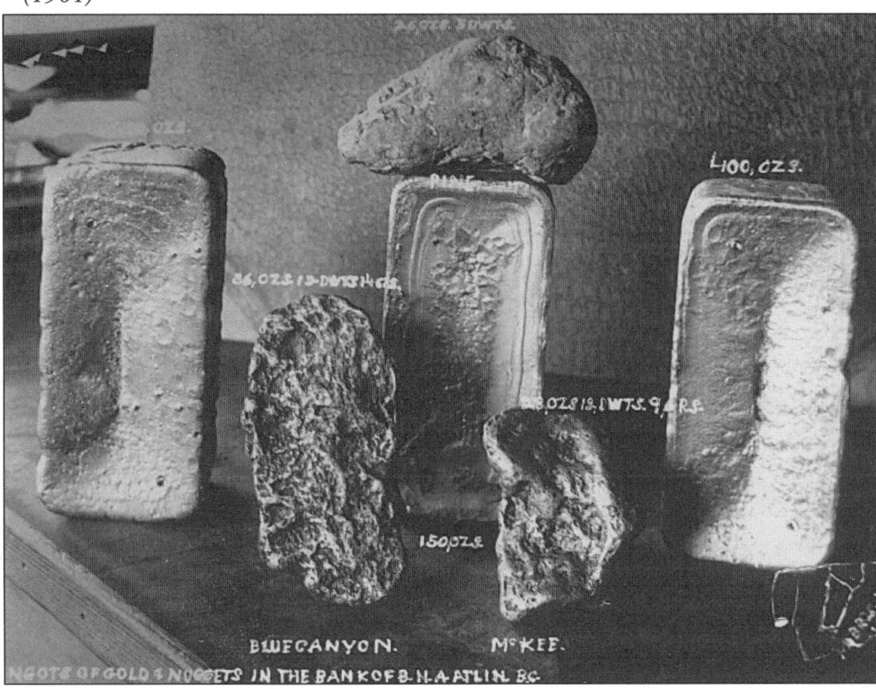


AHSP-2699

*Gold from Atlin Creeks was displayed at the Bank of British North America. (1901)*

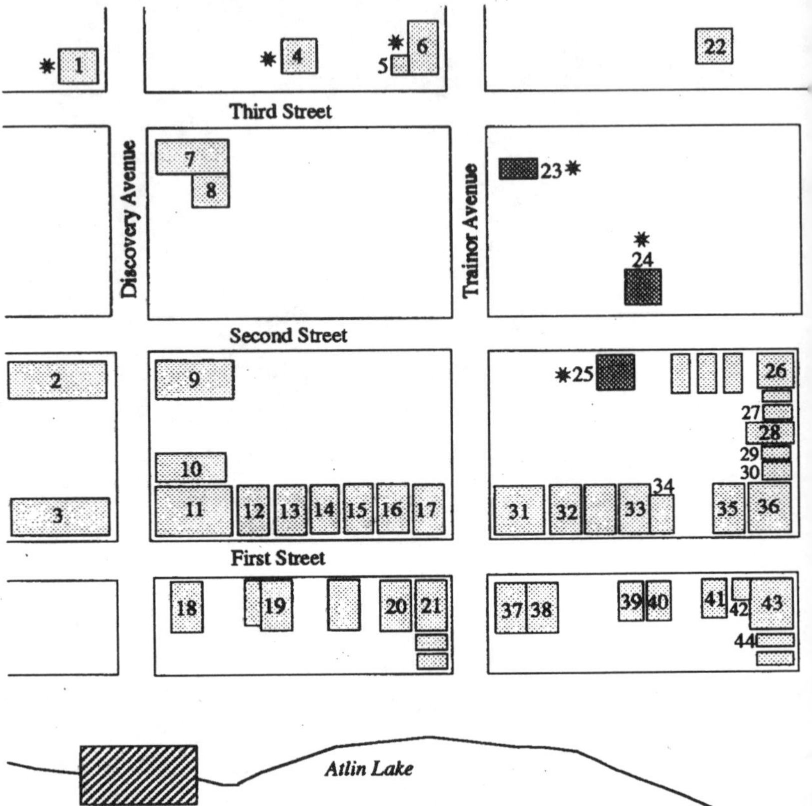


1. E. L. Pillman's House (1903)
2. Discovery Hotel
3. International Hotel
4. Rectory
5. St. Martin's Bell Tower (1902)
6. St. Martin's Anglican Church (1900)
7. St. Andrew's Hospital
8. Nurse's Residence
9. Opera House
10. Metropole
11. Grand Hotel
12. Mattress Factory
13. Quick Lunch Cafe
14. Pioneer Bakery
15. Jimmy Regan's Store
16. Vancouver Hotel
17. Vendome Hotel /later Clarence Sands Store
18. B & H General Merchandise
19. A Tent Restaurant
20. Kirkland Hotel
21. J. St. Clair Brackett's Store
22. Provincial Courthouse (1900)
23. present day location: School (#45)
24. present day location: Courthouse (#22)
25. present day location: Nurse's Residence (#8)
26. Post Office & Telegraph
27. John Irving Navigation Co. /later Globe Theatre (1917)
28. Leland Hotel/later John Garrett's Store (1917
29. Cafe
30. A.C. Hirschfeld Photographer
31. Kootenay Hotel
32. National News Depot
33. Northern Hotel
34. Northern Hotel Annex
35. J. R. Dockrill Drugstore /later Jules Eggert's Jewelry Store
36. Burns Meat Market /later E. L. Pillman's Store
37. Cummings & Richarson's Gents Furnishings
38. Fortin & LeChappelain Druggists
39. Godfrey Hardware

# Atlin Townsite circa 1899 - 1902

Many businesses changed locations a number of times. First and Second Streets, Discovery, Trainor, and Pearl Avenues were solidly built up. This map locates only some of the structures, as they are recorded in the Atlin Historical Society's Photo Archives. The town map was 'redrawn' with each fire and subsequent reconstruction. As more photographs and diaries are shared with the Archives, more pieces to this puzzle may be added.

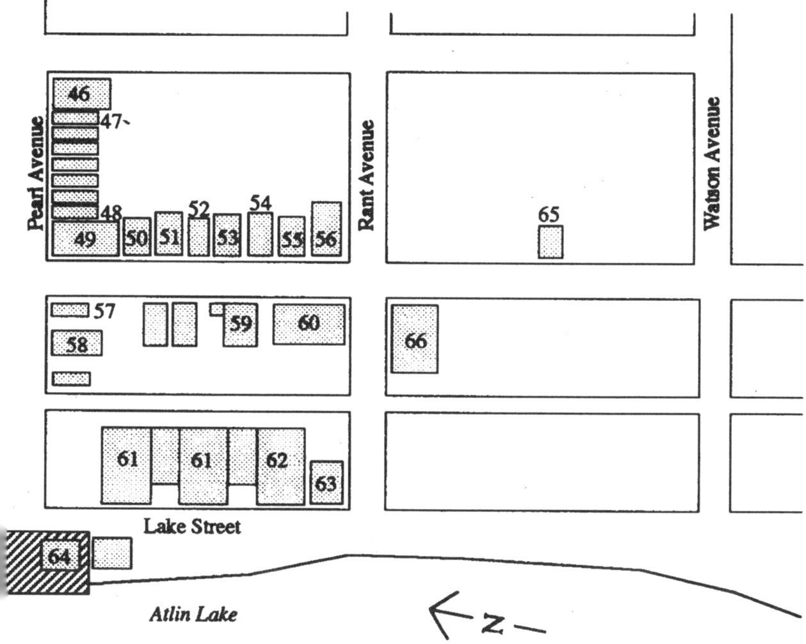


40. Parson's Produce Co.
41. OK Hotel
42. Assay Office
43. Bank of British North America
44. Atlin Claim
45. School
46. Bank of Commerce
47. Provincial Agents
48. News Depot/later addition of Royal Hotel, Louis Shultz's Meat Market
49. Bank of Halifax/later Royal Hotel
50. Assay Office
51. Jone's Restaurant
52. Captain Nickerson's General Store
53. Board of Trade Hotel
54 Queen's Restaurant
55. Barber Shop
56. McLennen & McFeely Hardware
57. Hotel Russell Annex
58. Hotel Russell
59. Olympic Hotel
60. Warehouse
61. Atlin Inn, first two wings (1917)
62. Atlin Inn, third wing (1928)
63. Atlin Club
64. White Pass & Yukon Co. Dock & Warehouses
65. Miss Simson's Women's Apparel
66. British North American Co. Hotel, Saloon &Warehouse

* Present day buildings dating from 1900-1903

ber was imported into Atlin only when a project demanded or could afford the cost of prime materials and the freight charges. For most purposes, local lumber was adequate. Atlin's first sawmill was owned by E. G. Tennant.

By chance Tennant had stumbled upon Miller and McLaren working on Pine Creek before their discoveries were widely known.[704] He recognized the signs of a major gold strike, and foresaw the area's future based on his knowledge of the demand for lumber in Dawson City. He anticipated the rush and the staggering demand for lumber it would generate. Seizing the opportunity, Tennant set up a sawmill near the slough in an area that become the south end of Atlin City. He and his partner, a man named Baker, logged and cut lumber throughout the winter of 1898-99.[705] They were well prepared when the stampeders arrived as expected. George Sinclair established a mill in early 1899,[706] and as the only suppliers, he and Tennant could ask high prices. For a time lumber sold for $150 - $200 per thousand board feet.[707]

Their financial success was a clear invitation to others. Rosselli and Ridd, two men who had worked for the Hastings Mill on Burrard Inlet, began cutting in the spring of 1899.[708] Their mill, owned by Rosselli, operated as the Atlin Lake Lumber Company. George Sinclair took on a partner, and they did business as Brown and Sinclair. At the end of the year the Atlin Lake Milling and Lumber Company, managed by Captain Westcott, joined the others. It could produce 20,000 feet per day. Eleven men were employed at the mill, and more in the bush.[709]

For two years, the mills were busy. Permanent buildings replaced tents and new structures filled town lots, and there was repair and rebuilding to do after the 1900 fire. Some local lumber was shipped to Bennett and Whitehorse.[710] All of the mines, small and large were steady customers for lumber. Nevertheless prosperity could not last for all the lumber companies, and by spring, 1901, the market was over supplied. Some of the smaller mills went out of business; others merged. To maintain profit levels, the remaining sawmills formed a combine, and adhered to agreed prices. To a degree, this move was self defeating in that it kept prices artificially high and slowed down construction.[711]

Lumber suppliers in Atlin were seldom long term propositions, and they tended to come and go with the demand for mining timbers. The Northern Power and Lumber Company had a large share of the market for several years,[712] but there were also some small mills operating near the working mines and other outlying areas. There was a mill on Taku Arm and two in the vicinity of Surprise Lake. For years the White Pass Company provided the mainstay for small portable mill operators whose main product was railroad ties.

Another high demand commodity, particularly when the area was overrun with stampeders, was liquor. Hard liquor was shipped in at considerable expense and effort. Beer was enormously bulky, subject to freezing, and would not bring in the same profit, relative to volume, as whiskey did. However, J. H. Russell recognized an excellent market for beer, if it could be stripped of transportation costs and risks, by producing it locally. After solving the problem of getting brewing supplies and equipment through Canadian customs,[713] Russell set up the Atlin Brewing Company in a small building near his hotel on Pearl Avenue in June, 1900.[714] Within a month, his manager, a man named Wehrfritz had produced the first Atlin beer. The new product was well received and pronounced by the *Claim* to be, "first class beer".[715] The temporary plant brewed both lager and "steam" beer in barrels and bottles. They advertised that they could fill orders for all parts of the Yukon.[716]

Russell located his permanent brewery at the north end of town, on a shallow cove, soon known as Brewery Bay. By fall beer was being bottled in the new plant, and production was to continue throughout the winter.[717] At Christmas, Russell complained that he could not supply the demand.[718] In March 1901 he obtained a contract to supply beer to all the White Pass boats on the southern lakes, and the Yukon River.[719] The Atlin Brewing Company was doing extremely well, or so it seemed. Actually Russell's profits did not cover the payment on his debts. A year after opening, a foreclosure forced the operation to a standstill.[720]

The brewing business was too good to ignore, and two years later a small local company acquired the remaining assets of the Atlin Brewing Company. They began production as the Northern Brewing Company in March 1903, with John Nicholl as manager. The brewmaster was a Mr. Youngmayr, an experienced brewer from the Red Cross Brewery in Vancouver.[721] Again the local product was widely appreciated, and this time the company did not overextend by attempting to supply more than local needs.

The brewery was threatened when a fire broke out in its stable in the fall of 1903. In the excitement Fire Chief Rosselli fell down in a pile of broken bottles, cut his hand and arm, and had to be stitched together at St. Andrew's Hospital. Still firemen managed to save the brewery building from any serious damage, and production continued without interruption.[722]

The second year, Youngmayr left. A Hungarian, who had worked in the brewing business for a number of years in Budapest, took over the technical

*The Atlin Brewing Company was established by J.H. Russell in 1903. Later owners named it the Northern Brewing Company.*

*Conrad Wawrecka, second from left, was the brewmaster for the Northern Brewing Company. Here he is joined by friends or family in quarters attached to the brewery.*

job of brewmaster. Konrad Wawrecka insisted on only the finest malts and hops for his main product, lager. Some stout was also brewed.[723] Deliveries to local hotels, and customers on the creeks were made regularly from a wagon driven by Bert Whitfield. In midsummer, when miners took little time off, and trips to town were infrequent, Bert and his cargo were always welcome at camps strung along the more remote creeks.

Wawrecka eventually took over as manager while retaining his responsibility for beer production. He remained with the Northern Brewing Company until it ceased operations about 1910.[724]

Sooty coal oil lamps and candles, both dangerous fire threats, had furnished flickering illumination for Atlin businesses and homes until electricity was available in September 1903.[725] P. F. Scharschmidt and G. H. Sproat bought a power generator in Skagway and shipped it to Atlin.[726] This was the seed of a new enterprise, the British Columbia Power and Manufacturing Company which was managed and operated by Bert Hartshorn, previously the engineer on the *Duchess*.

There was a customer fee of $3.50 for installation, and $3.00 per month for each 16 candle power bulb.[727] The opportunity to enjoy good lighting, especially in the winter, was seized by many businesses and residents.

In conjunction with the power plant, the company established a steam laundry, operated by Mr. and Mrs. Fenn.[728] This business offered a convenient service patronized particularly by the hotels and bachelor miners who had little patience with scrubbing on a wash board.

British Columbia Power and Manufacturing merged with the Northern Lumber Company[729] to become the Northern Power and Lumber Company, and several months later, in the early winter of 1904, built a large new power house.[730] Early in 1905, there were signs that the company was not doing well.[731] When its holdings were destroyed in the fire of 1905, company heads decided not to rebuild the generating plant.[732] It was a bitter blow to citizens who had grown to appreciate and rely on electricity.

For many years electricity was not generated on a community basis, but individuals sometimes were able to make their own arrangements, and those living on the creeks were often able to arrange a connection with a system being used to provide power for a mine. Many years later the Board of Trade acquired a generating plant, and once again there was a community power supply for domestic and business use.

Perhaps the most unlikely industry for Atlin was the brickworks established in 1900,[733] by building contractor, W. H. T. Olive. Clay of excellent quality was found in the area, and Olive had a use for the bricks, since he had just been awarded the contract to build the new Provincial Government building and courthouse. He located the brick plant on the Watson Avenue extension east of Atlin. While the building boom lasted the bricks were used for foundations and chimneys. Olive also put them to industrial use when he built a drying kiln for lumber in 1901.[734]

Olive thought he could export bricks to Dawson City, but transportation was expensive, and the trade did not eventuate. It appears some of his bricks were used in the boiler rooms of the White Pass vessels. His success, however, depended on the local market, and when the initial flurry of building was over, the bricks were no longer needed.

In 1904, when the Northern Power and Lumber Company built their new plant, they required a brick foundation for the boiler. Their plans also included a brick sawdust burner. Thomas Kirkland, a brickmaker by trade, contracted to supply 20,000 bricks for the project.[735] He hoped the contract would extend into an ongoing business,[736] but as Olive had discovered the demand did not exist.

Two energetic young Vancouver newsmen, Harry Cowan, and W. J. MacKay answered the call of the gold rush.[737] They arrived in town in the spring of 1899, and started the town's first newspaper, the *Atlin Claim*. They set about immediately to gather news professionally, and to represent the views of all segments of the community. They also vowed to have nothing to do with "booming."[738] This was the unethical practice of provoking a rush to some area solely on the strength of unfounded stories about new discoveries. Atlin had been plagued by "booming" on the part of Vancouver and Victoria businessmen, eager to tout a gold rush for British Columbia.

The first edition of the *Atlin Claim* appeared on April 29th, 1899. It was published every Saturday thereafter. In true newspaper tradition, nothing was allowed to delay an issue. When it was discovered that the type brought for the advertising columns did not include any spaces, the newspapermen improvised by using sawdust to separate the words in the copy.[739] From the start the four-page weekly assumed an important role in the community. Its offices were in a large, rough building at the corner of Rant and First. A few months later they moved to more convenient quarters on Pearl Avenue.[740]

Cowan and McKay were prominent in local affairs, as well as being deeply involved with the *Claim*. The hectic pace was perhaps too much. In the summer of 1900, Cowan left the paper.[741] That fall MacKay closed the office indefinitely to return south for a holiday.[742]

Publication resumed just before Christmas under new management. A. C. Hirschfeld, a staunch "Atlin-believer" viewed the newspaper as an important asset to the community. He was, however, deeply involved in his own photography studio, and a number of mining properties. He employed D. Todd-Lees, as sub-editor, and the production of the *Claim* was largely Lees' work.[743] James Simpson, with help from Alfred Carmichael,[744] worked as sub-editors during 1904.[745] They entertained themselves and readers during one uneventful winter with tales about a fictional miner.

Hirschfeld, anxious for the *Claim* to succeed, tried several innovations. When telegraph service reached Atlin in September, 1901, he began publishing the *Atlin Telegraphic Bulletin* twice weekly. It was a supplement to the *Claim*, including the latest national and international news, direct from Skagway.[746] Unfortunately the cost of publication could not be met by the returns from a small population. Later, in 1903, he tried expanding the *Claim* from four to eight pages. The extra space was filled with short stories and serials.[747] This failed to increase circulation, so it too was dropped.

Hirschfeld sold the paper in the spring of 1905 to W. Pollard Grant,[748] a lawyer and active political leader in town, and the best days of the *Atlin Claim* were over. For the previous six years, the paper had stayed clear of involvement in local political issues, but Grant was moved to use the *Claim* as a voice for his own political opinions. Dissenters canceled their subscriptions and withdrew advertising. Grant's local following was too small to support a partisan publication. The final blow came over the Japanese issue, when miners were angered by Grant's pro-Japanese stand.

In March 1908, Hubert Faulkner took over the *Claim*.[749] His earnest effort to regain lost readers and advertisers had no significant results. The damage done by Grant could not be quickly mended. The population was shrinking and he could not reestablish the newspaper on a profitable footing. The edition of April 11, 1908 was the last.

The *Claim's* only competition was a short-lived threat, the *Atlin Globe*. It was first published on August 30, 1899, and appeared intermittently until the spring of 1900.[750] The owner and publisher was William Baillie, a competent journalist, whose opinions were often quoted by other newspapers and journals. Baillie, in the style of a frontier editor, could do zero-hour improvising if necessary to get his paper on the streets. When a shipment of newsprint failed to arrive, he printed on wrapping paper. It was distinctly pink in colour, but served the purpose.[751]

Despite Baillie's talents and acumen, there was not enough business in Atlin for more than one paper. The *Globe* changed hands twice, and in May 1900 the plant was sold to the *Whitehorse Tribune*.[752]

In 1936, A. E. Windle started publishing the *Atlin Nugget*. Nine or ten issues appeared before the paper's offices went up in flames, destroying the press and all other contents. The incident had suspicious overtones since Windle had suffered a similar misfortune once before in Burns Lake.[753]

The need for a community newspaper was addressed by the Board of Trade in 1938. They arranged with Horace E. Moore, editor of the *Whitehorse Star* to publish a weekly paper for Atlin. The publication was named the *Atlin News Miner*.[754] Moore was assisted by an Atlin committee appointed by the Board of Trade. The committee's purpose was to funnel local news to the Whitehorse publisher. A young miner named Tom Kirkwood, was asked to be the reporter for Spruce Creek. He regularly covered the local news on his beat around the mines, and in the small community of mining families who lived there. Unfortunately the *Atlin News Miner* was sprinkled heavily with news from Whitehorse, Dawson City and Mayo, and Atlinites seemed to have trouble identifying with it. In turn the editor complained about the lack of subscriptions and advertisers.[755] World War II brought a halt to the publication in December 1943. Almost 30 years would pass before Atlin had another newspaper. In 1972, the same Tom Kirkwood became editor of the second *Atlin News Miner*.

A weather vane of a town's financial health is its hotels. They, more than most businesses, are sensitive to the vagaries of economic change, because clientele must come from outside as well as within the community. A quiet town attracts fewer visitors and house receipts fall as operating expenses remain constant. A thriving hotel, on the other hand, is an open invitation to new competitors.

Andy Galarno opened the Atlin Hotel on Watson Avenue in 1898.[756] Before the year ended four more licences were issued in the area, and Galarno's brief monopoly was broken.[757] By 1899 there were at least 19 hotels in Atlin, and 10 in Discovery. There were hotels at Surprise City, Spruce Junction, and Taku Landing.[758] There were also any number of unlicensed, makeshift saloons on the creeks doing good business, and being heartily protested by the hotel owners.[759]

For the most part, the first hotels were small, dingy, and devoid of embellishments. They were built at a minimum cost, and offered the barest essentials: liquor, smokes and bunks. Their advertisements in the *Claim* boasting otherwise were grossly misleading. In May 1899, the newspaper announced the opening of the Exchange Hotel, which offered the "choicest liquors and cigars in the city", and "table and beds unsurpassed"[760]. A week later, the Exchange, along with five similar places, were closed by the police for noncompliance with building standards.[761] The Exchange was a tent.

A definite improvement over the tents were log and frame hotels such as the Discovery, the Board of Trade, and the Vancouver. Most of these started business as little more than barns. Inside their arrangements were similar. The street level floor was the saloon. A long bar, perhaps with a mirror behind, stretched as far as space would allow. A decorative touch was often a gilt framed painting. The most popular subject was a voluptuous woman languishing on a divan, provocative in scanty gossamer scarves, with lowered eyes fixed on patrons across the counter. Her mere presence probably accounted for a small but steady per cent of the daily receipts. At Smith and Creed's Metropole Hotel, advertised as a family establishment, there was no bar room nude. Instead graffiti decorated the walls - "God Bless our Home", and "Leave your Dog Team Outside."[762]

Brass spittoons were placed strategically on practical sawdust floors. Secluded at the back, but known to all, including the police, were the gambling rooms. Upstairs there were one or two large rooms lined with iron bunks.[763] These dormitories were the credentials which distinguished hotels from common saloons. Basic frontier hotels such as these were usually retreats for men only and few women would have ventured inside.

Part of a hotel's appeal was the personality of the owner. A successful host was everybody's friend. He had a ready ear and his advice was usually good. He looked after his patrons and they were loyal. One of the most colourful was John Kirkland, sometimes called the "Mayor of Atlin". He was a rough diamond but honest and kind-hearted.[764] He was once described as a "burly salonkeeper who loved nothing better than an argument, unless it was a fight".[765]

Dance hall girls and hurdy-gurdy girls were not part of the Atlin hotel scene, but most owners promoted gambling, even though it was illegal. Faro, roulette, black jack and poker added to the profits, and hotel keepers were not easily dissuaded by the officials. The police raided frequently, laid charges, collected fines, and the games went on. It was often the gaming tables which kept hotels solvent during the winter.

As the smaller, shabbier hotels closed, others began settling into permanence and improving their amenities. The Kootenay installed billiard tables,[766] the Vendome added a bowling Alley,[767] and the Hotel Russell had a fine dance hall upstairs.[768] Other establishments improved their dining and sleeping accommodations. The Gold House in Discovery, owned by Letherdale and Burton was renovated thoroughly. When done, it featured an elegant staircase and entrance hall, a dining room with high backed chairs, and stylish table settings. The guest rooms were named, starting alphabetically, Atlin, Bennett, Cassiar, Dawson, Egypt, and so on.[769]

In the spirit of competition, hotels offered enticements to hotel patrons. The Leland Hotel advertised "A Boon for the Thirsty - Two Drinks for a Quarter."[770] Dining room menus were sophisticated by any gourmet's standards. One visitor commented, "You may have any delicacy New York could afford."[771] A Kootenay Hotel menu attested to that. It included lobster, broiled ox tongue with mint sauce, baked trout or chicken with oyster dressing. There was prime rib beef au jus, creamed potatoes, French peas, sliced tomatoes, all of which could be topped off with lemon custard pie, banana fritters with rum sauce, or English plum pudding with brandy sauce.[772]

The hotels staged entertainments throughout the year, and particularly at the close of the mining season. The Balmoral Hotel hosted the typically rowdy affairs so popular in Discovery. One celebration in 1903 featured wrestling between champions from Butte, Montana, and Belfast, Ireland. There were two boxing matches with local contenders. Less athletic was a competition between two patrons - a race to see if one could drink a bottle of whiskey before the other smoked two entire cigars. The evening included howling bagpipes and highland sword dancing. Patrons wended their way home as the sun rose over Spruce Mountain.[773]

There was no question that the premier establishment was Atlin's Grand Hotel. It belonged to George L. Rice, a Juneau businessman who had a number of other hotels and restaurants throughout the north, and David Hastie, who also managed the Grand. One of the first permanent buildings in Atlin, the hotel was built on the corner of First Street and Discovery Avenue during the winter and spring of 1898-99[774] at a cost of $14,000.[775] It was Rice's intention that the Grand would be the finest house in the north, and certainly there were few rivals.

The splendid ballroom was used for concerts and gala events. All rooms, public and private, were comfortable and tastefully decorated. The busy dining room, which had a number of private boxes, featured French cuisine. Lobster, crab, shrimp, caviar, lamb, imported cheeses and ice cream were served regularly.[776]

By 1902, there were only eight hotels in Atlin and six in Discovery. The Spruce Junction Hotel had burned to the ground, and the proprietor, W. Joyce, injured and apparently on fire, had thrown himself into the creek, and drowned.[777] The Surprise Lake Hotel and the Dawson Hotel at Taku were still in business.[778] The next year the numbers had dwindled further. There were still six hotels in Discovery,[779] but only five in Atlin.[780] At Spruce Junction, Jim Clark was preparing to build a new road house adjacent to the original Half-way House, which he felt was past its usefulness.[781]

In 1902, David Hastie turned management of the Grand Hotel over to A. R. McDonald,[782] who owned both the Irving and the Royal in Discovery. In 1904, the Grand was sold to Fred Wrong, who ran it for less than a year before it was closed for good.[783] The best hotel in Atlin then became the Royal.

The Royal opened in 1901 in the abandoned Bank of Halifax building. Dr. F. T. Troughton, the owner, had transformed it into a reputable and comfortable hotel, the chosen accommodation for many visiting mining men.[784] After several owners, it was purchased by Dixon and Schulz in 1905.[785]

Their partnership soon dissolved, but Louis Schulz retained ownership, of the hotel. He made the Royal the centrepiece of his many Atlin holdings. The well maintained premises did good business throughout the busy tourist era. Competition was, by then, from only two other Atlin hotels, the White Pass Inn built in 1916, and the Kootenay.

The Kootenay survived to be the longest continuous hotel business in Atlin. Jack Byrom, a popular and jovial host, built the first Kootenay House in 1899.[786] On Christmas Eve 1900, a fire destroyed the second floor. Damage was extensive, but Byrom and his friends hauled out furniture, billiard tables, liquor, and cigars before they were ruined. Because the Christmas trade was too good to miss, Byrom was back in business within three hours of the blaze. He simply moved his rescued assets to the nearby Hotel Northern building, which he held under lease, and opened the doors to his customers.[787]

Before reconstruction of the Kootenay was complete, Byrom sold it to George Hayes, who also owned Discovery's new Nugget.[788] Hayes made a number of improvements, adding a well equipped gymnasium, a dining room, and kitchen.[789]

The Kootenay was one of the few hotels to hold events designed to entertain the ladies.[790] It hosted a clientele drawn to an interesting variety of entertainments. Hayes sold out in 1904, and went to Oregon.[791] There were several other owners until, in 1910, the Kootenay came to rest with John Roxborough.[792]

AHS, Henning Collection P-1250

*The Merchants Bank of Halifax building on Pearl Avenue and First Street was remodeled into the Royal Hotel in 1901. Later, when Louis Schulz owned it, he added the far left wing for his grocery store and meat market.*

AHS P-168

*The Grand Hotel was the premier establishment in Atlin during the gold boom.*

AHS P-1080

*The Hotel Russell has a fine dance floor upstairs.*

BCARS HP 72680

*The Hotel Northern on First Street was designed by architect Edward Garden in 1900.*

Roxborough was an Irishman from Dublin. He joined the gold rush in 1898, and landed in Atlin in 1899. He owned and operated the Surprise Lake Hotel until 1906.[793] There was a post office at the hotel, and it served as the terminus of the stage route between Surprise City and Atlin. Roxborough was the manager, and perhaps the owner, of the Gold House in Discovery for a while also.

Fire struck the Kootenay again, when it was leveled along with the rest of the business district in 1914. Another hotel was built over the ashes. At the Kootenay, Sunday meals were served in a sunny dining room on tables covered in white linen. It became the favourite retreat of tourists who returned to Atlin year after year. The Roxborough family remained prominent in the hotel business in Atlin for nearly 60 years.

Charles Oubridge, a young Englishman, was one of the first to establish a business in Discovery.[794] Early in 1899, he opened the Pine Tree Hotel, which was destined to be the longest operating hotel on Pine Creek. Unfortunately Oubridge did not survive to enjoy the success of the Pine Tree. In September, 1899, he was stricken so suddenly by an illness that he fell from his horse, as he rode up to the hotel and died just a few days later.[795] His death shocked the entire mining community, which had dubbed him, unofficially, the mayor of Pine City.

Edward Sands, another Englishman, acquired the Pine Tree, and made numerous improvements. He built a large extension,[796] and turned it into a solid, spacious structure. Its soaring false front dominated Discovery from its position at the east end of Pine Avenue. The Pine Tree was not the grandest hotel in town. It lacked the architectural refinements and decor of the Gold House, the Balmoral and the Nugget but it was well patronized. Its gaming tables were an irresistible enticement for men from the creeks, and gold flowed with the liquor. The law caused Edward Sands some bother, but the gambling room remained a busy place.

The Pine Tree was taken over in 1912 by Clarence Sands, Edward's nephew, who had already been involved with its operation for several years. In 1923, Clarence closed the doors when there was no longer enough business in Discovery to support even the one remaining hotel.[797]

Over the years the merchants proved to be the real strength of the business community. In the beginning there were too many retailers for the market created by the stampede, and many amalgamated with their competitors. Large established companies which sent in agents to open branch stores never lasted long. Parson's Produce from Winnipeg encountered problems when they launched a scheme to transport fresh produce from the prairies to Atlin and Dawson City. After a year of big losses, they withdrew from both markets.[798] The Atlin business was purchased by two former employees, Nat Wheeling and J. A. Fraser. Another large company which soon sold out the Atlin branch to an employee, was Thomas Dunn, a Vancouver hardware merchant.[799] The purchaser in this case was Alex Godfrey, who decided that Atlin had enough hardware stores, and he shipped his stock by barge to Dawson City.[800] When the boom faltered, other companies, such as the British America Corporation, McLennan & McFeeley, and Kelly Douglas, all vanished like phantoms.

A gold rush partnership, whose influence would be felt throughout the Yukon, and particularly in Whitehorse for more than 75 years, was that of Isaac Taylor and William Drury. In time, these two Englishmen would have 18 stores across the Yukon, but their first opened in Atlin in the very early Spring of 1899. Taylor had $200.00 earned working on the track gang for White Pass. Drury had a 3.5 x 4 metre tent. In the spring of 1899, the pair made the best possible use of their meager capital by buying outfits from discouraged stampeders with Taylor's money, and selling the equipment from Drury's tent. Theirs was a common business all along the gold rush routes and in the mining camps, and it was a lucrative one, until merchants set up a steady source of supplies. Taylor and Drury made substantial gains, and Taylor left for Victoria to buy some new stock. On his return in July, he decided they should move their enterprise to Bennett City. The railroad had been pushed through to that point, and there was frenetic boat building activity along the lake shore. He sent word to Drury, who wound up business in Atlin, and left to join him.[801]

Patrick Burns, the "Western Beef King" won big profits selling meat in the north after gold was struck in the Klondike. He was quick to open a store in Atlin, and supplied it with cattle driven in on the hoof, and slaughtered on arrival.[802] When Burns decided to withdraw from Atlin in October 1901,[803] he gave the store to a 21 year old employee, Louis Schulz.[804] This generous gift served to keep Schulz in Atlin, a happy circumstance for the future of the town.

Young Schulz converted the butcher shop to a general store, and soon built a branch in Discovery.[805] With an eye for a good proposition, and a good head for business, he added to his holdings year by year. Thirty-five years later, Schulz still operated his store, but he also owned the Royal Hotel and dining room, a fleet of sightseeing cars, and the Chevrolet dealership. He owned two sawmills, he was a big game outfitter, and kept a hand in the fur trade.[806] His freighting business included local hauling, and he ran winter cat trains be-

AHS P-1251

*In Discovery City the Commission and Storage business dealt in the second hand goods of men who came to the Atlin gold fields during the rush, but left soon after for other camps or to return home.*

*The Nugget Hotel, owned by Jack Byrom, was on Pine Avenue in the center of Discovery City's business district.*

BCARS #91369

tween Atlin and Carcross. He carried passengers, and provided accommodations for them at Moose Arm for one dollar a night. Until airplanes took over the mail service, Schulz often held the winter mail contracts, and he was an active investor in mining.

Although he was generally believed to be a hard businessman, he was also known to be exceedingly fair and generous. Seldom did he turn away a miner in need of a stake. Many men could never have pursued their dreams or tested their hunches without Louis Schulz's support.

It could also be said of Schulz that he kept Atlin going during the lean years. Sometimes he invested everything he had to bring in winter supplies. He paid wages all winter to men cutting lagging[807] for the underground miners, knowing he could not recoup his investment until summer.[808] Schulz spent the best and most productive years of his life in Atlin, engaged in a variety of enterprises, exemplary in his commitment to neighbours and the community. He retired and left Atlin in 1941.[809]

Another of his ilk was Edwin Pillman, one of the most astute and enterprising businessmen Atlin knew. He was a born entrepreneur and the gold rush provided a backdrop against which he could test his talents. He arrived at the head of Lynn Canal in 1898 with his wife, Elizabeth, and his daughter, Ethel. He opened stores in Dyea and Skagway, and introduced that city to its first steam peanut roaster.[810] Ever an opportunist, Pillman headed for Atlin in 1899, and established a grocery business which was one of the first in town. It is probable that he had no previous experience as an undertaker, but he placed an advertisement in the first issue of the *Claim*: "E. L. Pillman, Mortician. Orders on short notice."[811]

As the miners spread along the creeks, Pillman was close behind. Soon he was selling groceries in Atlin, Discovery, and on Spruce Creek.[812] On Spruce Creek he offered a courier service to Atlin.[813] When the miners stampeded to the O'Donnel River in 1912, Pillman set up a store there. It was a source of amusement among his customers, although probably not all of them, that his faithful horse Bess pulled a wagon which served both as hearse, and for grocery deliveries.[814]

Not satisfied with two or three lines of business, Pillman started market gardening on a small productive plot in town. He harvested hay north of Atlin on a hillside first used by the Shirley brothers, but soon called Pillman's Hill. He also rented out bicycles,[815] and often rode one himself in the course of his business day. This provided entertainment for two young lads who liked to "borrow" his bike after Pillman went home, enjoy a spree, then quietly return it to his place on Pearl Avenue. Pillman was not fooled, though. Well aware of these escapades, one night he removed the bike's chain, and then waited. Out of the darkness came young Jackie Williams and Willie Campbell. Willie hopped on the bike, and headed down Pearl Avenue. In seconds, he was rocketing, out of control and heading straight for the lake. Helpless, Jackie waited for the inevitable splash that put a spectacular stop to their after hours pranks.

Elizabeth Pillman was an active partner in her husband's ventures. Her support made it possible for him to diversify their business. She worked in the Atlin store and took charge of their large circulating library. For a while, she leased the Hotel Northern, and rented the rooms.[816]

The Pillmans lost two stores at Pearl Avenue and First Street. Both burned in the fires of 1914 and 1916. Each time they rebuilt,[817] and after the second fire they also built the Globe Theatre, a few doors up on Pearl Avenue. Pillman brought the first silent movies to Atlin. To ensure a full house for each showing, he sent a young boy around town ringing a bell as a reminder the movie was about to begin. If the attendance was low, he sent the bell ringer around again, while patrons waited patiently. The bell ringing would earn the lucky lad a free seat to the show.[818] Pillman's cumbersome carbon arc projector whirred and flickered as he turned the hand crank, and scenes danced across the screen. It was tiresome work, and if Pillman got bored and sleepy, he sometimes nodded off, and the action ground to a halt. The audience would shout its displeasure, the projectionist would jerk to, and soon actors would be bolting across the screen at high speed. Later the projector was adapted to electricity, and in 1931, Pillman brought Atlin movie-goers their first "talkie", just six months after Vancouver viewers first enjoyed them.

Long past retirement age, the Pillmans, known as Pills and Pilly to Atlinites, retired from their only remaining business, a magazine and candy store, and moved south. Many stories were told at Ed Pillman's expense, because he forged his own busy path, and settled problems in unique, often comical ways. A cleft pallet made speech difficult for him, and invited imitators. Pillman was nevertheless a generous man, and his enterprising nature furnished the Atlin people with services they would not have enjoyed without such an innovative pioneer in their midst. Pills and Pilly lived and worked in Atlin for nearly half a century, a kind of frontier conglomerate in themselves.

Jules Eggert's fine jewelry stores lent an air of refinement to the business districts of Atlin and Discovery for more than 20 years. He came to Atlin in 1899 from Switzerland where he learned the trade of watch repair and jewelry making. At first he had two other competitors, but he took command of the market and the others disappeared. Eggert settled on First Street, but by 1902 he had replaced his first shop with a smart new store complete with oak cabinets and show cases.[819] He sold fine china and silver, and his own artfully crafted gold nugget jewelry. When his son, Paul, joined the business in 1903 another store was built for him in Discovery. The jewelry business sagged when the rest of Discovery's economy faltered, but in Atlin the lively influx of tourists meant a good trade for Eggert. He was a dedicated community man, always active in the Board of Trade, and, in 1907, he was made a Justice of the Peace.[820] In 1922, he erected a clock on the boardwalk outside his store. Originally its huge face was lit from inside, and at night it resembled a full moon hanging low over First Street. Jules died that same year, and Eggert's Jewelry business passed to his son Paul.

Many merchants in Atlin and Discovery provided long, faithful service. H. E. Brown had clothing businesses in Atlin and Discovery for 14 years. Isaac Matthews, a miner, also ran a grocery store on Spruce Creek. Margaret and James McIntosh were in the grocery business for 30 years. James sold hundreds of cords of wood each year, all cut by hand. Margaret helped fill the need for a bakery. The inviting aroma of fresh baked goods always brought customers to their store. Charles Gairns lost a leg in a mining accident. For a new line of work he bought the B.C. Hotel in Discovery, and eventually had a small store on Discovery Avenue in Atlin.

When Clarence Sands closed the Pine Tree Hotel, he opened a general store on First Street and Trainor Avenue. He also skidded a section of the old hotel to Atlin, and remodeled it into a home. When Sand's Store burned down in 1929, along with the Kirkland Hotel, and the Government Liquor Store across the street, he rebuilt. After Clarence died in 1954, his wife, Signe, stepped in as proprietor. The next year the store was destroyed again in a fire thought to have been started from a cigarette tossed carelessly on the boardwalk. Her elder son, Leslie built another store for her on the same site. When Signe finally retired and moved from Atlin, at about age 75, the Sands family had been in business there for over 60 years. Only a few years would pass, though, before Clarence and Signe's younger son, Lyman, a retired Government Agent, returned with his two sons to reestablish the Sands name in the Atlin business scene.

While others were lured by the search for gold, the merchants and business people resisted shrewdly, and reckoned on other fortunes gained through providing goods and reliable services. Those who stayed beyond the first rush were motivated by a faith in Atlin, a commitment to her people, and a love of the north, and they formed the backbone of the community.

*In 1902, The B.C. Hotel was one of six hotels operating in Discovery City.*

CHAPTER 11

# Links to the Outside

*In a northern mining community, the dog plays such an important part that he ranks second only to humanity itself.*

- The *Atlin Claim*, May 26, 1900

"Outside" was a word of special significance to the people of Atlin. It meant destinations at the end of long, sometimes difficult, journeys by boat or sleigh, train and coastal steamer- those places a world away from the raw wilderness of the Atlin Lake country. "Outside" - somehow this single word characterized the extent of the isolation felt by those willingly held captive by the seclusion and harsh winters of the north.

For all its brash new independence, the welfare of Atlin was closely connected with the "outside," but distance and trying conditions could turn simply coming and going, bringing in equipment, supplies or a letter from home, into grand enterprize, which was sometimes cast with heroes and overshadowed by tragedy.

By the early spring of 1899, any businessman who hoped to make profits from the anticipated influx into the Atlin area, was engaged in serious preparations. Not the least of these were the transportation companies headquartered at Bennett City, which by then had a large population of stampeders turned boat builders, and a wide assortment of businesses to serve and supply them. Since 1897, Bennett Lake had been the head of Yukon River navigation with large sternwheelers steaming from that port on more or less regular schedules along with the motley fleets of individually designed rafts, boats, and barges.

The Bennett Lake and Klondyke Navigation Company had used its 18 metre steamer, the *Nora,*[821] on a twice weekly run to Taku Landing for the new Atlin trade in 1898. At that time travellers were left to their own resources to move their freight over the 3.6 km portage, and to find some means of transportation across Atlin Lake.

The B. L. K. N. Company, which had a virtual monopoly of transportation on the lakes, was organized by the notable Victoria architect, F. M. Rattenbury. He had every intention of cornering the lion's share of the Atlin business in 1899, a route his company had pioneered the year previously. Experienced competition, however, was prepared to challenge, and furthermore, to provide passage right to the wharf in Atlin.

The threat to Rattenbury's secure position was Captain John Irving, a man of imposing stature and personality, the consummate entrepreneur, and a man of the frontier. He was a veteran of hard fought competition on the Fraser, Thompson, and Stikine Rivers, and seldom had he come out the loser. John Irving was regularly centre stage in dramas of his own making. In 1898, he set a dazzling record at the helm of his steamer, *Yukoner*, on a speed run from St. Michael to Dawson City.[822] He operated with a certain authority and flare, and his influence as one of the sitting Members of the Provincial Parliament for Cassiar, added to his might.

He set out to capitalize on the gold rush when, in 1899, he founded the John Irving Navigation Company, under his own management. Familiar with northern conditions, and aware that timing was important, he managed successfully to have suitable vessels built, launched and in service, at the very beginning of the season when the demand was greatest.

Irving commissioned James Carroll at Bennett City to build a small sternwheeler, the *Ruth*. He must later have entertained doubts about her size, 32 tonnes and only 15 metres long, because in January 1899, just a month after she was registered, the *Ruth* was sold.[823] However, by spring 1899, Irving had two other vessels under construction.

Irving himself, supervized building the *S. S. Gleaner* at Bennett City. It was 241 tonnes,[824] 34 metres long, and 7.5 metres wide.[825] She was launched on May 2nd, 1899,[826] and put immediately into service between Bennett and Canyon City, just up-river from Whitehorse, until Tagish and Atlin Lakes were free of ice.[827]

At about this time, the *S. S. Scotia* was being assembled on Atlin Lake near the Atlintoo River. Apparently prefabricated, the 24.5 metre long vessel had been hauled in over the ice in the early spring, and made ready to sail in less than two months.[828]

Some doubt surrounds the origin of the *Scotia*. One source claims she was built in Chicago and shipped west, where Irving saw her on the wharf in Seattle, bound for St. Michael. He purchased her after persuading her owners that she was unsuitable for service on the Yukon River.[829] A contemporary report in the *Atlin Claim*, however, states the materials for the *Scotia* originated in Portland, Oregon.[830]

Wherever the design work and the materials for the *Scotia* may have originated, it is difficult to comprehend the problems of landing every last piece on the shores of Atlin Lake. Long, heavy timbers, the engine, and a nine tonne boiler were brought to the summit of the White Pass by train, then hauled across rough terrain and frozen lakes on horse drawn sleds.

The ice went out on Atlin Lake during the first week of June, 1899, and the John Irving Navigation Company was the first to offer a full service from Bennett to Atlin City. The *Gleaner* sailed to Taku Landing, where the company had a railway of sorts spanning the portage,[831] and the *Scotia*, under the command of Captain W. E. Spencer, provided the last transportation link in the service to Atlin. The compact little boat, loaded with freight, arrived on its maiden cruise on June 7. A grand crowd, including a string band belting out the "Star Spangled Banner," sent enthusiastic greetings from the shore. A young community was jubilant to finally have adequate transportation for passengers and freight.[832] The situation was further improved when, a month later, on July 6, White Pass Company ran its first train as far as the head of Bennett

AHSP-2006

*The S.S. Scotia was put into service in 1899 by Captain John Irving and reigned as Queen of Atlin Lake until 1918.*

Lake.[833] This provided Atlin with a convenient transportation link to the "outside".

John Irving dominated Atlin's transportation scene for several weeks amid loud complaints about high fares, freight rates, slow service and over loading.[834] Deeply resenting his intrusion, the Bennett Lake and Klondyke Navigation Company viewed the general dissatisfaction as an opportunity to retaliate against Irving for so cleverly breaking their comfortable monopoly. With little delay a parallel service at competitive rates was initiated. The company even offered to pay railroad freight from Skagway, and duty at Bennett, for merchants shipping in goods.[835] The *Olive May* made the two day trip from Bennett to Taku Landing, and the company employed a horse drawn conveyance for crossing the portage. At Atlin Lake two steel hulled sail boats, the *Nellie* and the *Ellwood*, plied the distance between Scotia Bay and Atlin.[836]

Irving did not take kindly to the competition, and countered by laying claims that the *Nellie* and the *Ellwood* were carrying passengers which they were not licensed to do. When authorities upheld this complaint, the navigation company had to make other arrangements.[837] The *Ruth*, meanwhile, had been bought by Captain C. E. Bowden, and he agreed to bring his little sternwheeler to Atlin Lake to replace the sailboats, so the company could maintain connections with the *Olive May* and the fast *William Ogilvie*, owned by the Teslin Yukon Steam Navigation Company,[838] and advertized as the "North American Greyhound".

Bringing the *Ruth* to Atlin was no easy task. It took three days to line her metre by metre up the turbulent waters of the Atlintoo River.[839] When she finally arrived at the wharf at the foot of Discovery Avenue, which had been built by public subscription, Captain Irving stepped forward to inform Captain Bowden that the wharf was under his control. Furthermore, he would be charged a $5 docking fee and wharfage on every ton of freight he unloaded. Irving would brook no competition, and had wielded his considerable influence to gain control of both of Atlin's wharves.[840] Understandably Bowden was furious, but there was little he could do. He had to tie up to a log boom at Rosselli's sawmill at the southern end of town.[841]

The war between John Irving and an alliance of the other companies raged for several weeks. A charge was laid against Irving claiming the *Scotia* was operating without a qualified engineer or sufficient life jackets. It was thrown out of court on a technicality, but the Judge noted that the charges were nonetheless true.[842] By mid September, the struggle appears to have ended when the

*Scotia* was laid up for the season because of some disagreement between Captains Irving and Spencer.[843]

Throughout that first season Irving was also plagued by disputes over his right to access across the portage from Taku Landing to Scotia Bay. The Atlin Short Line Railway and Navigation Company was incorporated by a number of Victoria businessmen in February of 1899.[844] Their aim was to build a tramway or railway across the isthmus, and they "acquired powers to construct telegraph lines, to utilize the Atlin (sic) River for electric power purposes, and to build wharves and other facilities for steamship service."[845] When the line was officially opened on June 6th, 1899 by manager, J. H. Brownlee, the railway was actually just a wagon road.[846] For the first year goods were horse drawn along the road paralleling the Atlintoo River.[847]

Throughout the summer Irving made various attempts to gain the portage. He even tried to close down the wagon road, by pre-empting the land, and fencing it off, thereby blocking off the route,[848] which put him in control of it for a time. Repeatedly though, the Minister of Lands, F. Carter Cotton, requested Gold Commissioner Graham to intervene.[849] By fall, Irving had lost the struggle, and the right of way had reverted to Brownlee's control.[850]

The following March there were further disputes as Irving began to build his own railway.[851] In anticipation of emerging victorious from the struggle, he had, the previous season, brought in the area's first steam locomotive. The Duchess was built by the Baldwin Locomotive Works in 1878, and Irving purchased her from the Wellington Collieries on Vancouver Island. She was shipped north on the *S.S. Danube*, hauled by train to Carcross, and barged to Taku Landing.[852] However, there were further complaints from the management of the Atlin Short Line Railway, and in May, all work was suspended on Irving's project.[853]

The whole matter was resolved in June when the combined interests of the White Pass Company and the Canadian Development Company acquired the assets of the John Irving Navigation Company and the Atlin Short Line Railway.[854] This development marked the beginning of a long period during which most of Atlin's transportation services were controlled by the White Pass.

Work on the railway proceeded quickly, and on 18 July 1900, the *Duchess* pulled its first load across the portage.[855] Despite the fact that, at $2 for about four kilometres, the fare was probably the most expensive in the world, the ride was far from comfortable. The first accommodations for passengers were simply the flat cars where they rode among bundles and crates of cargo,

NA #2114-22 Glenbow Archives, Calgary

*The locomotive, Duchess, was purchased by Captain John Irving for use on a rail portage between Taku City and Scotia Bay. (1899)*

brushing off cinders, and jumping down to stamp out small fires that started along the tracks. Later a car with seating for 48 passengers was added. It was a large open air coach and along each side was a row of windows which were never intended to hold panes. Instead, roll down curtains were provided, but flying sparks, soot and nesting swallows that fluttered along beside the car, continued to plague customers. At first there was no turntable or round house, so the train travelled forward in one direction, and backwards the other way. The little *Duchess* had to make a run for it in order to crest the seven percent grade, and passengers were sometimes asked to help push.[856]

The White Pass was now firmly in control of the entire route from Vancouver to Atlin, and making travel arrangements was a simple matter. Although the distance was great, it could be covered in reasonable comfort, and passengers found considerable enjoyment in the picturesque beauty of the route. Its popularity would grow.

On Atlin Lake, the *Scotia* captured most of the passenger business, but there were other vessels used for a variety of purposes. The steamer *Blair Athol* was a 10.5 metre tug designed to tow scows and carry passengers. She was built in Atlin by Joseph Morrison, an innovative engineer,[857] who used hickory from old sleighs for part of her framing. She was launched in August 1900, and her first charter was to the copper mines at the south end of Atlin Lake. She was operated by a number of owners until 1904 when she burned to the water line.[858] Destruction by fire had also been the fate of the *Ruth* in 1901.[859]

William J. Smith had ventured without notable success into the hotel business, auctioneering, and blacksmithing in Atlin. His Hotel Metropole was a popular, if somewhat rustic establishment, and he was widely known as Metropole Smith, long after the hotel ceased to exist. In 1904, he turned his attention to the lake, and built an unattractive, but powerful side-wheeler steam tug with two engines.[860] The *Atlinto* was used mainly for towing freight scows, and ferrying prospectors. Although most of Smith's financial ventures seemed fraught with accidents, and the *Atlinto* had its share of mechanical and structural problems, this enterprize appears to have succeeded, because in 1906 he added 5 meters to the length of the boat. The changes to her super structure significantly improved her lines, and she was renamed the *Glengarry*.[861] Metropole Smith also owned a large sail boat, the *Lady of the Lake*. This vessel was used mostly for pleasure, and suffered severe damage on a couple of occasions.[862]

The Atlin vessel in use over the longest period of time was probably the *Gladys*. Perhaps built as early as 1899, she was used on the mail run, to haul freight, for fishing parties, and finally to carry ore from the Ruffner Mine on Fourth of July Creek.

Another common sight on Atlin Lake until about 1908 was the *Tagish* which was a 6.5 metre gas driven vessel used both by the British Columbia Provincial Police and the Gold Commissioner. She was commonly known as either the "Tag" or the "Government Yacht."

The unchallenged Queen of the Atlin fleet, however, was the *Scotia*. In addition to scheduled service between Scotia Bay and Atlin, she made special deliveries of livestock, machinery and dynamite.[863] Most fondly remembered were the many sightseeing excursions she made to remote beauty spots on the lake. Nosing up to an inviting beach, passengers spilled ashore with picnic baskets and Kodaks. On some special occasions a barge was towed along, and attractive tables were set under a canvas canopy festooned with swags of greenery, where passengers enjoyed an elegant wilderness repast.

In 1915, the White Pass faced the need to increase its carrying capacity for both passengers and freight. That winter pre-cut timbers for a new vessel were brought to Atlin over the ice.

During the winter, a ship 24 metres long with a 6 metre beam took shape in front of the townsite, and on June 19, 1916, she was named *Tarahne*,

*The M.V. Tarahne, built for the White Pass Company in 1917, carried cargo and passengers between Atlin and Scotia Bay.*

and launched.[864] She was the first gasoline powered propeller-driven vessel in the White Pass fleet, an experiment for a rapidly growing company. All of the other vessels on the Tagish Lake and Yukon River system, including the *Tutshi*, built for Tagish Lake the same year, were steam-powered and fuelled by thousands of cords of wood.

White Pass ran both the *Scotia* and the *Tarahne* until, in 1918, it was decided that only one vessel was needed, and the *Scotia* was retired. During the next decade, tourism blossomed in Atlin and mining strengthened. The *Tarahne* could not meet the demands of increased freighting and the large influx of travellers. In the spring of 1928, she was lengthened by a 9 metre addition amidships. With a significantly enlarged hauling capacity, she served the community well until she was abandoned in 1936, a casualty of the depression years.

The *Duchess*, originally a coal burner, was converted to wood for service on the Taku railway. Later she was refitted for fuel oil.[865] The little locomotive served at the portage until 1919 when a larger engine, #51 was brought in to replace her. In 1932, locomotive #52 was working at the portage. Both of these veteran engines were built in 1881 by Brooks Locomotive Works, and used during the construction of the White Pass Railway.[866]

The high cost of freight was a constant reason for complaint in a community reliant on transportation companies for delivering all manufactured goods and most of its food supplies. The viability of a new mine often hung on the scale of costs and service. For years representations and delegations argued and pleaded with the White Pass for lower rates. From time to time some minor adjustments were made, but the powerful company took advantage of its monopoly, and Atlin miners and businessmen were left with little alternative but to accept the asking price.

It was frequently pointed out that if the Atlintoo River could be made navigable, freight rates could be reduced considerably. If one steamer from Carcross could carry goods bound for Atlin all the way, the transfer and handling of freight at Taku and Scotia Bay would be eliminated, saving time and money. This possibility was investigated early by the Provincial Government,[867] but was soon rejected because of the enormous cost of dredging a river with a fall of 14.5 metres in five kilometres.

In 1902, Dr. Troughton, a sawmill owner, took it upon himself to employ a gang to dynamite some of the more prominent obstacles out of the river.[868] Even such a show of sincerity was not enough to prompt government action, and the Atlintoo River remained unnavigable for large vessels. Local businessmen were reluctant to give up the idea, however, and continued to lobby both provincial and federal governments, frequently until about 1907.

Each October when the last boat steamed away from the town wharf, it offered the final easy escape for those planning to spend the winter "outside". Its sailing prompted farewell celebrations, and the inevitable gambling which began in the final hours before departure time. More than one man lost his entire summer earnings to a turn of the card, and then stood bleakly on the shore as his boat sailed away without him.

While winter reigned over the next seven or eight months, contact with the "outside" was limited, but there was a regular horse and sleigh service between Atlin and Carcross, run by the White Pass Company. The $15 fare was expensive,[869] but comfort was a priority, and passengers rode aboard well-sprung sleighs, which carried twelve people and some freight. Buffalo robes and charcoal footwarmers were provided. Warmly dressed travellers could contemplate a comfortable journey, unless the temperature dipped low, or a northerly wind blew.[870] If the weather was good, passengers often found it a pleasant change to walk part of the distance. There were several roadhouses along the way where meals, beds and fresh horses were provided. Sleeping arrangements were often primitive, but the rooms were warm, and the food was welcome.[871]

In 1928, when Doris Nichols left Atlin after a year of nursing at St. Andrew's Hospital, she noted that the roadhouses on the stage route, which had once been rustic but adequate accommodations, were by then in poor repair. A midday meal was eaten in a small cabin, where she and other passengers sat on boxes and drank from saucerless cups, with no handles. That evening, Miss Nichols retired at 5 p.m. in a cabin at Squaw Point, near Moose Arm, slept in a bed without sheets, and listened to pack rats scurrying in the ceiling, while the mice ran around the room and over her blankets.[872]

Trail conditions were the test of men driving the six-horse teams. Passengers, tucked comfortably into the sleigh, were usually unaware of many of the dangers encountered, as drivers skilfully guided teams up icy inclines, through drifts, and over dangerous frozen lakes. Still accidents were bound to happen, and occasionally a sleigh would break lose, and fly backwards down a slippery hillside. As spring approached teams travelled at night to take advantage of the frozen trails. Too often, though, treacherous ice gave way to dump horses screaming and flailing into deep water with little hope of getting out.

Dozens of powerful teams and drivers played important roles on the Atlin transportation scene, and helped significantly to alleviate the many problems of isolation. The hardworking animals deserved the good care given them by most of their owners.

Teams formed relays to move freight and passenger sleighs over a length of trail. Huge loads and heavy weather could exhaust the toughest animals, and drivers were relieved to reach the freighters' camps where shelter and feed awaited their spent teams, and fresh horses were available for the next lap. Even dedicated care could not prevent some loss, and each spring a few animals succumbed to ruined lungs, burned too often by the sub-zero air.

White Pass had competition in the freighting business from a number of independent teamsters who worked all over the district. Joe Brooks, the "Old Reliable Packer" who had worked more than two hundred mules and horses on the Chilkoot Trail before the railway was opened,[873] knew the area, and was one of the first to establish a business. White Pass hired him as a teamster for a time, but for the most part he operated independently. By 1904, when he sold his business, he had made his fortune.[874]

The Dixon Brothers began their operation early in the spring of 1899. Within a year they had a virtual monopoly on the passenger service between Atlin, Discovery and Surprise City. During the mining season they had six stages making scheduled runs. They developed a profitable and growing business, and in 1903, took in Louis Schulz as a partner. The next year they bought out Joe Brooks which extended their monopoly to the freighting business.[875]

The dogs of Atlin, despite their wolfish delight in late night serenading, which brought cries of protest from sleepless residents, still gained solid respect as they became an integral part of the transportation network.

For many residents, a dog team was cheaper to buy, house and feed, and more convenient to use than horses. The ubiquitous dogs were the only working animals that required no reining, and worked from voice commands alone. Their harness evolved from horse rigging, but was fashioned on a smaller simpler scale. Brawny Malemutes predominated in the canine work force, but St Bernards, Newfoundlands, and dogs of mixed blood were also used. The priorities were large stature and a strong tolerance for cold weather.

Summer working dogs were regular companions and valuable pack animals for wandering prospectors. In good physical condition, they were capable of carrying up to half their own weight. They were used for a variety of jobs at the placer mines, and for numerous chores around town. The fish monger delivered his fresh catch from a dog cart. Gentle animals pulled baby carriages, and citizens rode in dog drawn carts which resembled small buckboards. For many years various canine waterworks companies delivered domestic water to householders around Atlin from tank carts that were filled from the lake.. This unique service intrigued and delighted summer visitors, who thought of dogs as pets only.

When winter came, dog teams proved their worth because of their speed and ability to get through when travelling conditions were at their worst. The delivery of the mail, without doubt, prompted a spirit of adventure and daring among men and their teams, and epic adventures were logged by those in the service. The importance of the mail was never forgotten by drivers who regularly faced hazardous, lonely trails and brutal weather. The run between Atlin and Log Cabin on the Fantail Trail usually took two days under fair conditions. The run from Atlin to Carcross in the fall of the year was considered the hardest mail route in Canada.[876]

Winter trails were often obliterated by whiteouts of blowing snow, and drivers were left guessing about directions, which their teams often knew instinctively. Sometimes men tramped for hours on snowshoes, breaking new trails where old ones had been erased by the wind. One miserable winter, an overnight snowfall of two metres stranded a driver and his team at Teepee on the Fantail Trail. For several days he could neither break trail nor move his team in the shoulder deep snow. His destination, Log Cabin, was only a few kilometres away, and he had no alternative but to wait.

Spring brought its own hazards to mail drivers. When lake levels began to rise from melting snow, pressure ridges became dangerous open leads, and overflow lurked everywhere under a disguise of thin ice. Some drivers carried small canoes so they could paddle across the open stretches of water, and they tested suspicious stretches of ice before venturing upon them. Still men and dog teams lived on the keen edge of danger in the spring, and despite due care and precaution, even veterans of the trail made fatal mistakes.

Late in November of 1902, Joseph Abbey and John McIntyre were on the trail to Log Cabin, with a load of mail, and some sacks of ore. The ice was thin and treacherous, but the code of the mail drivers demanded that the mail must get through.[877] Both men were experienced; Abbey knew the Fantail Trail well, and McIntyre had carried mail between Whitehorse and Dawson.[878] Ability and experience did not form a perfect armour against accident, however. At Golden Gate the unstable ice collapsed under them, and men and team slid into the dark water. Later, when searchers located the mail, sled, and team at the bottom of the lake, the dogs were still in harness. They discovered a fur cap frozen into the ice, but the bodies of Abbey and McIntyre were not recovered until the following May. Joe Abbey was found in an upright position with his legs in less than 20 cm of water. He was only 6 metres from the shore. McIntyre had drowned not far away.[879] They had probably struggled futilely to reach solid ice or land, before exhaustion set in and exposure took over.

Norman Fisher and Captain Alexander arrived in Atlin one day in May 1907, and told about the perilous journey they had just made to Log Cabin. They had set out with 22 kg of mail pulled by a team of seven dogs. They travelled on a deteriorating lake, and after breaking through "a couple of times," they abandoned the sled, and transferred the mail into the dog packs. At Taku Arm the ice was too unstable to cross, so they made camp and spent five days building a raft. Made from green logs, their raft was not really buoyant enough to carry the combined weight of men, mail and dogs, but they started out anyway with the top of the raft submerged in knee deep water.

Fisher related that "a big wind came up and (they) fought ice all that night. In fact, (they) bid good bye to each other two or three times."[880] At one point, the ice closed in to maroon them for a time on an island without food of any kind. The rest of the trip was spent wading in the shallows, and urging the makeshift raft along or swimming in glacial water. Thirteen days after leaving Atlin, they arrived within five kilometres of Log Cabin. Alexander swam across

AHS P-1253

*The Royal Mail wagon collected incoming mail at the White Pass wharf and delivered it to the Atlin Post Office and to the Discovery Post Office. It may also have delivered to Surprise City.*

the Tutshi River, and returned the next morning with help. The mail was exchanged and the two men started out on the return journey.[881]

Fisher said that Alexander was "a damn good swimmer," or neither of them would have survived the ordeal. Years later he recalled, "I should have had the sense to turn back, but in those days, it was kind of a disgrace to start on a trip, and then turn back."[882]

Jack Fox was not actually an official mail carrier, but his heroism and devotion to duty were in the best traditions of that brotherhood. In December 1918, he was carrying important legal documents from the Engineer Mine to Atlin. South of Golden Gate, the ice gave way and he began to sink. Before he surrendered to the icy waters, he managed to tie the crucial envelope to his snowshoes, and slide it some distance away on the ice. A search party later found the letter, along with a hat, sun glasses, and the snowshoes.[883]

In 1905, a mail route was opened between Telegraph Creek and Atlin, along the 375 km Telegraph Trail. This established a roundabout mail service to the river settlement during the months when travel on the Stikine was impossible.[884] The run was made once a month, and usually took six days. Norman Fisher remembered it as a difficult trip, because if there had been little or no traffic to keep the trail open, breaking trail became a major part of the job. At the end of the day he sometimes made a rough camp, and was glad to curl up with his dogs, using his furry lead dog, McSweeney, as a pillow.[885]

Mail routes were awarded on contract basis. In 1899, the Canadian Development Company held the contract for the service between Atlin and Log Cabin.[886] When White Pass bought out the Development Company, the mail contract was included in the transaction.[887] In 1903, John Kirkland,[888] a hotelkeeper, and himself a mail driver, won the contract. The Telegraph Creek contract was at first held by a Tahltan named McClusky. In 1907, it became part of Kirkland's contract. Kirkland was a controversial character. While he ran the mail, there were frequent complaints about poor service, he was accused of using poor equipment, and having too few carriers. Most of the complaints, though, were aimed at the length of time the mail was held up on the trail. In answer to one such complaint published in the *Atlin Claim*, and in defence of Kirkland, Henry Esson Young described a trip he had made with the mail driver in January 1906. On the first day, there had been a storm on Atlin Lake. The next morning they had wandered for two hours in a blizzard. That afternoon they encountered rotten ice around Golden Gate, and could proceed only by testing each step of the way with an axe. Later there was another storm on Otter lake. The next day they floundered in soft snow, waist deep, the trail completely obliterated. At this point they had to abandon one sleigh and still they arrived in Log Cabin on schedule.[889]

John Kirkland lost his life in the fall of 1907 when he drowned while searching for a sack of mail lost by one of his drivers. After his death, the contract went to Dixon and Schulz.[890] When their partnership was dissolved the contract was retained by Louis Schulz for as long as mail was carried by dogs.

Over the years, dozens of intrepid mail carriers risked their lives to maintain Atlin's links with the outside during the winter months. So thorough was their knowledge of the trail, and so trusted were their instincts, that travellers would pay $25 for the right to follow behind the mail sled.[891] Norman Rant, Malcolm McRae, Herb and Earl Faulkner, Jack McMurphy, Glen Hope, Henry Johnny and Leo Jack, Paul Tingley and George Ball were among the dedicated men who delivered the Royal Mail. Then in a mere blink of time, the mail drivers and their dog teams became a thing of the past, when in 1933, a giant leap in transportation technology, replaced them with the airplane.

The sight of the first automobile arriving in Atlin from across the frozen lake created a sensation. The tiny vehicle belonged to Count Emile Janne de

Lamare, the flamboyant Parisian newspaperman, and irrepressible show off. He was an enthusiastic autoist, and president of the Auto Club of Paris. De Lamare was well aware that introducing the first automobile to the north would create a whirl of publicity that could only help increase the circulation of the publications for which he was a correspondent.

The Count landed in Skagway in April 1900 with a three-wheeled auto-cycle and an automobile, machines guaranteed to draw crowds. Both vehicles were gasoline driven, and each had two complete sets of spare parts, pulled behind on a sled.[892] After reaching Bennett, a party of three, which included the intrepid Mrs. Mary Hitchcock, travelled on the slushy ice of late spring toward Atlin. The auto-cycle made the trip with little difficulty - its three horse power engine producing the unbelievable speed of 58 km per hour.[893] The five horse power automobile, however, experienced considerable difficulty, and managed a speed of 26 km per hour. The burner underneath was almost entirely submerged in water for the entire journey. In order to make repairs, driver and mechanic had to dig a hole in the ice.[894] The trip took five days, but de Lamarre was undaunted, and after proudly displaying his mechanical wonder to the citizens of Atlin, and giving rides to any who dared, he set out in his car for Dawson City. The trip proved extremely difficult, and there in the Klondike, the first car ever to visit Atlin and the Yukon was abandoned.[895] Nevertheless, the fact that he was successful in travelling more than 1000 km over territory where there was nothing that vaguely resembled a road, remains a marvel.

Routes used by horses and pedestrians did not lend themselves well to automobile travel, so in the early years automobiles were not really practical. However, men of vision soon recognized the transportation of the future. W. J. Robinson of the British America Dredging Company brought an automobile to Atlin in 1904.[896] A little later the Guggenheim Corporation had a battery-powered Oldsmobile.[897] In the years following World War I, it gradually became feasible to drive cars on improved wagon roads to Surprise Lake, Indian River and the O'Donnel River.

The fastest way to bridge the distances between towns and camps in the area was by telephone. The first line was installed in August 1899, between Atlin and Discovery, and connected two popular hotels, the Grand in Atlin and the Gold House in Discovery.[898] Frequently the line was out of order, and customers complained about the 75 cents per minute rate,[899] but eventually another line was installed to connect the telegraph office in Atlin with the Halfway House at Spruce Junction, and the Discovery Post Office. The rate

BCARS #18659

*Count Emile de Lamare, Paris newsman and Atlin gold miner, brought this three-wheeled motor vehicle and an automobile to Atlin in 1899.*

AHS Sheldrake Collection P-1254

*Nate Murphy's 1927 Ford and his dog team share the O'Donnel River Road.*

for this service was a more reasonable 25 cents per minute.[900] Through the years, citizens strung telephone lines on white porcelain insulators attached to convenient tree trunks, whenever they felt the need. The service was so efficient that trappers or other travellers who arrived at the O'Donnel River from the Telegraph Trail, could use Nate Murphy's phone to call an Atlin taxi to come out to carry them and their load the final stretch of their journey.[901] Fallen trees, or moose entangled in the line were the usual causes of telephone outages. When this happened, Murphy or another resident along the line would simply track down and repair the damage.

After gold was discovered in the Klondike, citizens in Dawson pressured the Dominion Government for telegraph service to connect them with the outside world. The government resisted due to the cost, but the founding of the Atlin camp and its need for a communication service lent some weight to the arguments from Dawson, since both camps could be served with one line. In 1899, a telegraph line connecting Dawson, Tagish, Atlin, Bennett, and Skagway was complete. On October 4, the first message for Atlin was received by J. H. Brownlee.[902] At this time, messages leaving the north could be telegraphed as far as Skagway, and then carried by steamer to its first port of call.[903]

It had originally been the intention to link this northern line with the southern system by means of a connection between Atlin and Quesnel. This section was 1600 km long, and the northern reach was an unmapped territory of dominating mountains, low lying swamplands, huge and dangerous rivers, and a tempestuous climate. It was not surprising that the government had initially balked at the expense. However, there were problems and delays in the transfer of messages from Skagway to south bound ships then back to the telegraph system again, so, in December 1899, the Dominion Government succumbed to pressure and agreed to build the connecting section.[904] The projected cost was $225,000.[905]

In March 1900, crews were sent to Quesnel, Hazelton and Atlin. The construction men were largely French Canadian, a fact strongly resented locally.[906] J. B. Charleson, who had overseen the construction of the northern section, supervized installation of the Atlin leg of the line.[907] Between Quesnel and Telegraph Creek much of the route was that blazed by the aborted Western Union project of the 1860's. However, the work was far from easy. Very little was known about the terrain, and there were no surveys. Foremen simply walked ahead of the workmen, marking a trail they hoped would eventually meet the crew building towards them from the south. In fact, at the end, the two ap-

AHSP-277

*The Dominion Telegraph provided a communication link between Atlin and the southern province. Men who served on the line lived in rustic isolation along the Telegraph Trail. This station is on the Nahlin River. (1932)*

proaching crews did select different mountain passes, and overlapped each other by about 110 km.[908]

The final connection was made in September 1901, at a cost well over $410,000, far in excess of the original estimates. It would be the last project of such magnitude in the North American wilderness until the advent of the Alaska Highway some 40 years later, but Atlin and Dawson at last had direct contact with the outside world. Government, business connections, families, and the Boer War no longer seemed so remote. There were many problems, however. The line was down as much as half of the time during its first winter of operation.[909] Falling trees, fierce winds, and heavy snows did most of the damage.

Thirty-seven cabins were built along the telegraph line, spaced a summer day's walk apart,[910] or 50 to 80 km. Each was home to a telegrapher and a lineman, who operated and maintained the line while living lives of unusual isolation. They kept busy beyond their duties, raising gardens, hunting, trapping and playing cribbage, chess, and other games over the line. Some became expert photographers, others took study courses or painted. They faced dangers from animals, hazardous rivers, deep cold, extreme snowfalls and slides. Supplies arrived once a year by packtrain.[911] Holidays came only every three years. Accidents happened and lives were lost. Still the lonely life fascinated many men, who often talked of quitting the service, but stayed until they retired.

The first telegrapher in Atlin was a Mr. Mansfield, whose quarters were in the Post Office building on Pearl Avenue.[912] This building had previously been rented but was purchased, renovated, and another story was added for staff accommodation.[913] At the time, it was considered Atlin's most imposing structure, and a reassuring symbol of the community's permanence. It also housed the customs office.

For almost 40 years, the Dominion Telegraph line provided the north's only direct communication link with the south. It was an expensive system considering the damage done by uncompromising weather, the great empty distances it spanned, and the amount of manpower required to maintain and operate it. The Dominion Telegraph Service gladly abandoned sections of the line as radio links became feasible. The line between Atlin and Telegraph Creek, was closed down in about 1942. The line between Atlin and Whitehorse remained in service until the early 1950's, because the mountainous terrain made early radio transmission impossible over that distance. Ted Smyth was the last telegrapher to tap out messages from Atlin. He continued to provide the link to the outside by simply switching his operations to radio.[914]

The first mail addressed to Atlin was delivered by the North West Mounted Police,[915] but the great influx of miners prompted the postal department to extend its northern service. On 1 May 1899, the Atlin Post Office was opened and E. C. Mahoney was named postmaster.[916] At the same time an office was opened at Log Cabin. Although the new service was welcomed there were many complaints during the first year.

When the office opened it was short of stamps. While postage on a letter was two cents, the only stamps available cost two and a half cents, and customers were, "forced to tip the Dominion Government to the extent of a half cent on every letter." [917]

Mail arrived twice a week throughout both summer and winter, but frequently there were delays. The dissatisfaction, however, does not appear to have been directed against the local postmasters, whose efforts were always appreciated, but rather the contractors and the distant post office bureacracy were blamed for the poor service. In 1902, J. A. Fraser became the postmaster, and when he was appointed Government Agent for Atlin, his place was taken by John Williams, who remained until 1910.[918]

The Discovery Post Office opened on June 26, 1899, in the store of Herbert Cancellor, who was the postmaster.[919] This arrangement did not last long, though, as Cancellor soon announced his resignation. He did this by flying a banner, which read:

> "Owing to the munificent salary of $25 a month the Postmaster has decided to resign forthwith."

His complaint was that the post office business was so great that it caused him to neglect his own business.[920] With the opening of the post office, Discovery became the official name of the town first known as Pine City and Pine Town, though residents were slow to adopt it.

On April 13, 1928, the nature of the mail service, and indeed the nature of all transportation in the Atlin area changed irrevocably. It was on that date that the *Queen of the Yukon*, a Ryan monoplane, piloted by T. G. Stephens, landed on the ice of Atlin Lake, with the first air mail delivery to the community.[921] Many residents had never seen an airplane, nor did they realize it signaled the end of the dog team mail service.

The *Queen of the Yukon*, which was the sister ship of Charles Lindberg's *Spirit of St. Louis*, belonged to Yukon Airways and Exploration Company. Initiated by Clyde Wann and James Finnegan, it represented the first commercial flying venture in the north. They did not have a contract to carry mail, but they were able to offer a service with the permission of the Post Office. This was rather a unique arrangement whereby anybody wishing to send mail by air, would purchase the usual two cent stamp, plus another 25 cent "stamp" issued by Yukon Airways.[922]

Less than a year later, on 28 February 1929, John Patterson flew Yukon Airways' second plane, *Northern Light*, on the first official air mail flight from Whitehorse to Telegraph Creek.[923] In March 1929, *Northern Light* made its first scheduled flight to Atlin.[924]

Clearly the age of air travel was well launched. Trappers, prospectors and businessmen, who had been watching the barnstorming and free rides offered by airline companies across the north, were quick to see the advantages of flight in a land where roads scarcely existed. Throughout 1929, there were three, and sometimes four, planes flying all over the Yukon, and there seemed to be plenty of work to keep them busy. There was an obvious opportunity for a company to operate out of Atlin.

AHS P-36

*Atlinites gathered around the first airplane to land on Atlin Lake. The Queen of the Yukon. (1928)*

*By June 1928, some mail was delivered to Atlin by plane.*

AHS P-2134

In May 1930, that opening was filled by the Air-Land Manufacturing Company of Vancouver, which sent the first Junkers F-13 to fly in British Columbia, with their chief pilot, E. J. A. (Paddy) Burke, and a mechanic, Emil Kading.[925] Burke was a genial Irishman, and one of the most experienced pilots in Canada at the time. He had flown with the Royal Air Force during World War I, and subsequently in the Middle East. In 1926 he had immigrated to Canada, and flown with the Aerial Photographic Survey of the Fraser Valley, as an airline pilot for International Airways, and a test pilot for Boeing.[926]

During the summer of 1930, he came to know the rugged country of northern British Columbia, and the southern Yukon, as his flights took him to traplines, mines, and the few settlements and distant outposts of the region. He loved flying, and enjoyed the frontier lifestyle. Within a few short months he had become a popular figure in Atlin.

On 10 October 1930, Burke, Kading, and trapper, Bob Marten left on a flight from Atlin to Fort Liard, the Hudsons Bay Post. All went well until shortly after becoming airborne on their return trip, they flew into a blizzard, which forced Burke to land on the Liard River. Although it was still fall by the calendar, the temperature dipped, and a blizzard howled. There they remained

AHS Burke Collection

*Paddy Burke, left, was an aviator of vast experience. He flew this Junkers in the Atlin area. (1930)*

overnight, and the next morning they took off trying to reach Atlin. Immediately though, another shocking unseasonal storm turned visibility into a blur of white, and forced them back onto the river. This time one of the plane's floats was damaged when it struck something submerged in the water. The three men were stranded, and poorly prepared for their situation.

They waited for help while food supplies ran low, and the early onslaught of winter did not let up. On October 17, they left a message with the plane, and set out in search of a cache which Burke had made earlier in the season. Unfortunately, the desperately needed food was farther away than they thought, and moving through the bleak country became even more difficult as they slowly weakened. Feet encased in leather boots were vulnerable to freezing, and they had no snowshoes to keep them from foundering in the deepening snow. They made slow progress at the cost of their waning energy. Game was scarce and their food disappeared. Temperatures continued extremely cold until at last they made camp, and waited for either rescue or death.

There was some delay before searchers began to comb the area from the air. Far to the south people followed the news of the search with nervous fascination. One plane, and three lives were lost in the quest, and eventually most fliers gave up as winter advanced. One man, Everett Wasson, a young American pilot from Juneau, believed Paddy Burke could somehow survive, and with Joe Walsh as his guide, he stubbornly persevered. On December 6, Wasson located the victims' camp, and found Kading and Marten, haggard and emaciated, hardly recognizable as the robust men who had flown from Atlin nearly two months before. For Burke, who was a diabetic, the success of Wasson's single minded dedication to the search came too late. He had died on November 20.[927]

Paddy Burke's death pointed strongly to the danger that accompanied early flights in the north. With no radios, and only the sketchiest of weather reports, flying over thousands of square kilometres of uninhabited wilderness placed aviators and their passengers in constant peril. The advantages of air travel were obvious though, and especially so in vast roadless regions, so there was no shortage of customers. Neither was there any lack of daring young men willing and eager to fly those early machines.

Throughout the depression, the flying business survived in the north, but it did not flourish. It was as a the result of a "get-rich-quick" scheme, generated by the depression, that two young fliers came north to stay. In 1932, James Eastman and Frank Barr joined the much publicized Mitchell Exploration Company on its prospecting expedition to the Finlayson River and Frances Lake area in southeastern Yukon.

Thomas Mitchell was a Detroit promoter, who attracted a number of investors and adventurers with the promise of remote creeks laden with gold. Jim Eastman was the vice president of Mitchell's company. He was an experienced flier, and the designer of the Eastman Flying Yacht. Mitchell and his company planned to rely heavily on five Eastman aircraft for ferrying supplies from their base camp in Atlin to Frances Lake. Frank Barr followed the expedition for the adventure, and in the hope of flying one of Eastman's innovative amphibians.[928]

It was only a matter of two months before Mitchell was discredited, the expedition abandoned, and most of its members returned south. Both Eastman and Barr, however, stayed; one bitten by mining, and the other by bush flying.

Eastman built a successful mining company on Spruce Creek, and his flying boats were used throughout the northwest for a number of years. Although they were unable to carry a large payload, the four-seaters proved useful.

Barr flew for Bill Strong, a trader on the Taku River, who planned to increase his business by adding an aeroplane. His plan did not materialize, so within a couple of years Barr started his own business. His company was North Canada Air Express, and the stockholders were dozens of Atlin businessmen and miners who felt Atlin needed Barr and his planes. He flew passengers and every conceivable kind of cargo all over northern British Columbia, the southern Yukon, and southeastern Alaska.

For a short time, Barr ran a scheduled service between Atlin and Juneau, to the chagrin of the White Pass Company, which by then had a fleet of planes, and viewed Barr's operation as a threat to its long time monopoly. Barr's passenger plane was a Pilgrim monoplane, manufactured by the Fairchild Aircraft Company. It could carry almost a 1000 kg of freight, or nine passengers, and Barr worked it hard. During his years in Atlin, he became a legend for his canny ability to emerge triumphant from forced landings, mechanical failures and impossible weather conditions. The general attitude was that, although Frank might be late returning from a charter flight, perhaps as long as a month, he would eventually fly home. He never lost or injured a passenger during his long flying career. In 1937, when business fell off, he left Atlin for Alaska.[929]

Flying had become an important mode of transportation around the Atlin area, but after 1937 the aircraft used were based in Carcross, Whitehorse or Alaska. In 1938, Northern Airways, a Carcross company, owned by George

Simmons, began a scheduled service from Dawson City to Vancouver, touching down in Atlin en route. They carried the mail, as well as passengers and freight, and provided faster transportation to the south than ever before known to Atlin citizens, who at last felt closely in touch with the rest of the province.[930] Air service was particularly appreciated in the spring and the fall when the community had endured days or weeks of complete isolation during freeze up or break up, which rendered lake travel impossible.

The lake ice provided a natural landing facility for nearly six months of the year, and a base for amphibians in the summer, but a landing strip was needed for year round use. In 1929, Atlin volunteers, without benefit of outside funding, built a 550 metre dirt strip east of the townsite. It was maintained and later upgraded to a length of 1000 metres by the people of Atlin. In 1941, when the Provincial Government granted $1000 for improvements on the strip it was the first outside money ever received for the facility.[931]

Air links had relieved many problems, by providing faster mail service and easy transportation for passengers, but there remained a vital need for ground transportation to carry heavy freight such as machinery, building supplies, groceries and fuel. After the large vessels were retired from Atlin Lake, freight was handled by smaller boats towing barges. The *Norgold*, did much of the work after it replaced the Tarahne in 1937. On one occasion this sturdy little craft towed a large diesel-powered shovel, so heavy it had to be off-loaded onto rails specially constructed at water level on the lakeshore.[932]

The faithful draft horses of winter were replaced about 1935 by cat trains. A common sight was a long string of heavily loaded sledges, hauled by one tractor equipped with modified steel tracks, for maximum traction on ice. The carrying capacity of the cat trains was enormous, and those run by Louis Schulz, Bob Roxborough, and George Simmons[933] were of tremendous value to the mining companies.

Generally, heavy equipment was hauled to Atlin in the spring while the lake ice was still solid. Goods were delivered to Carcross by the railroad, and the loads for Atlin assembled there. Cat trains proceeded along the lakes to Moose Arm, where an old roadhouse provided shelter and food. They arrived in Atlin on the second day, if there were no mishaps or breakdowns. One of the difficulties of the trip was manoeuvring the sledges, sometimes loaded with up to 36 tonnes of cargo, from the lake ice to the rocky shore.

Louis Schulz ran the largest freighting operation. His loads included stock for his own store as well as machinery and heavy equipment for several mines. Bob Roxborough hauled for several large mining companies, and George

*Robert Roxborough provided a vital freighting service for Atlin. Here he hauls heavily loaded sledges on the lake ice.*

Simmons did ground freighting in conjunction with his Northern Airways service. He carried parcel mail to the Engineer Mine and to Atlin. His deliveries were eagerly awaited since they often included mail order goods, surprise gifts from family and friends, and new reading material for the long winter evenings.[934]

Bob Roxborough, an innovative Atlinite, built a snow machine in 1925, long before any factory models were available to the consumer. It was made with a variety of automobile parts, and powered with a Model T engine. Driven on both tracks and wheels, it was capable of traversing not only ice, but slush and bare ground. He used it mainly for hauling up to half a tonne of mail or food supplies between Atlin and the Engineer Mine.

Handling of cargo, the need to shift goods from one conveyance to another during the long haul to Atlin, was a factor that kept freight rates high. Over the years community leaders pursued various ways to overcome the problem, and one particular scheme most often led the list of solutions. Proposals for a year round overland route into Atlin were reviewed regularly. At first railways were considered, and once the White Pass route was completed, it seemed no obstacle should stand in the way of making a connection with its service.

The first such plan involved the Atlin Southern Railway Company, incorporated in 1899, which proposed building a railway from Log Cabin to Atlin, and thence south to Telegraph Creek.[935] The Pacific Northern and Omineca Railway,[936] and the Coast Yukon Railway[937] both planned lines originating in Kitimat, and travelling northward through Atlin, and terminating in either Teslin or Dawson City. These were all-Canadian routes, and they were promoted vigorously, but in fact there was never enough business to support such lines.

Other shorter lines, originating in Alaska, were also considered. The Bennett Railway Company had a wild, if somewhat vague plan to start a railroad near Dyea, Alaska, tunnel beneath Chilkoot Pass, and proceed via Lake Bennett and Atlin Lake to the 60th parallel, where they hoped another company might pick up the route.[938]

The idea of a line originating in Taku Inlet was first brought forward in 1901,[939] but it was not until 1914 that the idea was actively pursued. Preliminary surveys were made, and there was some excitement among mineral claim owners in the area to the south of Atlin. However, the outbreak of war interrupted the work, and it was never resumed.[940]

A railway was the fastest and most efficient transportation option, but railway building was expensive and required a great volume of business to make it pay. A wagon road connection was not the preferred solution, but at various times the people of Atlin could see that it might be a feasible and attainable goal.

Between 1903-08, there was much discussion about the roads, and where the connection should be. The business community favoured a road to Juneau, despite the difficulties of the terrain. This would provide direct access to the Alaskan capital and a salt water port. In 1905, this proposal had a good deal of support both in Atlin and in Juneau. The route considered most likely was along the existing McKee Creek Road, and from there south and through the Silver Salmon, Nakina and Taku River Valleys to Taku Inlet. There seemed to be no problem with getting the Juneau portion built, and in Atlin, there was enough enthusiasm that money was raised privately to hire an engineer to prepare a feasibility report.[941]

The support for this project could not be denied, so Henry Esson Young secured government money for the preliminary survey and feasibility study.[942] A survey party of three, including Road Superintendent Hugh Molyneux, set out with a train of pack horses in June 1905.[943] In less than two weeks one of the party returned to hire an Indian guide and a canoe, since the terrain was unsuitable for horses.[944]

The report was complete by January 1906, and although its tone was optimistic, there was an unsolved problem created by the Taku Glacier. Furthermore, it was by no means certain that such a route could be kept open all year round. Most disheartening were the indications that the cost would be prohibitive.[945]

At this point Henry Esson Young conveyed the government's preference for a connection to the White Pass railway at Carcross. Not only was this a shorter and less expensive route, it provided access to the southern Yukon, it could be open all year, and it was all-Canadian. There was an existing road (perhaps trail is a better description) between Little Atlin Lake and the Tagish telegraph office, which had been used by the North West Mounted Police, occasionally by mail carriers and the men who maintained the telegraph line. Therefore, it only remained to construct a road from Atlin to the 60th parallel, and then to persuade the Dominion Government to complete the distance to Carcross.[946]

Both Young and Premier Richard McBride promised the people of Atlin that the Provincial Government would build the section of road within the province if the Dominion Government would build the rest of the distance to Carcross.[947] Such an agreement was never reached and so the plan never materialized.

The idea of building a road into Atlin did not gain serious attention again until 1939, when a Highway Commission was formed to gather information and public opinion about possible routes for a British Columbia-Yukon-Alaska Highway. Not surprisingly, Atlin opinion was unanimous that such a highway should pass through Atlin,[948] In 1942, the Alaska Highway, between Dawson Creek, B.C. and Fairbanks, Alaska, was built by the U.S. Army, and Canadian civilians as a wartime security measure. For Atlinites the highway passed a tantalising 100 km to the north.

The Atlin District Board of Trade recognized a new opportunity to promote a road link to Atlin, and after the war formed a committee to present a proposal to the Provincial Government. Tom Kirkwood, secretary of the Board of Trade, wrote suggesting some of the heavy equipment left idle after wartime road construction, and Canadian troops still stationed in Whitehorse be employed to build a road connecting Atlin with the Alaska Highway.[949]

At first the Provincial Government was reluctant to fund the 50 km of construction within British Columbia, but eventually they were persuaded to join forces with the Federal Government as a postwar development project. Construction began in June 1949, under the direction of the Canadian Army Engineers. As well as army equipment, contracts for machinery and operators were awarded to a number of private companies.[950]

This was the first major road-building project undertaken by the Canadian Army, and the job was placed in the charge of Lt. Bill Moore, R. C. E., and Sgt. Dan Cameron. A camp was established at Jake's Corner, about 80 km southeast of Whitehorse, where 130 military and civilian road workers were based.[951] Throughout the spring, summer and fall, they built through brush and bogs. Time and money limited the amount of rock blasting the crews could do. As a result the road wound around outcroppings, skirted rocky spurs, and passed over muskeg swamps, where vast amounts of fill was required. By October, a few vehicles, including a truck carrying the mail, were allowed through, but the road was still not fit for general use.

Just as the November completion date was approaching, and it looked as though the projected target date would be met, incessant rains turned the ground into a quagmire of heavy mud, and caused damage to the road bed along the route. Men and machines were stalled, and the crew moved into Atlin to wait for the weather to break. For a while anxious citizens thought they would not have a road until the next spring.[952]

The weather improved finally, and the Royal Canadian Air Force landed a final load of supplies on the Atlin strip. The road was completed in early December, on budget and more or less on time. Excited Atlin residents threw a grand party and dance at the Moose Hall to honour the road crew, and revelled in the knowledge that, after fifty years, their community was at last accessible by land.[953]

This change was to have a profound effect on the community. Many hoped that a link to the Alaska Highway would provide the boost that mining and business would need to keep the town alive. Others felt that with only 300 people, and with little hope that mining activity would return to prewar levels, the new road link was still not enough to maintain a healthy economy for the town. So a surprising and unwelcome spin-off came from the new access as many families used the road to move away.

Atlin did survive the exodus, and the road opened the gate to a new kind of tourism. Travellers who ventured up the Alaska Highway came under their own power, independent of schedules, and willing to forego frills. Those who took the branch road to Atlin were as beguiled by the little lakeside town and its magnificent surroundings as were their Victorian predecessors.

CHAPTER 12

# Keeping the Peace and Dealing Justice

Jefferson Randolph (Soapy) Smith, a smooth tongued master of deception and extortion, landed in Skagway late in the summer of 1897. Within days, his hoodlums had infiltrated the very framework of the town and the gold rush trails with their clever extortion schemes, brazen robberies and intricate scams. They did not flinch from murder either, if they deemed it necessary for the success of a profitable operation. All the while Smith, himself, paraded in high profile as an exemplary citizen who donated generously to churches and charities, and befriended the community's most needy citizens.

Smith and his gang turned Skagway into a black turmoil of lawlessness. At night the sound of gun shots mixed with the rowdy din of sleazy dance halls, saloons, and street brawls kept frightened citizens from their sleep. Rare was the stampeder who made his way through Skagway, and across the mountains without a losing encounter with one of Smith's gangsters.

Even though the only lawman in town was on Smith's payroll, there was an increasing number of citizens who understood that Smith was at the root of Skagway's corruption. On July 8th, 1898, an incensed citizens' committee drew him to Juneau dock in the Skagway harbour. There in a sudden close range gunfight, the notorious career of Soapy Smith was ended by the courageous Frank Reid, who lost his own life in the effort.

On the Canadian side of the mountains, the traditions of peace keeping became more firmly rooted because of Skagway's disastrous experience. The borders at the summits of the Chilkoot and White Passes were protected by the North West Mounted Police. The mushrooming town of Bennett City was kept free from any serious crime by two members of the British Columbia Provincial Police force. Soapy Smith's wary gangsters never attempted to penetrate the Canadian stronghold of law and order.

AHS P-1255

*At the Chilkoot Pass Summit, the North West Mounted Police post was buried in deep snow and surrounded by huge piles of supplies being relayed across the mountains.*

By the mid summer of 1898, Captain W. J. Rant, the Government Agent at Bennett, and Superintendent Sam Steele of the North West Mounted Police were keenly aware a rush to Atlin was imminent. They agreed the law must have a presence in the new camp on Pine Creek before word of the gold strike became widespread.

The two man force at Bennett was already over-extended, so Steele deployed two members of the Mounted Police, Sergeant Davis and Constable Woodhouse, to establish the Atlin Lake detachment.[954] In August 1898, the two men made their head quarters in a tent at Pine Creek where they could keep a close eye on the gold mining.

Steele had hoped his men would halt the spread of claim jumping on the creeks, but it was too widespread to curb. However, the mere fact that police were present lent an air of order to the camp. There was little violence, despite the volatile situation created by the many claim disputes, and the impatience of miners who had come so far for gold, and were being seriously hindered from finding it.

Police supervision was needed in another matter that summer, for although Alaskan law prohibited production, sale and consumption of liquor, the regulation was blatantly ignored and liquor thefts were common. Liquor shipments bound for Canada through the Port of Skagway were regularly stolen, or mysteriously watered down before ever reaching Canadian soil. Even after clearing the border, that particular commodity remained vulnerable to theft until delivered to its final destination. The need to protect these valuable cargoes taxed police resources, but two Mounted Policemen were assigned to escort shipments between Bennett and Atlin until the end of navigation in the fall of 1898.[955] It was this degree of vigilance, however, that helped keep Canadian settlements free of organized crime.

In September 1898, Superintendent Zachary Wood, who had replaced Strickland at Tagish, received orders from Ottawa to recall Davis and Woodhouse from Pine Creek.[956] It is not clear why the men were withdrawn, but it was likely in response to objections raised by the British Columbia Police who jealously guarded its right of jurisdiction over the new gold discoveries. In spite of that, their only representatives in the area were Captain Rant and his small staff who were fully occupied at Bennett, Log Cabin and Lindeman City.

In November, Supt. Wood learned that American miners were coming to Atlin via the Taku River. Since there was no customs station at the international border on the river route, they avoided paying duty on their foreign goods. Wood dispatched Sergeant Joyce and one constable to Atlin. The men set out by canoe, but winter winds had turned the lakes dangerously rough, and after several narrow escapes from the icy water, they were forced to return to Tagish Post.

After the lakes were solidly frozen, in January, four other policemen, a sergeant and three constables, make a successful trip to Atlin on the ice. Once settled they took charge of collecting taxes, and began to curb an unruly element that had developed in the absence of law enforcers. Theft of provisions and personal gear had become common, and no one who valued his possessions left his camp, a loaded sleigh, or any kind of cache, unprotected.[957] Crime had incubated quickly where there were no peacekeepers.

As soon as the Mounted Police were reestablished at Discovery, British Columbia's contentious Attorney General, Joseph Martin, wrote to the Prime Minister, Wilfred Laurier, concerning the Atlin district:

> We are sending in a Government Agent, Stipendiary Magistrate and a force of police. In the meantime the Mounted Police have gone into country. As soon as our police force and magistrate are installed there, we would be glad if you would withdraw the Mounted Police entirely. Their operations in connection with the Yukon country have been very unsatisfactory. We understand that nearly all of them

> are taking up claims in the Atlin District. Our policemen, and in fact all of our officials, are prohibited from having any interest, direct or indirect, in a mineral claim, and we feel that if the Mounted Police are retained in this country, there will be more or less clashing between them and our officials.[958]

The acrimonious Martin clearly had no appreciation of the sterling work being done under difficult conditions, by the Mounted Police, on behalf of both Canada and British Columbia. They made it their primary concern to ensure the criminal element, which held sway in Skagway and other Alaskan ports, did not move into Canada. Of equal importance was the onerous job, assumed by the Mounties, of ensuring that every goldseeker who entered their jurisdiction, was carrying adequate equipment and food to last for several months, and that they had a safe means of travel. The lives of countless novice gold seekers were saved because the Mounted Police safeguarded their passage.

Martin's allegations that police had acted in their own interest, did, in fact, have some substance. Strickland and eleven of his men had certainly rushed to stake claims on Pine Creek immediately after Fritz Miller had recorded the discovery at Tagish Post.[959] It was true that Sergeant Davis saw the real estate potential at Discovery, and made application to preempt 73 hectares of land which he planned to sell, lot by lot.[960] And there was evidence[961] that Norman Rant was involved, at least indirectly, with mineral claims, and that he too attempted to preempt land in the Atlin townsite, which he expected to have significant value.

Laurier took issue with Martin's statements, but assured him that there would be no interference from federal officials, just as soon as British Columbia took responsibility for her own territory. He also indicated that his government was prepared to co-operate in the demarcation of the British Columbia-Yukon boundary, to avoid future clashes of jurisdiction.[962]

Early in 1899, E. N. M. (Monty) Wood of New Westminster[963] was appointed Stipendiary Magistrate for Atlin. He was empowered to convict and sentence prisoners charged with minor crimes. On occasion, his impartiality[964] was questioned, and Gold Commissioner Joseph Graham considered him a "man of very little experience of life."[965] When he resigned in 1904, to go into private law practice in Atlin, he was succeeded by E. (Ned) Thain, who was well educated and earned the community respect never accorded to Wood.[966]

BCARS HP 13347

*The British Columbia Provincial Police displaced the Mounted Police in Atlin and Discovery. This is their Atlin barracks.*

About the time Magistrate Wood was sent to Atlin, the Provincial Police sent Chief Constable J. J. McKenna, and Constables E. W. Bickle, H. W. Heal, and G. H. Woodward to take up their duties in Atlin and Discovery.[967] Later in the spring of 1899, the police ranks in Atlin were further increased with the arrival of Constables W. H. Hodgson, E. W. Shaw, Walter Owen, W. H. Vickers, and H. G. H. Robinson.[968] With a contingent of 8 men,[969] the Province had clearly taken the responsibility for peace keeping in the district. At the end of April, the Mounted Police were withdrawn.[970]

The British Columbia Provincial Police was a small force of just over 100 men. They were not distinguishable by smart neat uniforms; in fact they had no uniforms, but they were a well disciplined and dedicated body. Locally, they appear to have been under the jurisdiction of the Government Agent, who represented their needs, dealt with personnel matters, and, on occasion, gave direction.

The provincial force was experienced in policing mining camps, and they posted officers to Atlin who were suited to working in a largely male community. It was necessary for them to be firm and authoritative, without losing the respect of the men they monitored. They maintained a fair level of law and order in spite of the prevalence of gambling, drinking, and prostitution, and the fact that a large amount of gold and money flowed through the camp.

In June, Chief Constable McKenna was transferred to take charge in the new Porcupine District to the west of Atlin. He was replaced by Chief Constable T. D. DesBrisay, a young Irishman, whose sense of self importance prompted him to report to his superior in Victoria:

> I shall endeavour to educate myself in this branch of the service by applying myself closely to the study of books that will give me the necessary information. I expect that at the end of the month I will have this district in good running order . . . [971]

About a month later, he reported to the Superintendent that he had raided the gambling rooms of the Grand Hotel, seized $169.50, and summoned the proprietors to appear before the Magistrate, who fined them $100. Thereupon he declared, "I now have gambling thoroughly under control"[972]

About the time DesBrisay was making this fatuous boast, J. D. Graham expressed dissatisfaction with his work, and questioned his suitability for the position.[973] Shortly thereafter the records indicate that the young man was demoted to Constable, and transferred to Bennett.[974]

There was no Chief Constable appointed for the district until 1902, when Walter Owen was sent to take charge in Atlin. Owen was a Englishman, who joined the force in 1899, and was first posted to Log Cabin. Later he was transferred to Atlin, and from there to Discovery in 1900,[975] where he had proved himself a competent policeman, even handed, and in charge. Drunkenness and fighting caused most of the problems, and it required a strong man, and constant vigilance to ensure that drinking sprees, and the unruly element did not get out of control.

As Police Chief, Owen was highly respected, and he served until January 1907, when his son's[976] illness forced him to resign, and move his family to Vancouver. He returned to his position in 1909, but in 1911, he left Atlin again, when he was promoted to Chief Constable of the Prince Rupert District. He was promoted several times, and in 1929, Owen was appointed Warden of Oakalla Prison. He was truly a distinguished member of the Provincial Police.[977]

W. H. "Billy" Vickers, who first went to Atlin in 1899, served for a number of years as both a special and a regular constable. When Owen left in 1907, he became Chief Constable, but shortly afterwards was transferred to

Prince Rupert to become Chief Constable in that city. His career with the B. C. P. P. was long and dedicated.[978]

Given the litigious nature of miners, and the particularly confused state of affairs on the creeks during 1899-1900, it is not surprising that the legal fraternity was well represented in Atlin. Some members of the bar had offices in town, and others came especially for sessions of the County Court. Among those established in Atlin were W. Pollard Grant, Frederick McBride Young, E. A. Jenns, W. G. Paxton, and C. W. Sawers. Some of those who made summer trips from their offices in Victoria and Vancouver became notable members of their profession. These included Richard McBride, Gordon Hunter, W. E. Fisher, A. L. Belyea, and A. Dunbar Taylor.[979]

The one missing component in the delivery of justice was the presence of a residential judge. Cases which could not be heard by the Magistrate, and these included mining disputes and serious crimes, had sometimes to wait months for a judge to arrive. This resulted in some mining properties lying idle for many months, and a lack of justice for some prisoners.

The first judge to arrive in Atlin was His Honour Judge A. P. Irving, who was in Atlin as a Special Commissioner with power to settle disputes which had arisen as a result of the first claims on Pine Creek being staked under the law of the Northwest Territories. When this task was done, and before returning home, he presided over a session of the County Court of Vancouver to hear a number of cases.

In subsequent years, Judges J. Martin, A. J. McColl, J. Drake, and A. Henderson held court sessions in Atlin. There was generally enough business to keep them there from two to three months, during which time they heard both Supreme and County Court actions.[980] Clearly there was a need, and gradually public pressure built for the appointment of a resident judge in Atlin. The *Atlin Claim* wrote editorials, and the Board of Trade lobbied the government.[981] Finally F. McBride Young was appointed in June 1905, to the newly created County Court of Atlin.[982] This new jurisdiction extended south to include Prince Rupert, Hazelton, and the Queen Charlotte Islands. Judge Young was a popular choice. He was well known in the district, as he had practised as a lawyer in the camp in 1899, and he was the brother of Henry Esson Young.

Atlin and Discovery were rough towns, but the police, for the most part, maintained a good degree of law and order. Misdemeanors and Saturday night rowdyism were the usual complaints; however, there were some notable exceptions when murders or robberies were committed, and sometimes the criminals got away.

In August 1899, the Discovery stage was held up by a masked man, who threatened the driver and passengers with a loaded Winchester. At least four passengers, who were carrying substantial amounts of gold, managed to scramble down a bank and disappear into the woods. Meanwhile the other passengers turned out their pockets and littered the floor of the coach with miner's licences, claim certificates, poker chips, jack knives and small change. The unlucky bandit fled the scene with only $4.70,[983] but he escaped police with his meager loot, and the crime was never solved.[984]

Partnerships were often formed out of necessity, with little thought given to compatibility. When the Baur brothers from California teamed up with a man named Garret Lieniga in Skagway, they had no idea that serious trouble would result, as it finally did in May 1899.

Nineteen year old Fred Baur and Lieniga were camped about 2.5 km from Atlin on the Watson Avenue route to Discovery. One morning Baur went to rouse Lieniga for breakfast, and playfully pulled his leg. Lieniga became angry and a few minutes later, he appeared at Baur's tent with a shot gun. Baur tried to run, but Lieniga fired, wounding the young man seriously. Strangely, Lieniga went for Dr. Lambert himself, reporting that Baur had been injured by a squirrel hunter who left after the shooting. When the doctor sent Lieniga for some medicinal whiskey he disappeared, but was captured that night at Otter Creek. Police were not able to hold him for long. A week later he escaped, still wearing leg irons, and brazenly sent his compliments to the police with a passerby. Baur eventually recovered, but Lieniga was never recaptured and brought to trial. In all probability he made his escape into Alaskan territory.[985]

The proximity of the Atlin district to Alaska presented a ready escape for lawbreakers who could reach Juneau or Skagway over established trails. In winter a desperado with a dog team could outrun police with only a one hour start. There was no way to signal ahead that a criminal was loose, and should be intercepted. Then in 1905, Police Chief Walter Owen requested that in the winter months a constable be stationed at Log Cabin, so that he could be notified by telegraph, if a wanted man was heading in that direction.[986]

Perhaps the best planned robberies committed in Atlin took place one night in 1903. The Atlin Mining Company was a profitable company, mining on a fairly large scale on McKee Creek, and near the end of August they were close to making a clean up, which was expected to be very rich. Just after midnight, work had closed down for the day, and a watchman was on duty. The area of the sluice boxes was flooded with light furnished by the generator located more than 2 km up the creek. When suddenly the lights went out, J.

Fetherstonhaugh, who was in charge of the dynamos, headed for the power plant to check for trouble.[987]

Having cleverly diverted the watchman, the thieves went to work to clean out one of the sluice boxes. This they did by neatly cutting away the side of box, and removing the riffles When the watchman returned, they were apparently surprised, as they left behind about $2000 worth of spilled nuggets, and a wide variety of tools. The exact value of their haul is not known, but it was thought they escaped with $15,000.[988]

That same night, 25 km away, 18 year old Bill Loveridge, the watchman at the Pine Creek Power Company's operation, was confronted by a large masked man and three accomplices. They seized and emptied Loveridge's gun, telling him to "get out," and threatening to shoot him. The young man ran for help. At Discovery a posse gathered with Constable Heal, and proceeded to the mine. The robbers had vanished with little, if any, gold.[989] There was, it seemed, a connection between the two robberies.

By morning, all routes of escape were patrolled, but Constable Owen believed the culprits were lying low in the area. Finally, in a spectacular saloon raid in Discovery, the police apprehended four suspects, all of whom had previous records. After several days, police neither located the missing gold, nor did they prove any crime against the suspects, so they were released. Despite the offer of a $1000 reward, the crime remained unsolved.[990]

From the beginning, Atlin and Discovery had a shortage of adequate jail space,[991] which caused a variety of problems. In June 1899, the police in Discovery were forced to shackle an inebriated prisoner to a tree, since they had no cells available. His plight caused much hilarity, and many of his friends came by to take photographs. Once sober, he was liberated.[992]

On another occasion the consequences were more serious. Harry Henning was arrested in October, 1901, for attempting to murder his wife for an imagined infidelity. He was a violent and disturbed man and there was no question of releasing him on bail. Unfortunately the cell building in Atlin was unsupervised, and totally inadequate for long term detention. A judge was not expected for 9 or 10 months and the situation prompted the Stipendiary Magistrate to telegraph his concerns about the welfare of the prisoner to the Attorney General. The plea was ignored. After more than six month's incarceration in a "pen", without any certainty of when he would be brought to trial, Henning managed to get Constable Heal's pocket knife, and cut his own throat. The Coroner's Jury roundly condemned the Government for such neglect.[993] While the government's neglect is indefensible, the question arises about the responsibilities of the usually self-reliant citizenry and the police who might have taken steps on their own, to ease the problem.

Police had to address the ongoing problem of illicit liquor trafficking between Atlin and Teslin. Liquor was readily available in Atlin, but the law denied the Tlingits and other native people the right to purchase it. This generated a profitable bootlegging trade between Atlin and the Tlingit village of Teslin, one that the authorities viewed seriously. In the winter of 1905, one party of native people hauled a great number of cases of liquor over the 120 km trail to Teslin. Complaints were subsequently laid by the Mounted Police in Teslin and prospectors working around Teslin Lake. Two Special Constables were sent to patrol the trail for a month, but the traffickers either found a new route, or stopped hauling, because no evidence of liquor movement was discovered while the officers were on duty there.[994] Police vigilance remained constant at both ends of the trail, and an occasional arrest was made.[995]

Probably because of the shortage of manpower, and declining population, the Provincial Police were not replaced in Atlin after Constables J. P. Scarlet and R. Webster enlisted for overseas service in 1916. This left Atlin without a policeman until 1919, when Sergeant C. S. Harper reopened the Mounted Police detachment. This occurred at a time when the force became responsible for the enforcement of Federal Statutes throughout western Canada.

Although stationed in Atlin, Harper was probably absent from his post a good deal of the time since he was responsible for a large portion of the surrounding area.[996] Bootlegging between Atlin and Teslin continued, and became one of Harper's major concerns. The Atlin detachment was closed again in July 1921 just after Corporal Tidd replaced Sergeant Harper. By then mining activity had declined, and Atlin's population stood at 375.[997]

In 1920, the Provincial Police reopened their Atlin office, and reinstated Bob Webster as Constable, but it was no longer the bustling gold rush post he had known in 1908. One solitary policeman, equipped with a dog team and a small boat, was responsible for a vast territory, and one quiet town. Discovery had all but disappeared, and just a few small mining operations worked the creeks. Only in summer, with the arrival of the *M. V. Tarahne* and its load of tourists, did the town take on an air of dignified bustle. Webster stayed in Atlin until 1929.

During the thirties mining enjoyed a resurgence, and the policeman faced a much more challenging task. A number of quite large mining camps accommodated several hundred men on Spruce Creek, and elsewhere, but still the policeman worked alone, responsible also for Game Department work.

Webster was left virtually alone by his superiors who trusted his judgment and ability to cope with whatever situation arose.[998] Equipped with only a radio for contact with his superiors in Prince Rupert, his jurisdiction encompassed British Columbia's vast northwestern corner, a formidable roadless expanse which covered thousands of square hectares.

The maintenance of law and order in Atlin and Discovery was never easy. Isolation, the proximity to a foreign border, the problems inherent in all boom towns, and the unique legal issues that beset the miners combined to create a stage for trouble. Occasionally policemen bungled and sometimes the dealing of justice was slow or inadequate, but the widely held respect for Canadian law and the essential integrity of the Provincial and North West Mounted Police were strong deterrents to criminal activity. Although the potential for trouble was always near the surface, the communities, on the whole, remained law abiding. A few bold and honourable men ensured a secure atmosphere because they watched closely and stood firm.

CHAPTER 13

# Three Men of God and Their Churches

The thousands who struggled north for gold imagined the rewards of their journey would far outweigh the trials and discomforts of the trip. Even so, the hardships were more severe, and the prizes less ample than most had anticipated. The church men who joined the stampede were not seeking riches; instead they sought to exercise their love for God and to help all who needed solace and direction. Fairly it should be said that for these dedicated men, too, the sacrifice was greater than ever imagined, but they gave willingly, and their unselfish efforts were deeply appreciated and remembered.

As "Klondike madness" raged in Vancouver during August, 1897, Rev. James Robertson, the Superintendent of Northern Missions for the Presbyterian Church of Canada, walked the teeming streets and watched. Good sense played no role in the picture he saw, as gold hungry men loaded outfits, sometimes representing entire life savings, aboard any floating hulk that had available space. A wild gamble was taking these frenzied men away from families, jobs and homes, and it was obvious most were ignorant of the hardships and many were unfit to meet the physical demands ahead. He recognized that these naive adventurers would soon need moral support and spiritual guidance. Then and there he resolved to send missionaries to the goldfields, without waiting to seek the approval of the General Assembly of Churches.[999]

Glenora, at the junction of the Ashcroft, Stikine and Teslin Trails was a growing encampment of gold hunters who were pouring in from the south and west throughout the winter of 1897-98. As well, the forces of Mackenzie and Mann were there to build the government road to Teslin Lake, and the railway construction crew of Dewdney Company added to the crowds of men and horses along the river banks. Neither the road nor the railway were built, and because of that many disgruntled stampeders turned homeward. Hundreds more decided to travel on, but were stalled in Glenora while they modified their outfits, and waited for the weather to allow passage over the existing trail. James Robertson wanted a missionary among those men too, and on 4 April 1898, Rev. John Pringle arrived in Glenora.

Pringle, the 45 year old father of four teenage children, was originally from Charlottetown, P.E.I.. His ministry had previously taken him to Ontario, Manitoba, and Minnesota. He could have had the choice of any number of comfortable urban parishes, but he chose the challenge of the north and the mission to miners. Full of lively enthusiasm, forthright opinion and spiritual purpose, he was an excellent choice for the job. He was a man of fine physical strength who stood over six feet tall. His imposing stature, infectious sense of humour, and companionable nature were arresting qualities that made him immediately popular with the miners.

His trek up the Stikine River to Glenora banished any romantic notions he might have had about the life of a stampeder. Harnessed to a sled loaded with 180 kg of supplies, he pulled through snow, rain, and slush for eight and a half long days. Wet clothing, blistered feet, and cold were the distress of every man on the trail with him that spring.[1000]

At Glenora, Pringle pitched a small tent, 1.8 by 2.5 metres, his new home. He was the only church man in the busy camp, or for that matter, in that entire part of the Stikine River, and converging trails. His first crowded service was in a bar room. In the congregation was James Lumsden, a singer from the D'Oyly Carte Opera Company of London. When he led the familiar hymns in his clear tenor voice, tears welled in the eyes of more than one lonely stampeder.[1001] After the Glenora service, Pringle hiked 20 km to Telegraph Creek for another.[1002] When river navigation opened up in the summer, his congregations numbered from 300 to 800, and services had to be moved outdoors.[1003]

Pringle understood the strain and hardship of loneliness that faced many of the men around him, and he knew the temptations of liquor, gambling and prostitutes, could easily lead them "to forget God, and serve the devil." He offered prayers whenever anyone was there to hear. To him it was of no consequence that, "my church has for a floor, the road; the sky is my roof, and two logs are my pulpit."[1004]

The furious winter of 1898 dropped a smothering blanket of snow on the Teslin Trail, piling it to depths of four and a half metres in some places,[1005] and the thermometer hit dangerous lows. Progress with heavy loads was nearly impossible, so travellers made camp in tents or cabins and settled down to wait. The delays taxed food supplies until the spectre of death haunted the trail with the menace of scurvy and starvation.

During those difficult months, Pringle and his dog team, passed up and down the trail delivering mail and good cheer, and, where necessary, caring for the sick, and injured. He gave healing doses of raw onions and potatoes to men suffering from scurvy,[1006] and if a man was very ill, Pringle would lash him to a sleigh, and take him back to Glenora, where he could be cared for.

Pringle had great stamina, and adapted to life on the trail quickly. In temperatures below -40° C., he could cook, sleep, and travel with the efficiency it took to survive. Enormously proud of his dogs, he cared for them well, and he could travel for hours to reach his scattered flock. He packed cornmeal and tallow for his dogs, and for himself, a few slices of fried bacon and a few ships biscuits tucked into his pocket.[1007]

In mid November 1898, Pringle advised his superiors in Toronto, that a rich strike of gold had been made in Atlin, and he thought the Church should send a man there the following spring. Many of the men he considered his "parish" were heading in that direction, and it was he the Church directed to follow. He arrived in Atlin on 25 March, 1899, and the next day he held his first service in yet another saloon.[1008] Just a week later, on Easter Sunday, prayers were offered in the bar room of the Metropole Hotel.[1009]

Pringle was the first resident clergyman in Atlin, but he was not the first to conduct a service there. On November 22, 1898, twenty-two people had gathered for a service. Among the congregation was one of the Kendall brothers, who noted the occasion in his diary.[1010] It is possible the minister was visiting from Skagway or Juneau.

Although Pringle was grateful for the use of any shelter for his prayer meetings, bar rooms, and all that they stood for, were distasteful to him. As a result he wasted no time in buying some property, and setting up a church on the corner of Third and Discovery Avenue. Like most others in town, the building was canvas- 7.5 by 18 metres,[1011] with a sawdust floor, and roughhewn benches for the faithful.[1012]

Before long a similar "church" was established at Discovery in the British Miners' Protective Association's assembly tent, but these roots did not confine Pringle or his mission. He continued to cover his parish on foot. Not only did he walk 10 km to Discovery and back on a Sunday, but during the week he would venture further to take his message onto the creeks. It was not uncommon for him to trudge over 30 km a day so that he could offer midweek services on Otter, Wright and Boulder Creeks, where the miners' tents spread for miles along their banks. Standing on a pile of dirt shoveled from a ditch, he preached to a handful of men and a few women, and their hymns echoed about the valley. Many nights he shared a miner's bed and, after a strenuous day on his rounds, slept soundly in spite of the racket made by a snoring sled dog.[1013]

John Pringle became a trusted friend and trail-mate of many miners, but he was distinctly straightforward. If he disapproved, he said so. Many a miner, especially the young and easily led, suffered the sting of his directness. He used the same approach when dealing with the government. He had lived and shared hardships with the gold miners, and when he felt they were being unfairly treated, he spoke out. He had not been in Atlin long before he was taking an active role in public affairs, and in June, he was elected to make representations to the Provincial Government on behalf of the miners.[1014] He chaired miners' meetings, and was involved with any matters concerning community welfare. When he was criticized for his political activity, he replied: "I shall meddle in politics as long as the devil meddles in politics." [1015]

His opinions were also heard very forcefully back in Toronto, when he wrote a barbed letter to the Superintendent of Western Missions about the lack of money to build a church. He was promptly reminded that it was his duty to get his church on a self-sustaining basis as soon as possible. He protested vigorously, contrasting his spartan church with the comfort of the bars and the saloons.[1016] Some funds were forthcoming, but never enough to build a proper church. Eventually the congregation found the money to buy the abandoned Arctic Brotherhood Hall on Second Street, but Pringle's ideal church never became a reality.

Even lacking proper facilities, the Presbyterian congregation was an active one. They had a well-organized Sunday School, a Girls' Society, and an excellent choir. The choir master was Postmaster J. Williams, who had himself been a choir boy at Westminster Abbey.[1017] One of the soloists was the accomplished James Lumsden, whose search for gold had led him to Atlin.[1018]

In the spring of 1900, Pringle went home on leave, after more than two years in the north. His brother George, a divinity student, replaced him for the summer. The next summer George returned to assist his brother, and in the fall Rev. John Pringle accepted a call to the community of Bonanza in the Klondike. Later he served with distinction as an army chaplain during World War I. He achieved the rank of Lieutenant Colonel, and was awarded the Distinguished Service Medal.

Rev. Frederick L. Stephenson, pioneered the Anglican mission in Atlin. He arrived within days of John Pringle. Stephenson had migrated to Victoria from England in 1883, and was ordained in 1889. He had worked for nine years among the Native people of the coast at Kitkatla and Port Simpson.[1019] In 1898, he was sent to Bennett by the Anglican Bishop of New Caledonia, Bishop William Ridley. In the Spring of 1899, he walked to Atlin, via Skagway.[1020]

Saloons, for all their evil connotations seemed ever available for church services, and Stephenson held his first meeting in one on Easter Monday. He described the situation with characteristic good humour:

> ". . . I was offered the upper floor of a saloon . . which was roofed over with rough boards and tar paper, but whose sides were additions of the future. Work was in progress during the whole service. Sawing, nailing, hewing, planing, above, and around us, beneath us a crowd in the saloon, drinking and gambling. Gradually, the large space we occupied was filled, some sitting on the floor or boxes, others standing, and slowly but surely, the noise around and beneath died away."[1021]

The next week, he used a large general store, but it was very crowded, and that created a problem. The pews were planks of rough lumber, supported by slabs at each end. At the end of a hymn, when a whole row of husky miners dropped to the seat simultaneously, it collapsed with a mighty crack. As they fell backwards, the next five rows toppled like a pack of cards.[1022]

Within weeks, the Anglicans too, had a church tent. An English missionary society, the Society for the Propagation of Christian Knowledge, had provided large tents for both the Bennett and Atlin Missions.[1023]

Bishop Ridley visited the Atlin mission, at the height of the boom, in July 1899. He preached to hundreds of miners at Atlin, Discovery and Surprise City. He was impressed with the size and dedication of the congregation, and the camp in general. He concluded the mission was important, and to carry on its work successfully, it would need financial support from the Church in England. Ridley bought property for a church building, and a small house for Stephenson, whose wife and four small children were due to arrive soon. In all, he spent 100 pounds.[1024]

It is evident that the Anglican and Presbyterian Churches viewed their work in the goldfields quite differently. Bishop Ridley's money had come from the Society for the Propagation of the Gospel, and Stephenson, himself, was already receiving money from the same body. Since it was never enough, Stephenson sometimes supplemented his small stipend by working as a carpenter. Still, raising funds to build a church was not quite the struggle for Stephenson that it was for Pringle.

The Anglican congregation was also blessed with a dedicated fund-raiser in Leonard de Gex, who was church warden, and a bank accountant. He was not only committed, but very persuasive. Stephenson, himself describes one of the warden's techniques:

> When we were raising the building fund my warden went to the office of another bank to interview the Manager. At the time everyone was on the *qui vive* to "get in" on some get rich quick proposition. Says Mr. Warden to Mr. Bank Manager, "Say G, do you want to "get in" on a real live proposition?"
>
> "Can you recommend it?"
>
> "Well I should say. The nearest thing this side of the grave."
>
> "Are you in it?" asks the Bank Manager.
>
> "Up to the eyes and then some."
>
> "How much is it going to cost?"
>
> "Fifty a share and sure interest on your money."
>
> "All right I'll flutter a couple of fifties." With that the Warden handed him the building fund list of subscribers. But Mr. Bank Manager stood to his guns and ponied up.[1025]

By November 1900, St. Martin's Anglican Church was completed on the corner of Third and Trainor. Built by W. H. T. Olive from locally milled timber, it was a fairly spartan but attractive building with a wood stove and sawdust filled walls, to help hold the heat for winter services. A year earlier funds had been raised to buy an organ which lent the final ecclesiastical touch to the new building.[1026] The little church was consecrated by the Bishop of Selkirk, the Right Rev. W. Bompas on 20 October, 1901.

Stephenson was respected and well liked by the miners, and the community at large. He was often called "F. L." by the locals who considered him a "good scout."[1027] He could race a fast dog team or keep up with the best of trail walkers. He was a good shot, and, if necessary, lived off the land. Stephenson and his family lived in a cramped 3.6 by 6 metre vicarage, two blocks from the lake. His slim income did not allow for extras, or even some essentials, and he carried all the family's water uphill to their cabin. He used a yoke on which hung two coal oil cans. The total weight would have been about 45 kg.[1028]

He had a good sense of humour, experience doing manual labour, and some medical knowledge which he was able to use on occasion. He was also more tolerant of the miners' way of life than was Pringle, and he was prepared to restrain his disapproval in order to gain their attention. A story is told of how he spent all night playing poker with a group of "tough characters", who agreed to have him close the game down with a few prayers. He frequently rode out to the creeks on horseback, was fond of hunting, and was an expert swordsman and boxer. Indeed, he was sometimes known as the "fighting" parson, and it has been suggested that not all of his boxing was done in the ring.

Stephenson's special appeal, brought him large congregations in both Atlin and Discovery. As the community became more established, he was able

AHS P-1257

*St. Martin's Anglican Church was build on Third Street in Atlin City in 1900. In 1902 the bell tower was added.*

to introduce improvements to St. Martin's. A bell tower was added in 1902, and a 90 kg bell was installed.[1029] That same year a new two story vicarage was built for the Stephenson family, which by then had five young children. Just two doors from the church, the new house was spacious and comfortable.

Stephenson and Pringle had a good deal in common, and got along particularly well. The two congregations celebrated the Christmas of 1899 together at St. Martin's. The ministers had gone together to cut a tree for the occasion. Similarly a concert to raise funds for St. Martin's building fund was held in the Presbyterian hall, and John Pringle was featured on the programme.[1030] Clearly the pair enjoyed a mutual respect and companionship.

When Pringle was sent to Bonanza Creek in the Klondike in the fall of 1901, his replacement was Rev. James Russell, who stayed for about a year, before the arrival of E. T. Turkington, a young and newly ordained minister, who kept the mission active until he left in 1905. W. J. Kidd arrived to serve for the summer, followed by Rev. M. Ross who carried on until sometime in 1909. Atlin's population had declined significantly, and Ross complained that progress was slow. He clearly preferred a pulpit where he could be more effective. For three years, though, the Presbyterian minister had been the only clergyman in Atlin.

By 1906, Atlin was a quieter town. Mining was in the hands of the large companies, and most of the individual miners had moved on. Fred Stephenson, too, would leave to begin a new mission at Aldemere (later Telkwa), in the Bulkley Valley. Because of increased agricultural activity, and plans to build a railway, the Bulkley Valley was receiving an influx of population at the same time as Atlin's size was dwindling.

Stephenson left for Aldemere in the late spring. He chose to go overland on foot with his dog team, a distance of 965 km. He wrote to a friend in Atlin that, " there were some very bad gulches and mountains to tackle, but I had 280 feet of good rope, and lowered away snubbed to a root of tree. . . "[1031] In his capable way he negotiated the long tough trail. His brother who had not seen him for nine years, walked north from Hazelton to meet him. Stephenson had abandoned his sled at the Nass River, and hiked along with his dogs running loose before him, packs on their backs. His brother barely recognized the bare headed man in overalls, burned dark by the sun, who strode down the trail towards him.[1032]

There was no incumbent in Atlin until July 1908, when Rev. Marcus Jackson arrived. He stayed for more than two years, and was replaced in March 1910, by Rev. E. P. Laycock who left six months later. It was 17 years before another Anglican minister resided in Atlin.[1033]

About the time Laycock left, Rev. A. C. Justice came to reopen the Presbyterian Mission in Atlin. It is unclear how long he stayed, but the building used by the Presbyterian congregation, the original Arctic Brotherhood Hall, burned in the fire of 1914. The minister at the time was a Mr. Batty. For a few months, he used St. Martin's Church on Sunday evening, but he left Atlin in September 1914, and St. Andrew's Parish ceased to exist.

St. Martin's continued to offer weekly services, due to the dedication of two faithful parishioners, Capt. C. S. Hathorn, and Major C. W. A. Nevile. Together they made sure the church never closed. During the summer, tourists often joined the local congregation. One was Miss Jesse van Brunt of Brooklyn, New York. She was so impressed with the pretty little church, that she returned home to design and execute a colourful painted glass window to be placed behind the altar. Installed in 1923, its symbols are traditionally Anglican, and uniquely northern, and it became St. Martin's decorative showpiece.

Over the years, the Bishop, or a minister paid periodic visits to St. Martin's. From 1927 to 1943, Reverends. Roy Manwaring, E. A. McCarthy, Gordon Bratt, and Richard Graham, continued the work started by Stephenson. Just before Rev. Graham left in 1943, Major Nevile died. Captain Hathorn had already moved from Atlin, and with these two faithful supporters missing, the church was closed until 1950.[1034]

Although Bishop Ridley had shown some interest in including the Tlingit people in his church's Atlin mission, and the Presbyterian Sunday School had included large numbers of Tlingit children on their picnics, neither Protestant church made a real effort to minister to the native community. There was a language difficulty, and the Tlingit people were not settled in Atlin all year round, as they traveled frequently to both Teslin, and the Taku River area. Some had, though, shown an interest in Christianity. Around 1899, and perhaps earlier, several had been converted in Juneau, where most Tlingit people were Russian Orthodox.[1035] It was not until 1907, and the arrival of Father Joseph Allard, an Oblate priest, that any significant effort was made to minister to the Native people of the area.

Before Allard's time Catholic priests had visited Atlin almost every year, but they too, were primarily concerned with the white population. Father Whelan of Vancouver, who was returning from Dawson City, spent part of the summer of 1899 in Atlin. He baptized babies, married some young couples on the creeks, and held Sunday services at the Hotel Russell. In 1900, Father Morgan, of New Westminster, held services in the post office.

In 1903, Archbishop Dontenville sent word from Vancouver to Rev. Bunoz, Superior for the Oblates working in the Yukon, to take charge of Atlin.[1036] From this time on, the Oblates, who had worked for many years among the Indian people in British Columbia, made the Atlin Tlingits their responsibility. A number of Indians already considered themselves Christian, and regularly said prayers together in their own language.[1037] Fathers Le Chesne, Corbeil, and Godfrey held occasional services in the summer months. In 1904, a church was built on the corner of 3rd Street and Trainor Avenue.[1038]

Father Joseph Allard, O. M. I. spent several years in the Yukon, and had been at Conrad City on Windy Arm of Tagish Lake before coming to Atlin in April 1907.[1039] The French priest was fired with missionary zeal. He was immediately captivated by the Tlingit people, and believed they were anxious to be members of his church. They showed immediate interest in his message about the Christian God, and when they expressed a desire to pray in their own tongue, he learned the difficult Tlingit language himself. None of the Tlingit children attended the Atlin school, so Allard determined to teach them to read, write and speak in English.

Never shy to ask for what he needed, Allard walked to Discovery, and asked the teacher if she would give him some pointers about teaching. Had they not been nailed down, he might have walked off with her blackboards, but instead he settled for some chalk. In Atlin Mrs. Henning gave him a small blackboard. For the next two weeks he held classes for the children in the meeting house at the Indian village, at the south end of town, until he discovered that someone actually lived in the place, and was objecting to the disturbance.

Right beside the village stood a large house owned by Mrs. E. Ridd, who was willing to sell it for $200. Allard headed for Spruce Creek and found Alex MacDonald, and nine other Catholic miners, who were unable to resist his pleas. Each surrendered $20.00, and the house was his. It was here that he set up a boys' boarding school, and looked after nine pupils while their parents were away hunting and trapping.

Father Allard had no real means of supporting his charges, but optimistically did not expect that any would go hungry. The boys' parents supplied him with meat, and town people and miners on the creeks were generous with donations of staple foods. He had a small fishing net, and with it he and the boys were able to supply some of their own food. Father Allard, himself, with help from the boys, did all of the household chores, cooking, wood cutting and the cleaning.[1040]

*St. Joseph's Catholic Church was moved from Third Street in Atlin to its permanent location in the Tlingit Indian village at the south end of town in the 1930's. At the right is the rectory.*

The school room was typical of its time, desks arranged in neat rows, maps and pictures decorating the walls.[1041] The students ranged in age from 6 to 16. Besides religious lessons, they were taught reading and writing, and how to mend their shoes and clothing. They learned church music, and their singing was impressive. Every two weeks, when Father Allard went to Spruce Creek to say a Mass, the boys went with him to sing, and the miners enjoyed their company.

Fourteen Tlingit children were baptized in May 1908, and, by August, 37 native people, some from Teslin, had become members of the church. Included in Father Allard's following were those who had been baptized by a Russian Orthodox priest.

By the next year, Father Allard was experiencing some financial difficulties at the school. Henry Esson Young, the M. P. P. for Atlin, and a Cabinet Minister, offered some assistance from the Provincial Government.[1042] With the support of the Atlin Board of Trade, and a petition from residents, Father Allard applied to the Dominion Government for funds. In January 1909, he was granted a teacher's salary of $75.00 per month, and with this he was able to carry on.

Because the girls of the village spent much of their time hunting and trapping with their parents, they did not remain in Atlin long enough to get an effective education, and Father Allard decided to start a boarding school for them too. He persuaded his sister, Bernadette Allard to come from France to help him. She arrived in October 1909, and although she was used to a comfortable life in Europe, and found the rustic conditions in which she was asked to live and work, a little shocking, she adjusted quickly. For nine months she taught the girls religion, reading, writing, sewing, cooking, and other homemaking skills.

In the summer of 1910, Father Allard's superiors sent him to Dawson City, and his sister returned to France. The Catholic Church in Atlin was left without a resident priest for the next 19 years. Only in the summer did clergymen make short visits.

In 1929, Father Allard returned for another term. He intended to reopen his school, but all of his potential students were away at residential schools. He remained in Atlin until 1936, and during this time he had the church moved from the centre of town to the Indian village, where it was enlarged and improved.[1043]

There is no doubt that Joseph Allard was a tireless worker and totally devoted to his people, but he was, to say the least, contentious. There had always been, until his arrival, the most cordial relations among all the Christian denominations. Just three years before he arrived, the Reverends Stephenson, Turkington and Corbeil had been hunting partners.[1044] Allard was intensely suspicious of the Protestant clergy, and their motives with respect to the native people. It does appear once Allard was settled in Atlin, the Anglican clergy in Carcross and Teslin did make some effort to convert the Atlin Tlingits. Similarly, Father Allard attempted to interfere with the Anglican mission in Teslin. The resulting strain on church relations ruined the friendly interchange enjoyed in earlier years. Allard also had disagreements with his superiors. He bitterly resented the fact that the Atlin mission was neglected for so long after his three year's of hard work from 1907 to 1910.

The distinctive personalities of Pringle, Stephenson, and Allard were as divergent as their backgrounds. Their approach to religion differed too, but each was equally devoted to the service of God, and his church. Just as the goldseekers had to face the difficulties of the trail, bear the disappointments of mining, and toil endlessly to gain their reward, these men of the cloth endured no less for the privilege of helping their fellow man.

CHAPTER 14

# The Care of the Sick

*It was a low log building sunk in the hillside overlooking the lake. Its shed roof was made of poles covered with "dirt". Its floor was six inches of sawdust, and as I stood my head was not more than six inches from the roof poles. Lying on a low cot was the man whom I had come to see, and on pole bunks around were five others injured and diseased. At the door was a rough box with a dead body in it, and outside was another . . . The only nurse was a so-called abandoned woman, who nursed, cooked and washed for the hospital without reward. God bless her for her work.*[1045]

Thus Rev. John Pringle described the distressful condition of the Provincial Government Hospital when he first visited in April 1899. The meager facility was put together with some government funding[1046] and local fundraising,[1047] under the direction of Dr. A. S. Monro, the Medical Health Officer.[1048] There was a desperate need to accommodate the growing number of patients from the camps, and those whose health had buckled on the trails. The winter of 1898-99 had been particularly cold, and a good number of those who arrived in the early months of 1899 were in a weakened condition. Some were suffering from scurvy, influenza or pneumonia, others from frostbite and a variety of injuries.

The crudely inadequate medical facility prompted Drs. Monro and F. M. Boyle to open a private infirmary on land made available by the government[1049] at the corner of Third St. and Trainor Ave.[1050] The first issue of the *Atlin Claim* proudly announced that the new facility would open on May 1, 1899,[1051] and advertised amenities including private rooms, a competent nurse, and moderate charges.[1052] There was even a room reserved for members of the Masonic Lodge.[1053]

Doctors Monro and Boyle and nurse E. Elliot cared for the community's sick through the summer of 1899 at the Atlin General Hospital, which was a spacious tent. Miss Elliot, who was from Winnipeg, was Atlin's first Registered Nurse.

Atlin General Hospital could admit paying patients only, because it was built with private funds, and was by no means a profit making venture. A Ladies Auxiliary headed by Mesdames Hitchcock, Ogilvie, and Galarno was formed to provide assistance.[1054] Government funding was available to help provide care for indigent patients. In just three months the hospital treated 118 patients.[1055] There is no record of it operating after the summer of 1899.

John Pringle's reaction to the shocking conditions at the dingy government hospital was to fire off an urgent request to his Church in Toronto for two nurses. Earlier in 1899, Rev. Robert Dickey had sent a clear message from Skagway that the best way to further the work of the Church would be to provide nursing care for the sick. Added to the pressure were articles written by Faith Fenton, a remarkable woman reporter for the *Toronto Globe.* She had traveled north with the Yukon Field Force and the Victorian Order of Nurses, which had set out in May 1898, on a sea and land route up the Stikine River, over the Teslin Trail, and down the Yukon River to Dawson City.[1056] Her first hand experiences on the trail and in the gold fields gave a clear ring of truth to the reports she filed with the *Globe.* Interest was aroused everywhere in the plight of sick goldseekers in the north, and Pringle's plea received an immediate response.

World focus was fixed on the great gold rush with the same fascination afforded a war. The stakes were surely different, but the hazards to life and limb were real enough. Those who dared not go, followed the drama vicariously. Although relatively few women were among the first stampeders who swarmed trails and crowded ramshackle camps, others were quick to organize support groups behind the lines when word filtered back that the goldseekers were in trouble. Women representing churches of all denominations raised money, gathered clothing, bedding, medical supplies and equipment and shipped it off to the north. It was the Presbyterian women in Toronto who organized the Klondyke Nurses Committee,[1057] and raised $1500. The need for nurses in the Klondike had been partially answered by the contingent of the Victorian Order of Nurses. As a result the ladies in Toronto turned their full attention to Atlin. They renamed themselves the Atlin Nurse Committee, and began looking for suitable nurses to fill Pringle's request.[1058]

Under the leadership of Mrs. Charlotte Macdonald, the wife of a Toronto judge, the search led the committee to two women, who at the time were right in their own city. Late in June 1899, Miss Elizabeth Mitchell and Miss Helen Bone accepted the challenge to care for the sick in a distant, isolated place called Atlin.

After a ceremony at the Toronto Presbyterian Church, which dedicated them as missionary nurses, they departed on 30 June, 1899, with six large bags packed with medical supplies, bedding and winter clothing.[1059] By July 22, they were in Atlin, where the hospital they anticipated did not exist, in fact, "was not even thought of."[1060] The original intention had been for them to go out to the creeks, from tent to cabin to administer care, but the time involved trekking far and wide to Discovery and the mines made this an impossible plan. Besides, their first patients simply turned up in Atlin looking for help, and they had to be accommodated in the nurses' tiny house. The first woman patient was a victim of typhoid from Discovery. Six men carried her on a stretcher for 10 km to deliver her to the nurses.[1061]

Helen Bone, whose home was in Paris, Ontario, was in her late twenties. Her training had been at Grace Hospital in Toronto.[1062] Her dismay at the harsh conditions afforded sick men in Atlin was evident in letters written to a member of the Atlin Nurse Committee. "Imagine being sick in a tent. One would scarcely believe that patients could recover."[1063] Nevertheless, she went quietly and capably about her duties, and improvements came. To begin with, Gold Commissioner Graham gave the nurses a hospital building. It had a mud roof and a sawdust floor, and very closely fits John Pringle's description of the first hospital.[1064] It may well have been the same building. It was only large enough for four cots, which were "made of canvas stretched between poles driven in the ground."[1065] There were two small windows which would not open. A tent was put up next to the cabin to provide more space, and the nurses improvised. They used the packing material that came around their cots for pillow cases, and stuffed them with hay. Although the tent annex was brighter than the building , when "it snowed outside, it rained inside."[1066] Just keeping patients and bedding dry was a nagging problem.

Elizabeth Mitchell was 50 years old, and the twelve hour shifts she and Helen Bone worked day in and day out would have defeated most women her age. However, she was strong, healthy, and deeply dedicated both as a nurse and a Christian. Her training in Philadelphia, combined with years of experience, made her invaluable to Atlin. Trim and efficient, with hair pulled back

AHS P-1258

*The Atlin Nurses Committee, formed by Presbyterian women in Toronto, sent nurses Elizabeth Mitchell and Helen Bone to Atlin in 1899.*

behind prominent ears, she made the most of the conditions, and poured her unusual energy into all her tasks.

Seven days a week, the pair laboured for long hard hours to cope with the sick. Helen Bone worked from 9 a. m. till 9 p. m., and Elizabeth Mitchell took over through the night and early morning hours. Meals for themselves and for the patients had to be cooked at home, and this was an extra chore for the "off duty" nurse. The pair remained surprisingly buoyant and felt far from oppressed. They wrote joyful letters home describing "glorious sunrises", and the magnificence of the northern lights.[1067] Their complaints were aimed at substandard facilities which caused discomfort to their patients. The ladies of the Committee in Toronto, whose support was crucial, never fully understood the magnitude of the work done by their nurses in Atlin. Through the Committee's efforts the nurses received salaries of $25 each month, furniture, food, fire wood, uniforms, and their small home, a 3.5 x 5.5 metre cabin.[1068]

The nurses soon attained a position of honour within the camp, where the miners regarded them as true "angels of mercy." There were few men who would not help either of them if he were able. Some brought gifts of wild game, and others called to see if odd jobs needed to be done. John Pringle considered their work was more beneficial to the church than any number of preachers. After observing their work for three months he wrote, "Thank God for such cheery, consecrated, efficient women".[1069]

One January night in 1900, Pringle left his cabin to walk the short distance to the hospital. The still cold air bit into his flesh, and as he released each breath it frosted his moustache. He looked about at the town shrouded in icy crystals, smoke rising in thin, straight plumes from a multitude of chimneys. Inside the hospital tent he found Elizabeth Mitchell, huddled in a fur coat. Three patients, buried in their covers, were bunched as close to the stove as it was safe to be. Their discomfort was so obvious that he decided, late that night as he lay awake, that a proper hospital had to be built.

The next day Pringle was scheduled to leave for Taku City and Tagish Post, where he would visit the North West Mounted Police, and Bennett City, to spend some time with Rev. John Sinclair. Before he left, however, he spoke to a stalwart member of his congregation, James Stables, about his hopes and determination to build a hospital for Atlin. Stables, a Scot, was a Stikine Trail mate of Pringle's, and a man of action. While Pringle was away he organized a force of miners, storekeepers, and professional men who were willing to donate time to build a hospital.[1070]

AHS P-1259

*St. Andrew's was the first Presbyterian Hospital in Canada. It was built by the miners for Reverend John Pringle in 1900. Two years later the nurses' residence, at right, was added.*

Ground breaking took place on January 13th,[1071] in the depth of winter. The cold presented some difficulties, but none that could not be overcome. The nurses warmed the frozen nails for the volunteer carpenters,[1072] and there was a great spirit of camaraderie amongst the workers, all of them happy to be part of such a worthy project. When Dr. Gustave Lewis crawled up on the roof to help the shinglers, they noticed he was not dressed for work, and gleefully mocked his formal attire by nailing his coat tail to the roof.[1073] When Pringle returned a month later, his hospital was almost completed at the cost of only $50, thanks to Stables' volunteers.

St. Andrew's Hospital faced Discovery Avenue at the corner of Third Street. Its numerous windows made the interior light and airy. It was finished with drop siding, and its outside walls were filled with the ever available sawdust for insulation. Inside there were three rooms; a nine bed ward, a bathroom and an operating room. A fine cooking range was provided by Mr. Charleson, who was in charge of construction on the Dominion Telegraph, but there was no room for it inside the new building and the kitchen was a tent behind the hospital. Patients were transferred to the new St. Andrew's Hospital on 24 February 1900.[1074]

The following year, an outbreak of scarlet fever made it necessary to provide an isolation hospital.[1075] The Provincial Government made funding available, and a separate building was moved adjacent to the new building. This was the building which caused Joseph Graham so much strife.[1076]

As the camp became settled and women came to Atlin to join their husbands, it was necessary to accommodate female patients. The arrival of new babies at the hospital had become commonplace, and often they and their mothers had to be cared for at the nurses' residence.[1077] In the summer of 1902, $1000 was sent from the committee in Toronto for the construction of a women's ward. It was named the Charlotte Macdonald Ward, in memory of the first President of the Atlin Nurse Committee. Mrs. Macdonald, who had been a tireless worker for St. Andrew's, died while the new wing was under construction. The new ward included six beds and a reception room.[1078]

The work of the nurses was so well appreciated that donations were constantly arriving from all over Canada. Blankets, sheets and pillows from a number of Vancouver Presbyterian Churches, books and a sewing machine from Ontario churches, clothing from Nova Scotia, and slippers from a manufacturer in Toronto.[1079] Nonetheless, maintaining a hospital was a constant struggle financially.

Local fund-raisers, concerts and balls, were always gainful, and the province provided a $500 grant to cover the care of indigent patients. Just the same, a certain amount of ingenuity was still needed, and in November 1900, John Pringle set out for the creeks selling hospital tickets for $1.00. The idea was that each $1.00 purchase represented a month, and that if a ticket holder became ill, he would be entitled to free hospital care for that long if necessary. This scheme, which had been used by other small hospitals, had all the appearances of an early medical insurance plan. It was simple, and it worked. On Pringle's first trip to the creeks he sold 300 tickets.[1080]

Before John Pringle left Atlin in 1901, he handed over financial control of the Hospital to a Board of Trustees. The first board was J. H. Russell, James Stables, Archie Shiels, and a Mr. Stewart. Soon, J. A. Fraser took Shiels place. The Board was responsible for maintaining all hospital property, and raising money for everything except salaries. At the earliest opportunity, they approached the Provincial Government and persuaded them to raise their annual grant to $1000. Then the innovative trustees planted a large garden behind the hospital to supply fresh vegetables for patients and nurses.[1081] The work in the garden was done by volunteers, and the well tended plot became a community showpiece.

The Board was also instrumental in building a new residence for the nurses in 1902. It was a two story frame building, attached to the west side of the hospital. For the nurses their new accommodation seemed palatial compared with their crowded cabin.[1082] Along with new housing, the nurses received a salary increase to $50 per month to help offset the very high cost of living in Atlin. Still they made only half the amount paid to a teacher, a policemen, or a cook.

Elizabeth Mitchell and Helen Bone had agreed to stay in Atlin for two years, and in the fall of 1902, after more than three years, they asked to be relieved of their duties. No one could have asked for more of these two women, who had worked long hours, under difficult conditions with few breaks. They were both in need of a rest, but in fact they had enjoyed their stay in Atlin. They had relished the challenges of frontier nursing, and both delighted in the northern environment.

By 1902, the hospital had become a busy place, the 15 bed were usually occupied, and there were also visiting patients to attend. Cases of scurvy and typhoid had declined, but pneumonia and frostbite were common complaints. Mining accidents were frequent and serious, and babies arrived routinely. Replacement nurses would find their hands full.

Kate Smith, and Kate McTavish were professionals who had chosen the Atlin assignment for the challenge and the adventure. Miss Smith had been in charge of the New Westminster Women's Hospital, and Miss McTavish, St. Andrew's new matron, had been assistant matron at the Toronto Home for the Incurables.[1083] Originally from Bala, Ontario, her working career had begun in the schoolroom, but she decided, at the age of 34, to enter nurses' training at Toronto General Hospital. She was about 47 when she came to Atlin,[1084] a woman whose firm confidence was backed by excellent schooling and experience, and whose words carried certain authority.

Life in Atlin agreed with these women too, and they engrossed themselves in their work, and the life of the community. The system of "selling tickets" was continued, and it was often the nurses who tramped to the creeks, met the miners, and requested their support. They had no trouble selling $5 tickets which guaranteed miners five months of free medical care. They particularly admired Rev. Turkington, and helped him with the Presbyterian Sunday School.[1085]

At times there was funding for a housekeeper, but usually the nurses did the cooking and chores themselves. In 1902, the Toronto committee allowed $25.00 per month for a helper, but it was difficult to keep anyone at that wage, when other such workers were earning three or four times as much. A Miss Dickenson, and then a Miss Douglas were sent by the Church from Vancouver, and the Atlin trustees added $10 to the salary paid by the Toronto Committee. However, the helper's salary was withdrawn in 1905, and afterward, when help was needed, the trustees paid the entire wage.

Kate Smith left Atlin in October, 1904, but Kate McTavish stayed until August, 1906, when she returned east to undertake further training.[1086] Before she left Atlin, the miners, who held her in high esteem, gave her a gold nugget watch chain, and a purse of money.[1087] For a time the Toronto Committee hoped that she would return to Atlin, but instead she became the Matron at the new Prince Rupert Hospital, where she was also responsible for training nurses.

A few years later another extraordinary woman became Matron of St. Andrew's Hospital. She was Rachel Hanna, a woman of great spirit and adventure. She graduated from the Toronto General Hospital School of Nursing, joined the Victoria Order of Nurses, and signed a contract to serve anywhere in Canada for a salary of $300 a month.[1088] "Anywhere" led her on a journey to Dawson City, with three other nurses, and the Yukon Field Force.[1089] Her service in the Klondike was exemplary, and during the height of the rush Nurse

Hanna had cared for thousands of sick, undernourished, and injured miners. When the Victoria Order of Nurses withdrew from Dawson, Rachel Hanna stayed on to work as a Presbyterian missionary nurse at the Good Samaritan Hospital.[1090] As matron of Atlin's St. Andrew's Hospital during 1910-11,[1091] she was irrepressibly cheerful, willing to tackle any task under any conditions, dedicated, hardworking and revered.

Over the years, a few of Atlin's nurses married and stayed in the community. Rachel Hanna's niece, Alma Hanna, married Sidney Smith of Atlin. They had two children, but in 1918, the entire family was lost when the *S.S. Princess Sophia* sank near Juneau. Mary Queen, a local girl, who nursed at the hospital for a short time, married Dr. A. K. Connolly.[1092] One of Atlin's nurses, Annie R. W. Moore, also taught piano lessons.[1093] John Pringle's tall and imposing sister, Lucy, served St. Andrew's for a short time in 1922.[1094]

By 1906, the Atlin Nurse Committee, now known as the Women's Home Missionary Society, considered withdrawing their support from Atlin, and there was considerable apprehension at St. Andrew's.[1095] Atlin was a well established community, and certainly no longer presented a missionary situation. Moreover, the Society was deeply involved with a number of other hospitals in Western Canada, and its resources were stretched. But Atlin was the Society's first project, and the first Presbyterian Hospital in Canada, so for little more than sentimental reasons, the Society continued to pay salaries for two nurses.

The nurses were not alone in their work. Usually there was at least one doctor. In the boom days there were dozens, but most preferred mining to medicine, and only a few were licensed to practice in British Columbia. This became a matter of contention between the residents who needed doctors, and the Government, whose regulations denied them the right to practice. Dr. F. T. Troughton came to Atlin in 1898, and became involved in a number of successful business and mining ventures. He often found himself awkwardly placed between his desire to help an ailing neighbour, and the threat of prosecution if he did. On one occasion, when there was no doctor in town, he received a telegraph message from the Premier asking him to attend an urgent case.[1096]

Dr. Gustave Lewis was the first doctor to practice at St. Andrew's Hospital. He stayed until the summer of 1902. His replacement, Dr. Lazier, stayed only a month, saying the country was, "too damned healthy".[1097]

Henry Esson Young was the doctor most closely associated with Atlin over a long period of time. He practised for some time without a British Columbia licence; then in October 1902, he was appointed Medical Officer of Health.[1098] A dedicated doctor, and an immensely popular man, Young was elected to the Legislative Assembly in 1903, and from that time a replacement was needed for the months he was in Victoria. Until he was appointed to the Cabinet in 1907, he returned whenever he could to care for the residents of Atlin. Even after he could no longer practice medicine in the community he represented, he worked hard to ensure that St. Andrew's was adequately funded, and that there was a resident physician.

In the last days of the summer of 1904, Fritz Miller was admitted to St. Andrew's, and on August 27 the young pioneer died there from "surgical shock" following an operation. The communities were distressed and saddened. As a tribute, flags were flown at half mast for the miner who was deemed "the father and discoverer of the Atlin camp." Tributes and condolences came from all quarters to his widow, Anna. Premier Richard McBride wrote that his death meant "the loss of one of British Columbia's most active and energetic citizens."[1099] Ironically, Miller's untimely demise could not be prevented despite good facilities and skilled help, because there were few effective medicines. Rev. Stephenson held the funeral service at St. Martin's, and Fritz Miller was laid to rest in the cemetery on the Discovery wagon road.

In 1919, while the population was still reeling from the ravages of World War I, the first fatal wave of the great influenza epidemic hit Atlin. It spared no corner of the globe and it attacked the young and robust, as well as the aged. The staff and resources of St. Andrew's were strained as they had not been since the early years. The first outbreak was in March of 1919, and there so much apprehension about the rapid spread of the virus that the school was closed.[1100]

It was not until the spring of 1920, though, that Atlin really experienced an epidemic. New crosses appeared at the cemeteries as 19 citizens died. At the height of the scourge, there were 50 people lying ill in both Atlin and Discovery. Worst hit were the Tlingit people, who lost 15 of their number.[1101]

Dr. Affleck and the hospital staff were driven to the point of exhaustion. A number of local people volunteered to help out, but the hero of this tragic time was Postmaster Thomas James. He was a dispensing chemist, and he worked day and night alongside the hospital staff. In time, he too fell victim to the savage virus, but he continued to help until just a few hours before he died.

Except for the frantic months of the epidemic, the smaller population during the War and through the twenties brought fewer patients to the hospital, and a lighter work load to the staff. Nurse Doris Nichols recorded an average of six patients at any given time during 1927-28. She filled her off-duty hours with a busy social life of teas, dinners, sing songs, boat trips, and hikes.[1102]

In 1929, after 30 years, the Women's Home Missionary Society finally withdrew its support from St. Andrew's Hospital. By that time Atlin was a destination for wealthy tourists, and by no means a suitable object for missionary charity.[1103] The community supported the hospital throughout the 1930's, while there was a resurgence in the mining industry in the north, and the community was relatively prosperous.

During most of St. Andrew's history it had a loyal supporter in Major C. W. A. Nevile. He was Secretary of the Board for many years, and important to its survival during difficult times. It was believed he was a remittance man, but whether real or not, he was considered to be a true English gentleman, by all who knew him. For a time he provided the water service to Atlin, delivering to households for 5¢ a pail. He lived in a tiny house on Third Avenue, and each evening he dined, appropriately dressed, at the Royal Hotel.

The hospital, built by the pioneers, was condemned by the Fire Marshall in 1940.[1104] By then the White Pass Company had eliminated Atlin from its tourist itinerary, and closed their hotel, the Atlin Inn. With Government funding, the hospital board bought the south wing of the deserted inn. The large three story section was dismantled and rebuilt at the edge of Brewery Bay. Ready to receive its first patients in November, 1942, the new St. Andrew's was an imposing two story structure, white with dark green trim. Its front windows looked out on the town and the old wharves and warehouses cluttered along the shore.

The new facility cost $21,000. Dr. G. W. Meyer was the physician in charge. Those who gathered for the opening dedication could not have guessed the new facility would have a shorter life span than Atlin's first crude government hospital. World War II caused mining to decline, and again the population eroded as men left to join the armed forces. The new hospital was a white elephant.

For 44 years, Atlin was served by dozens of remarkable doctors and nurses. With the help of a group of Toronto women, who never saw the results of their steady support, and with the generous work of local volunteers, through boom and bust the hospital had been a constant. In June 1943, only eight months after it opened, Dr. G. W. Meyer left St. Andrew's and its doors were closed, marking the end of the era of Presbyterian hospitals in Atlin.

CHAPTER 15

# Big Blazes and Bucket Brigades

On Sunday, 26th August, 1900, a lively breeze danced along the chilly surface of the lake and twirled down the quiet streets of Atlin. It whispered of the cooler days just ahead, as the shadows grew longer under a waning summer sun. Residents who planned to spend the winter in the area knew it was time to build wood piles and fill root cellars.

Then suddenly the wind carried another message, urgent and frightening. It brought the metallic clang of fire bells to the startled ears of citizens relaxing over dinner, and enjoying an easy day at the end of a hard week. It was 5:30 p.m. and the hotel portion of the British-American Corporation warehouse on Rant Street was on fire. Thick smoke churned from the building, and the wind chased it through the community, as everybody raced outside to answer the alarm.

Arriving first at the scene were J. L. Wallace and B. Dixon who attacked the fire with extinguishers. Their efforts were gallant but futile against the gathering blaze. Two barkeepers just barely escaped the burning building with their lives and the clothes they wore. As they reached the street several kegs of black gun powder exploded in the warehouse, and masses of flaming rubble were thrown across the street, onto wooden footpaths and trees. Suddenly many more buildings over a wide area ignited.[1105]

At the town wharf the *S. S. Scotia* started its pumps, and provided a non-stop supply of water to the bucket brigade, organized by Fire Chief John Kirkland. Dripping pails moved quickly along a human chain reaching from the lake to the burning warehouse. Icy water slopped over the hands of women, children of all ages, and dozens of men, including several lawyers, three clergymen, and a judge, all desperately intent on stopping the fire. The effort was impressive, but it is doubtful if even one bucket arrived at the scene more than half full.

Other residents and businessmen scrambled to save the contents of doomed buildings. One man jettisoned a trunk and a small dog from an upper window, then, in his rush downstairs, he tripped over two fur coats he was carrying and broke his leg.[1106]

The determined wind fanned the blaze and swept it to "Hirschfeld's Photo Studio, Miss Simson's old store, Fetherstonhaugh's Assay Office, Mr. Rose's dwelling house and several small cabins (which were) sheeted in flames."[1107]

Fire ran before the wind up First Street, where frame buildings and tar paper fed its hungry rampage. The OK Hotel, a log building,[1108] which also housed the Broder and Jones Restaurant and Bakery, was in flames."[1109] Then the Board of Trade Hotel building next door was consumed after a failed attempt to dynamite it in order to create a firebreak.

The fire roared along to Capt. Nickerson's General Store. As that building blazed, men stationed on top of the new Bank of Halifax on First and Pearl worked furiously at spreading wet blankets to save the roof from flying cinders. Two blocks away, at the grand new Government Building, officially opened just days earlier, other fire fighters struggled to clear away burning tar paper and other hot debris dropped by the wind on its steep shingled roof.[1110] All over town flying cinders ignited canvas and wood. Firemen were hard pressed to locate and extinguish isolated blazes.

The news that Atlin was on fire reached Discovery, probably by phone. Just over an hour after the alarm sounded in Atlin, a crowd of miners arrived from the creeks to join forces with the town people.[1111]

It was determined that the fire had to be stopped at the Bank of Halifax Assay Office and the Olympic Hotel, or all of Atlin would be engulfed. Fire Chief Kirkland, Tom Switzer and several other men stationed themselves atop the Olympic, where water from the bucket brigade was passed to them. Tom Dixon and Capt. J. Marten, who had raced in from Spruce Creek, manned the roof of the Assay Office, and held their post against terrible heat which singed the hair on their heads. Dozens of blankets were spread out and doused from barrels of water brought in carts by the Dixon Brothers and Joe Brooks.[1112]

Four difficult hours later, as night pressed down on the immense land, fire fighters gained control of the conflagration. In the end the "manhood of Atlin did not fail."[1113] The fire was squelched, no lives were lost, and the town still stood. "The battle had been fought and fair Atlin was once more safe."[1114] The fact that the wind died down with the setting sun might have contributed as much to stopping the fire as did Atlin's manhood, though.

While blood still raced and spirits soared because the danger was past, several hotel keepers, grateful that their businesses had been saved "kept open house during the remainder of the evening."[1115] Grimy firefighters took advantage of the hospitality to celebrate and sing into the early morning hours. As the victory party went on, the watchman stationed to report flare-ups, could not resist the free whiskey, and eventually it lulled him to sleep. By morning several other buildings had been destroyed.[1116]

The destruction done during the short life of the fire was devastating. Three hotels, six stores, houses and cabins, along with offices and warehouses, were destroyed totally. Buildings could and would be replaced, but personal treasures were gone forever. Forty cases of the very finest champagne, belonging to the flamboyant Count E. J. de Lamare, had added fuel to the fire.[1117] J. H. Rose lost three fur coats and 20 pairs of blankets. One storekeeper was especially chagrined since at the height of the fire he had turned down an on the spot offer of $1000 for his business which stood two blocks away from the centre of the blaze. A short time later, his building too was consumed.[1118]

The total damages were estimated to be $43,000 and virtually none of the property nor the business inventories were covered by insurance. Surely there was much sage head nodding by those who had recognized the importance of fire protection from the start. Although the lake offered an endless supply of water, conveying it quickly and effectively to the site of the fire posed a problem. In the winter the problem was more serious, since that vast reservoir lay beneath several feet of ice.

Stern warnings had been issued by the editors of the *Atlin Claim* in the spring of 1899 about the inevitability of a major fire.[1119] At first the business community was too busy getting established to be concerned, and it was not until July that a meeting was held at which J. H. Russell, Andy Galarno, and Jack Kirkland, all hotel keepers, were elected as fire wardens. A decision was made to establish three fire stations at the foot of Watson, Discovery and Pearl Avenues. At the same meeting over $600 was collected to purchase buckets, ladders, and axes.[1120] In August Tom Lillie was chosen as Atlin's first Fire Chief, but there was little interest in forming a trained Fire Brigade.[1121] While manpower committed to training was a problem, money and volunteers were plentiful.

In September a ball was held to raise more funds for the cause of fire protection,[1122] and Ed McFeeley donated a large fire bell.[1123] Although the community had felt ready to face a fire, when the time came, they realized their preparations were scarcely adequate.

*The fire that destroyed Atlin in 1914 started in one of these buildings on Pearl Avenue.*

Those who lost their property to the fire of 1900 had serious decisions to make. For a few, the ongoing frustration of bungled legislation, and a host of administrative problems, along with lower than expected gold production in 1900, left no reason to stay, and they decided to leave Atlin. Most were not disheartened though, and those who would rebuild also resolved to organize a well-trained fire fighting unit. Safety must come first, and it was clear that a Fire Chief working with sundry volunteers and a few pieces of simple equipment, was not adequate.

In January 1901 fund raising began for a proper pump truck, which it was estimated would cost $1000.[1124] By April, through the auspices of the Gold Commissioner Graham, a used, manually operated fire engine was located in Portland, Oregon. It was thoroughly tested by the Victoria Fire Department, and shipped north, free of charge, courtesy of the White Pass Company. Equipped with manually operated brass pumps, the wagon had a long water box, and carried 305 metres of hose. It weighed 364 kg,[1125] and would be pulled to fires by the Brigade members themselves.

There being no match for machinery to generate enthusiasm, the very sight of the bright little fire wagon drew a great turn out for the inaugural meeting of the Atlin Fire Brigade, in June 1901. Soon a reservoir was built behind the Government Building, and a fire hall erected by volunteer labour.[1126]

A month after the Fire Brigade was formed, and just a year after the big fire, the new team and equipment were put to the test, when a fire broke out at the Rant and Jones Building on Pearl Avenue. The wagon was quickly whisked down to the lake, where ready brigade members stood shoulder to shoulder, on each side, grasped the gleaming brass rails, and began to pump, one side up and the other down. Water gushed into the tank, and through the hose, already rolled out to the burning building. Soon flames hissed and steam billowed while a strong stream of water poured onto the fire. The structure sustained considerable damage, but there was a communal sigh of relief that no other buildings were involved.[1127]

Another important factor in Atlin's defenses against fire was the night watchman. Officially his job was to stoke the furnace at the fire hall, to keep the water holes in the lake ice open, and to sound the alarm at the first whiff of smoke.[1128] Unofficially, his duties included chasing children home to bed, keeping the "canine choristers" quiet, and keeping a wheelbarrow handy for pushing inebriated citizens home when they were unable to make it under their own power.[1129]

At first, the night watchman was employed by the British Columbia Provincial Police, but in 1902 the force decided not to continue with this arrangement. The Board of Trade assumed responsibility with occasional financial help from Victoria. During the winter of 1906, Fred Markus served as

AHS Heming Collection P-1261

*The stately Atlin Post Office / Dominion Telegraph Office, engulphed in smoke and on the verge of bursting into flames in this photo, burned to the ground. (1914)*

AHS Henning Collection P-1139

*Firemen continue to man the pump to squelch the last embers. The old Bank of Commerce building at centre background would become the Post Office. The Royal Hotel stands unharmed at far right. (1914)*

AHS P-2701

*Only a few business buildings remained following the fire of 1914.*

the night watchman, but he was the last town guardian, because at that time the Fire Brigade became fully volunteer.

Readiness, training and equipment meant that Atlin was better armed to fight fires and surely the minds of nervous citizens were eased, but no amount of preparedness would prevent fires from starting. Wood burning stoves were seldom allowed to cool during the bone chilling winters, so worn and soot-filled chimneys were a constant danger. The arid climate drew moisture from wooden buildings, and even the contents became tinder dry. When human carelessness was added to these conditions, there was a constant threat of a serious outbreak.

In the winter of 1902, a fire started on the roof of the Vendome Hotel, which was closed at the time. The alarm sounded, but at that hour of need the fire engine was being used on the other side of town to pump water on to the skating rink. Valuable time was lost and the Vendome was razed. The day was a chapter of accidents, and when the wind came up, the fire spread to the Vancouver Hotel. At this point the fire engine, operating at the lake became clogged with gravel, and the Vancouver was lost. Ironically, it was a bucket brigade which finally stopped the fire from doing any further damage.

In June 1905, a fire occurred that rivaled the big fire of 1900, in terms of the value of property lost. When the Northern Power and Lumber Company burned to the ground, its holdings included the town's power plant, a large steam laundry, a sawmill, a planing mill, a blacksmith shop, and a large stockpile of lumber. Every available man in Atlin, as well as many more from Discovery, were called in to fight the blaze. The effort was to no avail, since everything was lost, including Monday's wash which belonged to a number of citizens, and all the hotels. The power plant and laundry were not re-opened, and for many years, residents would not have electricity.[1130]

For several years Atlinites were spared any serious fires, but the worst was yet to come, and when it did, it left Atlin "a heap of cinders."[1131] One day in May, 1914, someone was heating a can of tar over a stove in a building on Pearl Avenue. Whether it upended or boiled over matters little, but it set loose a devastating fire. There was no stopping its rampage, and precious little was saved from the dozens of buildings devoured in its path. It gutted the centre of town from the north side of Pearl to Discovery Avenue, and from First Street to Third Street. The business district was levelled. Thomas James wrote that the fire spread so rapidly, many found it impossible to secure a place which was not eventually reached by the flames.[1132]

Once again bewildered residents surveyed the wreckage. The Kootenay and Grand Hotels, noteable establishments by any standards, were only memories. There would never be another Post Office and Telegraph Building to match the lofty three story structure built on the corner of Second and Pearl during Atlin's construction boom. At the site of Eggert's Jewelry Store only the huge safe was left cooling after the fire. Inside was Mrs. Eggert's butter, in perfect condition. Thomas James lost both his drug store, and his dry goods store, which were on opposite sides of First Street.[1133] A good deal of cash was also lost in the flames, because, it was close to breakup, the ice was unsafe for travel, and there had been no way of sending it out for a couple of weeks.

The fire damage was so great that, in the aftermath, it was easier to count the buildings that were spared. The Government Building with its important mining and land records remained to serve the community. The little school house at the very top of Pearl Avenue remained for the children. St. Andrew's Hospital survived for the sick and injured, and St. Martin's Church stood fast, offering solace to a downhearted congregation. On Pearl Avenue, right across from the source of the fire, Louis Schulz's retail store and meat market was apparently the only major business remaining. Total losses were estimated at $100,000.00.[1134]

That spring of 1914, Atlin had been poised on the brink of another busy summer season. The business community had enjoyed a growing tourist trade for several years, which provided a second economy for the district. Hopes for the coming season were dashed with the sudden destruction of hotels, restaurants and stores. Businessmen, acutely aware of the economic need for tourism, gathered their resources, and faced reconstruction. By fall, several imposing permanent buildings were taking shape along First Street, each, for safety's sake, set apart from its neighbour, rather than joined wall to wall as the original business centre had been.

By 1916, the tourist business had fully recovered. More tourists than could be accommodated, wanted to visit Atlin, and the White Pass Company was building a magnificent new hotel on the site of the Atlin Club. A large modern ship, the *M. V. Tarahne*, had been commissioned to handle increased passenger and freight traffic on the lake.

The economic future looked bright with tourism fast becoming a main stay - until fire struck again, and all of the new businesses on First Street were levelled. The White Pass Inn, still under construction, was spared, and rushed to completion. It opened its doors to the season's first tourists. With awesome tenacity, Atlinites put their faith in tourism and each other, and started rebuilding with little delay.

The little red fire wagon was retired after the 1916 fire.[1135] It had become difficult to prime, and often a bucket brigade proved faster. The old engine had served the community well, and citizens recalled with fondness all the times it had been pulled through the streets on either wheels or runners, depending on the season. It was now appropriate, however, to replace it with a more modern truck.

One more large scale fire struck Atlin. In 1929, on a stormy April night a fierce 50 km gale, blowing from the south, spread embers over a one kilometre radius from the midst of the blaze. The Fire Brigade saved all but three buildings. The Government Liquor Store, the Kirkland Hotel, and Clarence Sands' Store, all on First Street, were burned to a pile of ashes, $32,000 worth of damage.[1136]

Despite all of Atlin's trials by fire, Discovery, in many ways more vulnerable, never suffered a major fire. Fire wardens were elected from time to time, but there was no community equipment. Miners habitually switched the course of Pine Creek from one side of the valley to another. At times it ran along the opposite bank, far away from the townsite, so an adequate supply of water was not always close at hand.[1137]

Notwithstanding Discovery's experience, Atlin's fire record was probably closer to that of the average frontier settlement at the turn of the century. Building styles, heating methods, and the weather, contributed to the frequency of fires. Fire fighting equipment was inadequate against major blazes. Fire was an undeniable menace. It was the willingness of a town to rebuild, repair, and retrieve which determined the long term impact of each catastrophe. In Atlin's case each experience was not only salutary, but it served to strengthen the community, and confirm its determination to remain and prosper.

CHAPTER 16

# Mining Camp Schools

When the gold rush summer of 1899 ended, a significant population of miners and businessmen elected to stay in the area. Atlin had all the trappings of a blossoming frontier town. Rough false facades and tents faced streets from the lake edge to the top of the rise on which the town was located. The pioneer citizenry was offered an impressive roster of services. There were churches, a hospital, an assortment of hotels and saloons, restaurants, stores and a cadre of professionals at the beck and call of anyone needing consultation. Community leaders were making ambitious plans. First among their concerns, though, was the urgent matter of providing for the education of the camp's children. They also recognized that a school would be a valuable symbol of the community's permanence. In August a meeting was held with the Gold Commissioner. Three prominent businessmen: J. St. Clair Blackett, A. S. Cross, and J. H. Russell, were elected as temporary School Trustees.[1138]

As they discovered, establishing a school was no easy task. The Provincial Government placed the onus on the community to demonstrate the viability of a school, and to make all necessary arrangements. At first the Province promised to send a qualified teacher, and to pay the incidental expenses of running the school, if the community assumed the responsibility for providing a building.[1139]

Soon it became obvious a teacher would be hard to find, and to operate a school in the far north of the province would be unusually expensive. The cost of cord wood for heating, for example, was considered extremely high by the government. The initial offer was retracted and, in a new arrangement, $60 a month was offered for a teacher's salary. Hiring and all other arrangements were in the hands of the trustees.[1140]

For two months the trustees looked in vain for a teacher willing to accept the meager wage. Even when they decided to augment the Government's $60 with another $15, they had no better results. In the meantime, a class of

*The Atlin School was built at the corner of Pearl Avenue and Third Street in 1902.*

18 children had assembled with Rev. Fred Stephenson, who was teaching them on a temporary basis. Immediately there was a complaint from officials in Victoria, who pointed out that it was not legal for a clergyman to hold the post. Stephenson had been willing to work as a volunteer until a teacher could be found, but he was so offended by the official reprimand that he resigned.[1141]

The community, already chaffing from problems the government had thrown in the way of mining development, felt strongly that their efforts to establish a school were being foiled as well. Not only was Victoria quibbling over the cost of cord wood, but the promised grant was being held back. The reluctance to provide a reasonable teacher's wage had made it impossible to fill the position. Bitter resentment boiled around the fact that, although some $50,000 had been generated in government revenues by mining, land sales, and businesses in the area, it would be necessary to support the school as a local charity with dances and concerts.[1142]

Two new trustees, J. H. Brownlee, and D. S. MacDonald were elected to join Blackett on the board in November.[1143] Soon afterward, and mostly by luck, they finally hired a teacher. Young Hulet Wells, who had been mining on Spruce Creek, came to town with $100 in his pocket. These savings he meant to spend on his winter food supplies. Wells also had a teaching certificate and a year of experience in an Agassiz, B.C. classroom. Certainly his mind was far from teaching when he wandered into an all night gambling session. By morning he had only 50¢ left, no grub and winter was upon him. Needless to say, although he preferred mining, he was more than willing to accept the teaching position, and the salary of $75 per month.[1144]

Wells and his students met in the abandoned Board of Trade Hotel on First Street between Rant and Pearl Avenues.[1145] The student population had grown to 35. Once winter set in, teacher and class braced against the cold as wind and snow blew through the cracks in the walls of the ramshackle building. They gripped chalk and slates in stiff fingers, and squinted at their books in the gloomy room. When spring came, Wells said a final good-bye to teaching, packed his bed roll, and set out on foot for the mines on McKee Creek. Eventually he settled down to become a lawyer in Washington State.

In 1900, Miss K. C. Smith took charge of the students.[1146] That September, classes met in a spacious tent previously occupied by the Gold Commissioner and his staff. Half the floor was sawdust, the other half rough boards. Two wood stoves were kept burning, and when the first snow fell, it was carefully banked around the bottom of the tent to stop icy drafts.[1147] Still, in the short days of winter, when temperatures dropped low, the pleasant light that had filtered through the walls of the white canvas school, was replaced by a penetrating chill. The snorting wood burners warmed only those students sitting nearby.

Although comforts were minimal, and the noise of the tent snapping in the wind was a regular distraction, the dedicated Miss Smith offered students a full curriculum. Reading, writing, spelling, composition, arithmetic, grammar, geography, drawing, history, hygiene, and sewing kept them busy at their desks.

Attendance was irregular, particularly among children who came to school from out of town. Charlie Bruce rode the stage from Discovery until the weather became too cold, then lessons were set aside until the spring.[1148] Boys, especially, were absent when they were needed to work at mining or help with a family business.

When school convened in 1901, there was still no appropriate school building. Instead classes met in a building previously used as a hospital.[1149] Leonora Hughes from New Westminster was Atlin's third teacher. When she married Dr. Gustave Lewis in 1902,[1150] the regulations, which were clear, required that she resign. Married women were not allowed to teach.

After Atlin children had spent three school years in uncomfortable, makeshift facilities, the government finally provided $1600. Lobbying and complaints from trustees and parents paid off. In September 1902, classes met in a new one room school house at the top of Pearl Avenue on Third Street. Wide stairs led to an enclosed entry; there were large windows to brighten the classroom, and a stylish cupola perched on the roof. There was a basement for indoor recreation, but winter recesses could also be spent playing in the snow. Behind the building a steep slope became a perfect sliding hill from the school yard down to the alkali flats. Two outhouses, matching the school with drop siding and shingled roofs, were set discreetly at the back of the property.[1151]

Earlier that same year, the Board of Trade had successfully persuaded the Province to increase their annual grant, so a more highly qualified teacher could be hired.[1152] Miss E. I. Miller, a former Victoria high school teacher accepted the job, and became the first teacher in the new building.[1153]

During the first term, enrollment crept up to 37 pupils.[1154] Coping with so many students, comprising eight separate grades, was no easy job, but it was a common situation for rural teachers. The students themselves pitched in. Older students helped with the younger classes. Keen students often listened to lessons being taught to the class ahead. In the pleasant new surroundings, with a well qualified teacher, Atlin School began to develop a sound academic record.

The main school issues had been solved, but there would be others. One parent complained to the editor of the *Atlin Claim* that, on many mornings, no one started the fire at the school until after most students arrived. Lessons could not begin until about 10 am when the room began to warm. The editor suggested that the trustees have the night watchman start the fire instead of relying on a student to show up early to do the job.[1155]

When a young woman teacher came to Atlin, she was seldom aware of the stir generated by her arrival. School board members and parents were anxious to meet her to determine how effective she would be as an educator, but her presence created an even wider interest in the large population of single bachelors. As a result, many new teachers willingly traded their professional careers for marriage to a successful local suitor.

Miss Miller was a vivacious and energetic young lady, who soon captured the heart of local druggist, Charles Bourne. They were married during the summer following her first year at the Atlin School. Although the rules indicated otherwise, she was allowed to teach the next year from September until March, 1904. At that time H. C. Boothby, a retired army officer, and an honours graduate from Christ's College, Oxford[1156] arrived to teach for the remainder of the term. It seems that for once the children's education took precedence over stuffy regulations. Mrs. Bourne stayed in her classroom even though she was pregnant by the time Boothby took her place.

Malcolm J. R. Reid from Barkerville opened classes in the fall of 1904,[1157] and stayed in Atlin for three years. Under his experienced direction, the Atlin School further consolidated its good academic standing. In 1905, Ethel Pillman and Jennie MacDonald became the first Atlin students to write, and pass, the Province's high school entrance examinations, a mark of proficiency for both students and teacher.[1158] The following year, Maude Haslett, Nellie Stables, and Alan Fraser became successful candidates. The examiner commended their results, especially on the history paper.[1159]

In November 1905, the written work of every Atlin student was displayed at the Dominion Exhibition at New Westminster. The good quality of the work by students attending the most northern school in the province was reported by southern newspapers.[1160]

As in many other public schools at the time, school examinations were often held orally, and in public. Parents and trustees would visit for an afternoon to witness a display of the students' competence in such activities as reading, spelling, mental arithmetic, and recitation.

*Mr. Reid presided over a class of young boys in bow ties and girls with ribbons in their hair at the Atlin School. (1906)*

BCARS HP 45739

Dr. Henry Esson Young, Atlin's representative in Victoria, was particularly interested in the Atlin School and its students. His ongoing efforts brought increases to the funding allocated to the school. The government grant was raised in stages until, by 1906, it was $1140, which was approximately equal to the amount required to operate the school for a year.[1161] In 1905, Dr. Young, himself, offered a Winchester rifle to the boy making the best progress in all subjects.[1162] No record tells who won the rifle, but the competition would surely have been spirited for such an attractive prize. Young's concern and influential support of Atlin's students added to his well established popularity in the community. By 1907, he was appointed by Premier, Richard McBride, to the Cabinet as Provincial Secretary and Minister of Education.

Mrs. Young, formerly Rosalind Watson, had taught high school in Victoria. In Atlin she acted as sub-examiner for the high school entrance examinations,[1163] and reassured authorities in Victoria of the school's solid academic footing.[1164]

Students worked hard, but they were still children and when not in the classroom they found no shortage of things to keep them amused. In spring, bicycle riding on the lake ice provided hours of amusement, and in winter there were always plenty of dogs to whisk them away on exciting sleigh rides. Throughout the summer and fall there was exploring to do on the alkali flats, as well as berries to pick and fish to catch.

At the instigation of Herbert Cancellor, a Discovery merchant and first postmaster, discussion was held in 1899 about opening a school there. It was the fall of 1900, however, before classes met. By then a school board had been elected, Thomas Miller was hired to teach,[1165] and a school room was established in an existing building.[1166] There were only six students that year, and in June, it was decided to close the school.[1167]

For two years children from Discovery and the creeks attended the Atlin School when they could get there. In 1903, a new Discovery School Board was elected consisting of D. G. Stewart, Walter Owen, and a Mr. Reid.[1168] Mary Smith was hired to teach, and the Discovery School was reopened with 15 students. Enrollment remained fairly constant for the next ten years. Mary Smith taught at Discovery until June 1906. She left a community well pleased with her efforts and the results of her devoted work.[1169]

A new school house was provided by the residents. It was located amid the long row of buildings facing the creek on Main Street. Neighbouring buildings on either side housed stores, hotels and restaurants, all connected by a boardwalks. Because the school house was in the midst of the business district, with its door opening directly onto the boardwalk, it was not unusual for classes to be interrupted by an inebriated pedestrian who mistook the classroom for a saloon.[1170]

The curriculum at the Discovery School closely resembled that taught in Atlin. Students from the two schools competed aggressively in all contests, from spelling bees to hockey games.[1171]

*Atlin children enjoyed all seasons. Young mushers stopped to smile for a photographer at the corner of First Street and Trainor while two men enjoy a chat in front of the Vendome Hotel at upper right.*

AHSP-1239

Around the time of World War I, when the population in Discovery and on the creeks declined, the school struggled to remain open with fewer and fewer students. Jean Gairns remembered being enrolled in Grade 1, while she was actually too young to attend school, in order to add to the number of students, so the school could remain open.[1172]

Teachers sent to Discovery were generally inexperienced, and frequently stayed only one year. Rudyard T. Kipling remembered teaching in Discovery in 1922-23 as an adventure. He enjoyed the winter weekends spent racing dog teams over lakes and trails, with his friends. As a bachelor cook, he was pleased to receive any invitation to a good home cooked dinner at the home of a students. Walking there under a sky shimmering with northern lights made the occasion even more memorable.[1173]

The Discovery School could accommodate students from grades 1 to 8, as did Atlin. The small enrollment at Discovery, however, seldom provided enough students to represent every level. It was classified as an assisted school, meaning the government funded the teacher's salary and other expenses, but it was the community's responsibility to provide a building and maintain it. By contrast Atlin, after it had been open a few years, became a common school, and received government funding for a teacher's salary, the school building, and maintenance.

The last Discovery teacher was Matthew Buckpitt. His nine students came mostly from the surrounding creeks rather than from Discovery itself.[1174] When the children dispersed for the summer of 1924, the Discovery School was closed for good.

In Atlin, a procession of students and teachers came and went. At some point, when it was no longer practical for teachers to live in hotels and boarding houses, the school basement was converted to a teacherage. The school room, on the other hand, remained virtually unchanged as hundreds of young northerners, and their teachers learned from each other within its walls.

Many citizens, who built the foundations of their education in the Atlin and Discovery Schools, went on to higher learning and noteworthy careers in distant cities, or brought their knowledge back to the north. Others, keyed to the land, stayed to mine, prospect, trap, guide, or to become involved in local transportation. Some devoted their energy and business skills to promoting and stabilizing the Atlin economy. In 1975, Walter Owen, a former Atlin schoolboy, returned to pay his respects to the town of his childhood, and its residents. At that time he was the Lieutenant Governor of British Columbia.

CHAPTER 17

# Gold Camp Society

The tremendous energy needed to fuel a stampede, mine, and build towns, businesses and communities was generated by the youthful participants. A gold rush was no game for old men. Only a relatively few elders managed to keep up the pace. After the mines were working and the building going smoothly, citizens were willing to settle for other routines. There was time and energy to spare, and predictably they were channeled into a great variety of entertainment. The earliest events were almost always rough and tumble sports: wrestling, bare-fist boxing, horse racing, shooting, tug-of-war, jumping, vaulting. There was never a shortage of participants and raucous supporters.

Everyday life in Atlin calmed as wives and children came north to join their husbands, eager to reestablish homes. The unsavoury aspects of the boom period, such as all-night gambling, hard drinking, brawling and prostitution, became shadowy backdrops to community life. The social structure changed quickly. Dances held in 1899, overwhelmingly masculine occasions, had, in a year, gained a certain air of propriety and balance, with the addition of a few more women.

The population was an unusual mix of nationalities, educational backgrounds, professions and experience. Differences were of no particular concern, though, and a strong interdependence developed. Isolation was probably a significant factor. Early Atlin society was neither closed, nor overly cliquish. A rather strange reason for this may be that most people were not who they seemed to be. In the limited area of employment, a job rarely matched the employee's former occupation. Often there were not enough jobs to go around, and anxious men took any work offered, or tried to invent their own. As a result, a medical doctor might enter the lumber business, or an engineer open a store. Some ventured into a variety of unfamiliar enterprises. The average mine labourers, in particular, came from every sector of the working world, and a good number of them had never before paid any serious attention to the business end of a pick. New lives and titles were assumed, and not necessarily

*The Vendome Hotel, still in a tent when this photograph was made, was decorated for a summer celebration.*

by choice. A man would hardly have known where to seek his social level, had he felt so inclined.

Atlin's citizens mingled comfortably at town functions, and, when called upon, worked as a unit for the common good. Social layering was not much in evidence except for one notable exception. The prostitutes were pointedly shunned by any lady who considered herself respectable.[1175]

Spring sunshine, a relief from the bone chill of winter, fostered a temptation to throw responsibility to the wind, and become lost in the beckoning landscape. Instead, miners went to work on their claims. There were precious few weeks for mining, and it was the paramount concern. In the tradition established in 1899, however, there were two important exceptions - Queen Victoria's Birthday, and Dominion Day - when almost everyone found time to celebrate.

As each occasion approached, the *Atlin Claim* kept anticipation at a high level, with weekly reports on preparations at all locales, scheduled contests and prizes offered.

Queen Victoria may well have been appalled at the bizarre games held in celebration of her birthday. The sight of messy contestants trying mightily to shin to the top of a greasy pole, would have raised a royal eyebrow. Still, in the outrageous spirit of the gold camp, where life itself was a contest, the silly and preposterous seemed quite fitting.

In the early years, Dominion Day and Fourth of July were sometimes combined into one celebration. Miners in scrubbed clothing and trimmed hair were out in force to join the festivities. Stores and hotels were draped in billowing bunting. Flags flapped and streamers flew. Dozens of small spruce and pine were arranged along the boardwalks. In Discovery, foot racers tore down the main street, raising dust as they sped from the Pine Tree Hotel, past the Gold Pan Cafe, Moberly's store, the Gold House, the Nugget Hotel, and on to the finish line. In Atlin baseball games drew cheering boosters. Children excited to fever pitch, circled on their bicycles, or ran pell-mell among the onlookers. Ladies promenading in their finest summer gowns, twirling pretty parasols, lent style and colour to the lively scene.

Athletic events dominated the programs. The big attraction was the tug o' war, a contest for all able-bodied men. In 1900 the international melange was sorted to some degree when the teams formed along national lines. French pulled against Scots, and English against Canadians. In the big play off for the championship, Canadians triumphed over the French.[1176]

In the evening, celebrations were usually capped by a concert and a grand ball. Sometimes, in the wee morning hours, a photographer took pictures of the revelers, just for fun, and because it was light enough to do so at 3 a.m.

The festivals of summer, organized by the first Atlinites, became annual traditions that would always be recognized. The contests varied, but in the 1930's and 40's there were still races, and noisy baseball games. Eventually, though, an event was introduced that the first organizers would not have considered. A queen contest was added, and it became the highlight of the day.[1177]

Always included in these very British celebrations were any Tlingits who might be in town for the summer months. They particularly enjoyed running in the races, and frequently were among the prize winners. Mrs. Emma Jack, the wife of Chief Taku Jack, was a fine athlete who could rarely be beaten. From time to time local population and visitors watched with curiosity and wonder as the Tlingits hosted a potlatch for their relatives from Teslin and Carcross. Sleek canoes, carrying Tlingits resplendent in ceremonial garb, slipped majestically across Brewery Bay to the village, which became a scene of celebration and traditional rites. It was a compelling sight.

AHS Sheldrake Collection P-1263

*Atlin Lake was an arena for summer recreation which came in many forms. With hats firmly in place, Atlinites cooled their feet in the icy water. Circa 1910.*

AHS P-2008

*The S.S. Scotia was often used for picnic excursions to favorite inlets and island beaches at the south end of Atlin Lake.*

Whatever the occasion, social events were well attended in all cases, although one problem never ceased to taunt the young men of the community. There simply were not enough girls to go around. In summer that situation improved somewhat. Young women came with the influx of tourists, and many accompanied family and friends involved in mining. Olga Blunk was the particularly attractive daughter of F. Theodore Blunck of Davenport, Iowa, whose mining interests in Atlin included Pine Creek Power Company. Olga's arrival drew immediate attention from the bachelor population. Soon she was at the centre of a flock of ardent admirers. Her suitors showered her with gifts, although not the common tokens, such as candy and flowers. Atlin had no florist. Instead, Olga received gold nuggets. Although she did not find her true love in Atlin, she would not forget the generous young miners she knew there. Eventually she had the nuggets fashioned into a necklace, a tangible memento, and a tribute to her charm and popularity. It measured well over 4 metres in circumference.[1178]

For children, the Sunday school picnic was a summer highlight, equaled only by Christmas festivities. On one early morning in 1903, a large group of excited youngsters met at St. Martin's Anglican Church. Five stages stood ready to provide transportation. When the children were settled on board, nickering horses pulled away for the two and one half hour trip to Surprise Lake. After a picnic lunch, some exploration along the lakeshore, and races for all age groups,they headed for home. There was a stop at Swan Lake, near the mouth of Birch Creek, for an outdoor dinner. Then, with the kids singing at the tops of their voices, the swaying coaches rolled through Discovery, Spruce Junction and finally to Atlin.[1179]

Lake excursions on the *Scotia*, and later the *Tarahne*, were sometimes organized by the Arctic Brotherhood or the Masonic Lodge, but usually everyone was welcome. Destinations were choice bays and inlets on the lake, where a picnic meal, games and hikes, fishing or berry-picking filled the day. The late summer excursion to Ben-My-Chree, arranged especially for the people of Atlin, drew large crowds in the 1920's, and marked the end of each tourist season.

Perhaps the major sporting activity of the summer was baseball. In the early days there were clubs in Atlin, Discovery, Bennett, and Taku.[1180] By the late thirties the competition was chiefly between the creeks and the town. There were a number of teams playing in the league including: Noland's Placers, Noland's Work Dodgers, CMS Placers, Gold Nugget Cafe, and Vicki's.[1181] There was also some competition with teams in Carcross and Whitehorse.[1182]

In winter Atlin residents felt most acutely their separation from former homes. They drew instinctively closer. Time and darkness posed a threat to the morale of the isolated residents, but their exuberance and energy were a match for the challenge. Concerts and dances were a welcome diversion, and all special occasions were celebrated - St. Patrick's Day,[1183] Robert Burns Day,[1184] Valentines Day,[1185] and Christmas.

Yuletide celebrations, held in Atlin and Discovery in the winter of 1899, were especially significant to the young communities. Participation was wholehearted. The Christmas Concert at Discovery drew people from Atlin, Surprise City and Spruce Creek in -25° C. weather. Rev. John Pringle gave the opening address in a hall so crowded a few people fainted.[1186] Christmas was a time to feature the community's children, and they sparkled in recitals and pageants. In 1901, the Atlin youngsters traveled by sleigh to entertain the miners at MacDonald's Hall.[1187] For men separated from their own children, the occasion was filled with poignant moments.

To ensure the success of Christmas feasts, Burns and Company ordered 680 kg of turkey for the 1899 holiday.[1188] Many miners gathered in small groups to plan dinners. No single miner lacked an invitation. The community Christmas trees in both Atlin and Discovery held gifts for everyone, and were an important feature of the holiday celebrations.

Atlin's most sociable citizens were not willing to concede that their gold camp culture lacked any of the refinements and social opportunities enjoyed by their contemporaries in Victoria. To this end, during the winter of 1899, the Arctic Brotherhood contributed to the remarkable agenda of social events, by organizing a weekly concert followed by a dance in their own hall on Second Street.[1189] Throughout the week they also held whist tournaments, which were enormously popular.

The Brotherhood Hall was a hub of social activity; Mary Hitchcock lectured there, and Rev. John Pringle used it for Sunday services. Nonetheless, the Atlin Chapter soon folded. The Arctic Brotherhood building was on the block at a Sheriff's auction in the fall of 1900.[1190]

The upstairs hall of the Hotel Russell was the scene of crowded weekend dances for a number of years.[1191] The Kootenay Hotel was also the setting for dances which were sometimes hosted by the bachelors of the community.[1192] Billiard tournaments at the Kootenay were well attended.[1193] In 1901, the Dixon Brothers, who operated the stage service between Atlin and Surprise City, built a stable on Pearl Avenue. The upper floor, a spacious 9 x 12 metres, was opened as the Atlin Entertainment Hall, and made available to rent for social occasions.[1194] The ballroom at the Grand Hotel offered the finest surroundings for dances, receptions and special parties. Numerous organizations met there and among other things, the hotel sponsored ping-pong tournaments.[1195]

Long sleigh rides between Atlin, Discovery and the small winter community on Spruce Creek, were scarcely a consideration for young people out for a party or dance. Actually, the moonlit trip home in the biting cold of the small hours, could be memorable, and sometimes helpfully sobering. Couples with young children and babies simply took them along to most occasions.[1196] Cloakrooms were likely to be crowded with cradles and children sleeping blissfully among the parkas and mufflers. One evening a prankster switched the babies around in their beds, and caused hysterical commotion the next morning when parents awoke to find they had brought home someone else's offspring.[1197]

The educational and cultural health of the gold camp was nurtured from the beginning. In December 1899, a public meeting was held in the former Cassia Hotel in Discovery. James Stables was elected President of the Fine Reading Room Association. The hotel building would become a library, supplied with books and a selection of newspapers.[1198]

Two years later, the Atlin Operatic Society was formed under the leadership of Herbert Young, a clerk in the Gold Commissioner's office and Mrs. E. M. N. Woods, the wife of the Magistrate.[1199] If the organization's name was an indication, the group had grandiose ambitions, but there is no record to suggest that these were ever realized. In 1903, however, they presented Gilbert and Sullivan's "Trial by Jury" at Dixon's Entertainment Hall to good reviews. A few months later they changed their name to the Atlin Musical and Dramatic Society.[1200] Over the next couple of years members met regularly at the Grand Hotel. By early 1904, James Stables was the President. They acquired a piano and some stage props,[1201] and probably afforded more entertainment for themselves than the community at large.

The performing arts had a keen following during Atlin's first decade, in part because there were several exceptionally talented residents. Concerts marked every occasion, and the programs included outstanding performances by professional and near professional artists. James Lumsden was a former member of London's D'Oyly Carte Opera Company.[1202] His business partner and Stikine trail mate, James Stables often sang with him in concerts. Stables' forte, it was said, was humorous Scottish song.[1203] There were soloists for piano, flute, guitar, mandolin and concertina. Frequently an orchestra could be assembled from local musicians. L. C. Read, organist for St. Martin's Church, photographer

AHSP-1262

*The Bachelor's Ball was a popular annual affair in Atlin. In 1904 it was held at the Kootney Hotel.*

and guide, had previously directed a band of considerable renown in Vermont, and toured throughout Europe with his company.[1204]

Young Ethel Pillman, whose recitations delighted concert goers in Atlin, Discovery and Spruce Creek, was often in demand. In later times, Kathleen Keats White entertained with memorable violin recitals and readings, at times including her own poetry. Kathleen had entertained throughout the American Northwest before her marriage to MacLeod White, who was a mining engineer.

Sometimes a humble log cabin on a mining claim became the setting for a cultural event. F. T. Hamshaw and George Adams, hosted a popular weekly musical evening in their cabin on McKee Creek,[1205] and other similar events were not uncommon.

Occasionally, Atlin was included on the itinerary of a touring artist. In 1905 Marguerite Longacre, a "dramatic soprano and reciter", and Carrie Florilla Spalding, a pianist, gave concerts in the Grand Hotel Ballroom, and at Discovery.[1206]

The Atlin Club, well-appointed and fashionable, became the centre of social life for a few high flying years. It was situated on prime lake front lots, made available by the Provincial Government.[1207] The Club was the pride of a group of men who had been inspired by the proposal of W. J. Robinson to build a club which would rival any other on the continent. Amid the widespread commotion and optimism surrounding the installation of the dredge at Gold Run, Robinson was able to finance the Atlin Club's new building almost entirely from sources in Philadelphia, New York, and Boston. Shareholders in the British American Dredging Corporation were so positive of Atlin's rosy future as a major mining centre, that they did not question the advantage of membership in the Atlin Club.[1208]

The club had a tentative start in 1900, when a group of prominent citizens arranged with Mr. Byrom, one of the proprietors of the Kootenay Hotel, to use his premises for a gentlemen's club.[1209] It seems there was a desire on the part of at least a few citizens for a degree of social aloofness. The arrangement did not succeed at that time, however.

The Atlin Club was finally built by Marcus and Letherdale in the fall of 1903. It was a handsome log building, 12 x 38 metres, with large bay windows on three sides. Inside were several large finely furnished rooms, including a dining room, a library, a ladies' parlour, a billiard room, and a card room.[1210] Contributions of furnishings and equipment from Vancouver businessmen were especially generous. The grand piano, 100 volumes for the library, pictures,

*The Atlin Club, built in 1904, had every possible luxury.*

china, cutlery, a clock, kitchen and bar equipment, cigars and even insurance were provided at little or no cost to members.[1211]

The entire town of Atlin was invited to the Inaugural Ball which was held in August, 1904.[1212] Residents were immensely proud of the new facility, and very impressed with plans for the addition of a bowling alley and tennis court, the acquisition of a steam launch, and the construction of 30 cottages for visiting members and tourists.[1213] Such lavish improvements should of course, have been expected from an organization that included King Edward VII, Lord Roberts, the Governor General, Sir Henri Joly de Lotbinere, the Prime Minister, Sir Wilfrid Laurier, and the Premier, Sir Richard McBride, among its members.[1214]

For three years the Atlin Club thrived in a profusion of elabourate activities. A French chef was in residence,[1215] and each occasion was enhanced by the attractive surroundings and accoutrements of the club. Fine hospitality was a part of each gala dinner, dance or concert, game evening, and sporting event. Music was often provided by the Spruce Creek Orchestra. These musicians were a group of talented miners, under the direction of violinist, Jack Cartmel.[1216] On one disappointing occasion, the Queen's Birthday Ball, the orchestra missed its engagement at the Club. High water coursing down Spruce threatened their claims, and the musical miners had to fight to save the tools of their livelihood.[1217]

The decline in mining, and particularly the failure of dredging, must have shaken some of the absentee members of the Atlin Club who had harboured such confidence in the success of the mining district. Perhaps financial support started to dry up. At any rate the dominoes fell. Gold production dropped off triggering a decline in local business, and the slump affected the welfare of the Atlin Club. In 1907, the organization was disbanded, and the assets sold.[1218]

Winter weather complicated the business of everyday chores. Hauling water was a fight with slippery pails and frozen mittens. Animals needed extra attention, and there were the voracious appetites of wood burning stoves to appease. Even moderate cold made travelling a game of chance. On the other hand, winter set an irresistible stage for entertainment. Ice skating was the common theme for a remarkable range of festivities. The first makeshift ice rink was a flooded area on the lake at the foot of Pearl Avenue.[1219] It was well used but there were problems. The surface was often rough from newly fallen snow, and the slightest breeze coming down from the glacier and across the lake was a chilly deterrent. Gold Commissioner Graham provided a warming cabin for the ladies,[1220] but the fact remained that the fun of skating on open ice could quickly become an ordeal.

In 1901, Dockerill and Olive built a covered rink, complete with ladies' dressing room and lunch counter.[1221] It was located on the alkali flats[1222] just east of Third Street. That area had the advantage of freezing over earlier than the lake. Another decided attraction was a skating instructor from Toronto, who called himself Professor Barnes. He offered lessons on a daily schedule.[1223] In January that year, the skaters and the community at large, anticipated the arrival of John Philip Sousa whose band Barnes had promised to engage for an ice carnival.[1224] Unfortunately Sousa did not include Atlin in his itinerary. Skaters had to be satisfied with events which lacked the dazzle of such worldly entertainers. Not discouraged, they staged grand masquerades on ice, with the rink alight with Chinese lanterns, and hung with colourful bunting. Skaters circled in exotic costumes, and the most imaginative won the prizes. On-ice entertainment was, without a doubt, the most popular sport of the winter. Many young couples did their courting to the swish of the blades, if not to the oom-pa beat of a Sousa march.

Late in 1902, another new rink was built on the alkali by Lewis & Ward.[1225] This facility was larger than the last. In Discovery, skating also ranked high on the list of winter diversions. The first rink was built by a group of sportsmen, principally hockey players, in 1902.[1226] It was simply a flooded area, surrounded by a board fence. Basic and useful, it was, nevertheless, replaced by a rink under canvas. The huge cover was made by miners using a sewing machine.[1227]

Hockey games became weekly events, and the competition between Atlin and Discovery was keen. Large crowds of supporters traveled with their teams, and post game business at the saloons was brisk. In 1902, a cocky team of both Atlin and Discovery players sent a challenge to Whitehorse; then they mushed 112 km to Carcross, where they caught the train to Whitehorse, only to be soundly trounced. It was a severe blow to the confidence of the local sportsmen.[1228] When Whitehorse players came to Atlin, the game results were no different, but the elabourate dinners, concert and ball arranged for the visitors' benefit must have made the event memorable for all.[1229]

Curling was first initiated as an open air event, and players toughed out uneven ice and piercing cold on the lake. When Ward & Lewis' rink was built in 1902,[1230] indoor curling on the skating rink became possible, and the enthusiasm spread. In 1904, the Curling Club bought the rink, and set about enlarging it.[1231] The *Atlin Claim* filled weekly columns with the results of games

*Half the work and twice the fun on a bicycle built for two.*

played in Atlin and Discovery. Male teams dominated the sport until, in 1908, the *Claim* reported, with obvious surprise, that some women were "playing as lead and doing well as deliverers and sweepers".[1232]

On rare winters, weather conditions met in perfect combination at freeze up time, and Atlin Lake became a flawless icy spread. Free of snow and drift, it presented a superb sports arena. The winter of 1904 -05 was one of ideal ice, and a new sport, ice boat racing, became the local passion. Groups of enthusiasts became intensely occupied with building faster and better craft. The boats were little more than a beam supporting a seat and outriggers fitted with skis. The sail was rigged overhead. A well designed ice boat was capable of incredible speed over frictionless ice, and some of the Atlin boats were high fliers. The first ice boat was built by M. V. Trotman, and sailed successfully on several trial runs by Messrs. Moberly and Wallace. The *Third of December*, the *Flyler*, *Windbuck*, and the *Idaho* were the most prominent members of the fleet and everyone had their favourites.[1233]

Atlin's children, robust offsprings of the frontier, took naturally to ice and snow, but one young lad, who may have been inspired by a verse of doggerel published in the *Atlin Claim*, found himself, literally, in over his head.

AHS P-916

*Snowshoe hikes and picnics in the snow were winter entertainment for Atlin's hardy population.*

AHS P-151

Oh, the sleigh, the merry sleigh
As down Pearl Avenue we scoot away
Thumping and bumping on roots and logs
Running into and over the Malamute dogs
Pleasantly pass the November day
Who would be a kid without a sleigh ?[1234]

A few days after this verse was published, 10 year old Albert Durie was streaking down Pearl to the wharf at the bottom of the street. The thrill of the run must have numbed his better judgment, because he miscalculated his speed, and zoomed off the end of the dock into two metres of ice cold water. The astonished speedster clung instinctively to his sleigh, which acted as a sort of life preserver. Henri Maluin, who happened to see the spectacular splash, leapt right in and retrieved the bewildered Albert. The *Claim* reported the unscheduled polar dip, which they may unknowingly have instigated, and declared, "Mr. Maluin is deserving of all praise for his manly action."[1235] Certainly Albert agreed with that.

After the town was levelled by the fires of 1914 and 1916, Edwin L. Pillman built a theatre on Pearl Avenue. The Globe was a simple frame building, but it had an architectural feature usually reserved for the more pretentious structures of the period. There was a vaulted ceiling, which probably lent something to the quality of the acoustics. The Globe, usually just called "Pillman's Hall," and Pillman's cumbersome carbon arch projector spun out silent movies for rapt Atlin audiences. After a stage was added, dramatic productions and concerts were held there along with dances, musical evenings, magic lantern shows, boxing matches, and the occasional funeral.

In the years prior to World War II, during a time of significant mining activity, the Atlin Athletic Club became prominent in organizing sporting and social events for the community. Under its auspices, volunteers built a covered ice rink for skaters, curlers and hockey players. Sometimes there were softball competitions with Whitehorse teams.[1236] In 1940, the Athletic Club reorganized as the Atlin Community Club and events such as debates, lectures, musical programs, whist tournaments, and outings dominated.[1237] Quieter activities reflected an aging population. Young men had gone to war, and the mines grew quiet. Post war years saw Atlin's young adults leaving in search of jobs. Finally an absence of youthful vitality had changed the social activities of residents. In no way, though, was there a change in the closeness that had always been an unshakable strength of the community.

CHAPTER 18

# They Lived Off the Land

The expanse of rugged land that stretches from the rain shadows of the Coast Mountains to the western rises of the Kawdy Plateau is isolated by massive ice fields on the west, and broad sweeps of sub-arctic forest to the east. Here hundreds of swiftly flowing creeks, and wide rivers collect in deep lakes which empty westward, eventually to form one dynamic waterway, the Yukon River. These vast lakes, although dangerously cold, have a certain beneficial influence on the climate of the basin, and average temperatures are perceptively higher than in the country beyond it.

During short summers the lakes reflect benign skies, and a warm sun, while forests flourish as plants and creatures thrive and grow. When winter snuffs the life from this two-faced country, the heart of the basin lies still, storms pound from above, and ice and snow seal the landscape in frozen limbo.

For uncounted generations, the Tlingit people lived in and around the huge basin. They were masters of their environment, skilled at coping with erratic weather patterns, and able to survive successfully because they understood their harshly beautiful homeland. They lived off the land in their own economical way, and the settlers who followed benefited from their experience. When white men came to mine the valleys around Atlin Lake, they discovered that their long supply lines from southern cities could not provide for all their needs. It was not long before they turned instinctively to the land, and it had to surrender generously to sustain them, but resources were abundant.

First harvests included trees for building, and fish and game for food. Business minded pioneers turned nature's crops into profit as they built saw mills, sold fire wood, or hunted for the marketplace. Men who could not contend with the uncertainties of gold mining, but who had grown to like the region, looked for still other ways to make a living. Although it was not abun-

*The Tlingit Indians were masters at living off the land. Their incredible knowledge of the country and its uses was passed to eager young learners.*

AHS Henning Collection P-1266

dant, there were pockets of arable ground where vegetables such as potatoes, carrots, beets and turnips flourished. Other hardy crops such as cabbage and cauliflower also fared well, in spite of predictable summer frosts. Indeed the long daylight hours of June, and July induced rapid growth that amazed experienced gardeners used to southern conditions.

Beginning in 1899, fresh vegetables were available at Parson's Produce on First Street in Atlin. Prices, however, were high, and freshness, doubtful, after the long haul from the prairies, or southern British Columbia. Many miners could not afford common vegetables which had become luxuries in the camps. The opportunity for local gardeners was obvious, and during 1899 a number of would-be farmers made application for crown land.

A certain amount of trial and error preceded successful production, and one early experimenter was Robert Grierson who located his vegetable gardens near Discovery.[1238] Lee Garden started a farm at the mouth of Fourth of July Creek where he grew hay and other forage crops.[1239] Apart from the human population there were many horses and other livestock to feed. George Shirley grew fine quality rye, barley, timothy and clover on a warm south facing slope, known as Shirley's Hill, just outside of Atlin.[1240]

Not all the crops grown for the Atlin market were raised close at hand. A Manitoban named H. M. Wooldridge noticed the lush growth of wild hay in the country around Caribou Crossing. The soil was rich and fertile, and in 1898, Wooldridge paid $3.00 an acre for a tract of crown land he named Ten Mile Ranch for its distance from the Northwest Mounted Police Post on Tagish Lake. Four years later he was raising oats, turnips, potatoes, carrots, celery, and a small variety of other garden vegetables for the Atlin market.[1241] He also cut well over 90 tonnes of hay, 70% of which, was marketable for $53 per tonne in Atlin.[1242]

The Butler Brothers, Jeff and Claude, who ran a popular roadhouse on the south shore of Graham Inlet, raised hay and root crops for Atlin buyers. The Butler's most celebrated crop, though, was large and luscious strawberries which grew in great profusion at their farm, and were a special favourite among the miners.[1243]

About the same time as the Butlers started farming in 1902, E. J. Hughes started a farm on 65 hectares near Scotia Bay, about 6 kilometres away. An unplanned crop from that area grew along the Taku Railway. Over 13 tonnes of good timothy mixed with a large quantity of fine clover was harvested as a result of seed being scattered from the flatcars in transport.[1244]

In the early years, Atlin's most notable farmer was Elsworth P. Queen, who, early in 1903, applied for land on the northern edge of town. His application was granted because the land was swampy and considered unsuitable for building.[1245] After a good deal of drainage was done, he was able to plant 9 hectares in potatoes and oats.[1246] He was a perfectionist and an efficient farmer, and Queen's Ranch became a point of interest in Atlin.

Storing crops over the winter was always a problem, and during the fall and winter, it was not unusual for farmers to sell their vegetables, including turnips, carrots and potatoes, frozen solid.[1247] Robert Grierson's frozen turnips and carrots were said to be superior in flavour to any imported vegetables. Queen overcame the threat of freezing by building a large log structure that served as a root cellar. His crops were safe, and he had good produce for his customers throughout the year. 1907, the year that significantly marked the end of the mining boom, also saw the end of Queen's Ranch. Many years later the property became a site for picnics, softball games, and camping. The name, Queen's Ranch, was all but forgotten when it was officially named Tarahne Park.

Farming and gardening proved commercially feasible in the Atlin Basin, but growers who always had to make concessions to isolation and the weather, also needed to experiment and invent. During the twenties when Noel Lavadiere needed a better way to handle the hay crop on his Indian River Ranch, he and Nap Morrow, a blacksmith, collabourated to build an ingenious hay baler.

Strawberries were a successful crop on several area farms. They were perfectly suited to conditions in the Atlin basin, where hardy plants thrived. It is not known who developed the "Atlin berry" by crossing the wild strawberry with a tough cold resistant domestic variety, but the resulting fruitful plants bore delicious crops which became an important market crop. The largest producer was Fred Lawson, whose fields were at Hale on Taku Arm. He shipped his product in commercial quantities to Atlin and Carcross. Another big berry producer was Charlie Brown at Thrond Gulch. A good source of pocket money for the younger generation was to hire a horse and cart, go to old Charlie's place, and pick strawberries to peddle in Atlin.[1248]

In later years a farmer with the magic touch was Sam Johnston who homesteaded a large property south of Atlin, near the mouth of Pine Creek. Johnston brought water from the creek by a system of flumes to irrigate a large garden. Throughout the war and during the twenties, he raised literally tons of root vegetables each summer. He stored his crops in two large cellars built into a bank near his barn, and all winter he traveled to town each Saturday to fill the orders placed with him by Atlin housewives.[1249]

Over the years several others grew produce and crops for the local market - James "Cassiar" Cullen had gardens at Discovery, as did Hans Gunderson. Davy Hall farmed at a lake named for him 15 km north of Atlin. John Fink took advantage of the thermal waters that seeped from a hillside near the Warm Springs. By directing the flow of warm water he developed gardens which enjoyed a longer than average growing season. Joe Yonaites farmed near Surprise Lake, and Ed Pillman, who tried everything, raised crops on the old Shirley fields north of town. The area then became known as Pillman's Hill.

While farmers had little trouble selling what they grew, they were in competition with the established household garden which became a feature of Atlin's domestic life. In the towns and up and down the creeks most houses and cabins had a vegetable patch. Amateur horticulturists kept track of species which grew best, and over a period of decades, knowledge accumulated and experience established ways to successful growing. There was always a spirited competition between growers of carrot, cabbage and turnip who vied to produce the largest specimens. The backyard garden did not yield easily, though, and in most cases soil and water had to be hauled, and there was always the discouraging possibility of a frost or marauding wild rabbits to ruin a promising harvest.

Gardens were sometimes a community venture. The hospital garden which produced bountifully for the nurses and patients, was tended by volunteers. The potato patches at Warm Springs provided a staple crop for several families, as well as some enjoyable socializing. Planting was a rite of spring that, more than anything else, marked the end of winter.

Despite opinions[1250] that there was a potential for raising livestock in the area, none who attempted it was really successful on a significant scale. Charlie Doelker, who owned a butcher shop, tried for a while to raise chickens on First Island.[1251] Charlie Brown, also known as Billy Goat Brown, raised goats at his berry farm. To protect his garden and everybody else's, he kept the mischievous animals on First Island during the growing season. Although the goats were hardy enough to survive the winters, Brown's venture was never very profitable. Perhaps there was no ready market, and perhaps the farmer lacked business acumen.

Canned and powdered milks were a fact of Atlin life for decades, and had it been possible to keep a dairy herd, someone surely would have been successful. Many attempts were made, but keeping beef or dairy cows was

*The fish peddler travelled to Discovery with his cart and dogs to sell a fresh catch from Atlin Lake.*

never a sound proposition. Even for family use, the cost of feeding livestock was expensive for anyone who could not grow his own crops.

If domestic animals were not a major source of fresh meat in the area, then wild game certainly was. Hunting for meat was taken seriously by most residents. Moose were not a plentiful resource during the early years, because their browsing grounds lay farther south,[1252] but caribou, whose migration routes had previously crossed Atlin Lake, inhabited the district in significant numbers. The sale of wild game was legal, and it was the Tlingit people who brought most of the meat to market. They were experienced hunters and had established hunting grounds over a wide range. Over the years there were occasional complaints from both the white community and the Tlingits about the amount of game being harvested by the other group.[1253]

Fish was the one fresh meat available in all seasons, and nets were commonly set under the ice in the winter. Fresh lake trout (Arctic char) was featured on hotel menus, particularly when visitors from the south patronized the dining rooms, eager to sample a local speciality. Clearly there were few regulations or quotas to control commercial fishing. Tom White, who ran a cafe, set out 180 metres of net to catch greyling.[1254]

Fish stocks in Atlin Lake were healthy, and a commercial fisherman named Benfield returned from the north end of the Lake in the fall of 1899 with 2,500 kg of trout, salmon and whitefish. From time to time there were conflicting reports about the presence of salmon in Atlin Lake.[1255] Although it is doubtful that salmon was ever actually caught in the lake, it was sold in Atlin, and was probably brought from the Taku River by the Tlingit. They were the most proficient fishermen in the community and had sold Arctic char, whitefish, pike and salmon to the miners since 1898.[1256] Prices were cheap, and fish of all kinds sold for 20 - 35 cents per kg.

When mining was at its peak, and the towns teemed with people, A. Brunich operated a busy wholesale fish market on First Street.[1257] As long as the popular large hotels offered elabourate menus, he had a lively business. There were others too who made a living from Atlin's commercial fishery. The fish peddler, wheeling his catch on a barrow up and down the streets of Atlin and Discovery, was a familiar sight. Charlie Griffiths who took his catch directly to the miners on the creeks, used a cart pulled by a large dog.[1258]

After the rush Alex Marshall, who owned a small sail boat, fished systematically along the eastern lake shore from the moment the ice disappeared in the spring. Even without the demand from the large hotel dining rooms, which were closing one by one, he had little difficulty selling his weekly catch.

Furs were essential to the Tlingits for clothing and for trading, which had its roots in the Russian occupation of Alaskan Territory. The aboriginal people who circulated through the Atlin basin, supplied furs to their coastal relatives who dealt with the Russian American and Hudson Bay Companies. The varieties of fur bearing animals in the area included mink, marten, muskrat, beaver, lynx, coyote, wolf, and wolverine.

In time gold miners who needed an income during the off season, found trapping a natural recourse. It was also a lucrative spare time occupation for some of the men assigned to isolated duty along the Dominion Telegraph Line. They were perfectly situated to run traplines, and in the thirties, one lineman took $10,000 worth of marketable furs with him when he left his post.[1259] The financial benefits were unpredictable, but the challenge, the solitary lifestyle, the outdoor activity, and the companionship of a team of dogs appealed to many trappers.

AHSP-775

*Young Shorty Jackson and Tom Williams were on a hunt for live fox pups for one of Atlin's fur farms when a photographer took this picture.*

Although it is difficult to estimate how many trappers were active during any given winter, by 1906 there was enough fur trapped in the area to attract the attention of some major fur buyers who would send in their representatives from time to time. C. Goldstein from Juneau, and Louis Levy, representing the prominent New York house of Joseph Ullman both advertised in the *Claim*,[1260] and made regular buying trips for a number of years.

Trappers, however, preferred to sell locally, in order to receive payment immediately, and to have a hand in negotiating prices. Many were grubstaked regularly by local buyers, generally merchants, in which case they were obligated to clear their accounts with the furs they harvested. Since he supplied them with groceries and was generous with credit, Louis Schulz was able to buy most of the fur trapped by the Tlingit.[1261] Other buyers were Clarence Sands, John Garrett, and Bill Strong who operated a trading post on the Taku River. No doubt there was some sale of Atlin furs to fascinated tourists, but merchants who purchased in large amounts turned a profit by selling to professional fur traders, or by shipping outside to the auctions.

By 1900, the long history of fur trading was showing its effects on wild stocks throughout North America, and some species were seriously depleted. Trappers on the land could no longer harvest enough to meet the demands of an expanding world market. Although prices slumped during World War I, when there was no European money for furs, post war prosperity and new sources of demand changed the face of the fur trade.

In the early twenties the trapper found his profits profoundly influenced by fashion. Suddenly fox and mink garments were in style, prices soared, and supplies lagged behind demand. As well, the common use of the automobile created a lively market for fur auto robes, gloves and hats to comfort passengers in open vehicles. While new ideas and designs sent the industry spiraling upward, there also remained the long established need in northern regions where the insulating qualities of fur had no substitutes.

The time was ripe for fur farming on a serious scale. Not only could the domestication of fur bearing animal increase the supply, but the harvest could also be controlled to produce pelts in colours made voguish by fashion. Often higher values were placed on domestic furs than on their wild counterparts. Farms sprung up everywhere. All that was needed were the breeding stock and a steady source of protein to feed them. Atlin seemed an ideal place, and several farms were developed.

Domestic mink need a quiet environment, or the excitable females will kill their young, so Fred and Omar Lemieux built their pens at Gladys Lake on the Teslin Trail. Their breeding stock were trapped nearby, and the operation

AHS P-1148

*A silver fox dances for this man who, like many in the Atlin area, made at least part of his living from fur farming. In the background are pens and welping boxes.*

*Jerome Von Valkenburg raised prime silver fox at his fur farm at the mouth of Fourth of July Creek north of Atlin.*

AHS P-742

had few overhead costs, but it was work intensive. Cages had to be kept clean, the animals carefully monitored for signs of disease, and they had to be fed. The daily requirement of 50 kg of food was a ground mixture of fish and moose meat. To process the feed, the Lemieux built an ingenious grinder powered by a small Fairbanks Morris gasoline engine. The brothers hunted the moose, and caught and dried most of the fish during the summer. Each November they harvested 250 pelts. Along with the mink harvest, the Lemieux trapped more than a thousand muskrat each year from a population they maintained in nearby Fish Lake, by constructing a large dam to sustain sufficient water levels.[1262] The mink ranch was eventually sold to Henry Rudolph who operated it on a smaller scale.

Apart from Lemieux there were other mink farms. Ed Haennel established his at Pike River, Bruce Morton at Warm Springs, and Charlie Hill at MacDonald Lake. In addition there were fox farms. Noel Lavadiere raised foxes on his ranch at Indian River, and Jerome Van Valkenberg produced fine silver fox at the mouth of Fourth of July Creek.[1263]

In the spare words of his diary, Nate Murphy recorded his venture into fox farming at the O'Donnel River. In the fall of 1916, he and his partners, Tom White and Paddy Bar, built fox cages by taking time, when they could, from the everyday jobs of fishing and hunting for the dinner table, trapping, washing, cooking and a multitude of other chores. Logs were hauled with horses, for framing the cages which they enclosed with wire. Inside was a small dog house-like den, in which females could tend their litters. Wild foxes were placed in the pens, and once mating began in the spring, the fur farm was established.

For as long as fur prices were high, breeding fur bearers in captivity was a worthwhile and profitable venture. Certainly the sale of fur, wild and domesticated, was a significant factor in the economic life of the community.

By and large those who chose to earn their living from the land could not restrict themselves to one or two activities. To survive, such men had to be adaptable, willing to supply whatever commodities were needed, and on occasion, work for wages. One job which occupied many was woodcutting. Wood was needed to fuel a great number of steam engines. Hundreds of cords of wood were burned to propel the lake boats, to operate shovels, hoists and drag lines at the mines, to run laundry equipment, and to heat some hotels. Almost any man with an axe and saw could be a woodcutter, but keeping pace with Ole Skelle, who could buck up four cords a day by hand, was impossible for all but the very best. His strength and stamina made him a legend.

AHS P-2472

*The prospector and his dog were summer partners in the bush.*

*Mrs. Dunlop and Mrs. Tintinger were among the dozens of women who harvested the wealth of wild berries from the hills and valleys around Atlin*

AHS Henning Collection P-1264

*Dick Dunlop used a boiler from a drill rig to run his buzz saw. Wood piles indicate they were cutting cord wood for the steam boats. Circa 1906.*

*The Atlin City Water Works provided work for some Atlin dogs. Delivering lake water to households was a steady business in all seasons. (1906)*

The Tlingit, so adept at using the natural resources, were industrious people with an aptitude for shrewd trading. In the years following the arrival of the white miner, they were usually able to live well, since some worked for wages in the summer, and hunted and trapped the rest of the year. There was always a market for their furs, fish, and wild game.

The Tlingit's ability to win a living from the land depended on their ancient right and freedom to roam rivers, streams, lakes, woods and meadows. They had never shared the pioneer's interest in agriculture or horticulture. Until 1915, their freedom was unrestricted, but in June of that year the Tlingit people of Atlin met the McKenna McBride Commission. This commission was appointed by the Dominion Government, and headed by J. A. J. McKenna to investigate the welfare of the province's aboriginal people. Many believed that the Commission's chief purpose was to confiscate valuable Indian lands for the future development of the fastest growing cities in the south.

In the case of the Tlingit, however, the Commission had no such ambitions. These northern people had neither signed a treaty with the Crown, nor had they been granted reserve lands. Here, the Commission sought to regularize the band's relationship with the Dominion Government by recommending the establishment of a number of reserves in the area but they totally misunderstood the way the land was regarded and used by the Tlingits. When Chief Taku Jack was asked if there was any special land that he wanted granted as a reserve, he replied:

> "You got no land to give me. This land belongs to me....
> If you give us people a piece of land, we are not free.
> This is my country and I want to keep it."[1264]

Taku Jack, nonetheless, was aware of the economic advantages that white society had brought, and was adamant that he did not regret their arrival.

> "...we are just like brothers." [1265]

Influential members of the white community did not feel the same, however, and the Board of Trade made forceful representations to have the Indian village removed far from the surveyed townsite. With only the one dissenting voice of Captain Hathorn, their argument was that, although they did not wish to push the Indians out (they had in fact been very useful when

*Chief Taku Jack, when asked what land he wanted reserved for the Tlingit people, said, "You got no land to give me. This is my country and I want to keep it."*

AHS Henning Collection P-1266

the town was struck by a large fire), their population was very unhealthy, and dying rapidly of tuberculosis. The suggestion was that the new reserve should be at Five Mile Point, to the south.[1266]

The Commission was unable to grant these requests since most of the band's houses were located on the property of the Catholic Church, over which it had no jurisdiction.[1267] On the other hand ten other reserves were granted to correspond with some of the traditional hunting, fishing and trapping areas. Just how these areas were identified is unclear since Taku Jack remained steadfast in his determination not to request the gift of land he believed was already his. It is likely, though, that Scott Simpson, the Indian Agent for the Stikine Agency, negotiated with the Commission in what he considered to be the best interests of the Tlingit.

While these changes were academic at the time, over the ensuing decades they had a striking impact on the way the Tlingit were able to pursue their traditional lifestyle. Inevitably the sense of pride and security displayed by Taku Jack suffered as they struggled to reconcile the demands of white society, and their own inherent identification with the land.

CHAPTER 19

# Some Women of Atlin

The all-male character of gold rush photographs portrays the acute shortage of women in early mining camp societies. The few who bravely struck out on the trails to the raw country of gold mining shared a sense of adventure seldom evident in women whose dull roles had been fashioned by the era. It was fortunate for communities like Atlin, that these spirited women came north with the stampede, because eventually they became a compelling force in the orderly development of the settlements.

Almyra Short and Amelia Greenslate came with their husbands from Juneau to join Miller and McLaren on Pine Creek before any claims were staked, and before news of the strike became public.[1268] For Amelia, Atlin was undoubtedly the safest place to be, since her husband, James, had annoyed Soapy Smith, and he needed to avoid both Smith and his gang.[1269] Almyra and Al Short staked on Pine and Otter Creeks, and decided to stay in the area through the winter of 1898-99. Good-natured Almyra was known as the "Mother of the Camp". Her yeast bread, was highly prized by miners when sourdough and inexperienced cooks were the norm.[1270] It is believed that she was the first white woman to come to Atlin.[1271]

Among the Pine Creek claims recorded a few days after the discovery group was one for Mary Turner, the wife of a dentist who quit his Juneau practice to search for gold.[1272] Mary had more nerve and grit than physical strength, and like some of the men, she could not endure the privations of camp and trail. On her second trip to Atlin in April, 1899, the 32 year old Mrs. Turner became sick with pneumonia, and died at Discovery the day after she arrived.[1273] A month later, a similar fate befell Amelia Greenslate, also in her thirties.[1274]

Mrs. John Merrifield accompanied her husband into Atlin late in 1898, as a working member of the expedition. The attractive black-eyed woman toiled side by side with the men, and took hardships in stride. By the time the group

*Berry pickers, Anais Roxborough and Rachel Hanna at the Discovery Avenue Wharf. Circa 1910.*

AHS P-1267

arrived in Atlin, she was expert at guiding a loaded hand sled on narrow packed trails, awkward slopes, and in tangled frozen bush. When she was later interviewed by a reporter in Juneau, she had good advice for other women contemplating a trip. She recommended they dress in men's suits or wear trousers under knee length skirts lined with buckskin or chamois. Long dresses were entirely out of the question. It often happened, though, that a woman dressed in appropriate mannish clothes was mistaken for a boy.[1275]

If Martha Grace Watson wondered why she was floundering in ice and snow on the Chilkoot Trail with four children aged two to thirteen, no one ever knew. If she had serious misgivings about the strange journey she was on, she hid them beneath a remarkable sense of humour. In the spring of 1899, it seemed that half the world was migrating north, and Martha, her husband Matthew, and their children were a small party in the sea of gold hunters washing over the Coast Mountains.

The Watsons were going to Atlin. They had the required mountain of supplies, a horse and three dogs, as well as Matthew Jr., aged 13, Bruce, 11, 6 year old Bill, and little Grace, a 2 year old toddler. The bulk of their supplies was transported to the summit of Chilkoot Pass on the aerial tramway, while the family hiked the busy trail. At last they faced the grueling 300 metre rise from the Scales to the top. They found the section used by pack animals, the Peterson Trail, so steep that Matthew Sr. had to stop every 20 meters to rest their horse. At each pause, Martha and the boys manned the sled brake to keep the entire outfit from sliding backwards down the trail, taking with it baby Grace who was tucked into the sled with the baggage.

Before long the weary horse began stumbling and falling down until he lost his purchase entirely, and started rolling down the slope. With the strength born of impending disaster, Matthew Sr. held the floundering animal with a rope while Martha rescued Grace, and Bruce threw something under the runner of the loose sled, keeping it from escaping. At that crucial moment two strangers appeared, righted the frightened horse and led him to the top. Matthew carried Grace, while Martha boosted little Bill up the last steep stretch, and the other children gathered up the spilled supplies. Before Martha could catch her breath or breathe a sigh of relief, she found the whole situation working in reverse as they began the descent on the other side.

Martha and the three youngest children camped at Lake Lindeman for a week while Matthew Jr. and his father ferried their supplies from the tram terminus. One evening while they waited, the Police Commissioner and his men invited all the residents of the camp to a social, at which there was mandolin and guitar music for dancing. The irrepressible Martha had a most enjoyable evening, contributing to the entertainment with a couple of songs.[1276]

When their supplies had been cleared through customs, and the duty paid at an average of 20% on American goods,[1277] they were ready for the next leg of the journey to Atlin. Trail conditions had deteriorated dangerously and for the safety of all, especially the children, the remaining distance had to be covered quickly. They cached two loads, and pressed on with just their food and bedding. To save time, Martha spent her evenings in camp cooking enough food to last the next day. On a portable stove Matthew had built to her liking, she stewed dried fruit, rice, beans and moose meat. She baked bread and large batches of cookies, and she always cooked dog food. One morning they were about to set out with two loaves of bread, just enough for lunch, when one of the dogs stole the largest loaf. That meant an unscheduled stop in the afternoon to make twelve loaves of baking powder bread.

In spite of the upsets and rigors of the trip, Martha kept her spirits intact, and after 24 days en route, she wrote, "We are all enjoying our camp life very much and we are having lovely weather."[1278]

The rest of the trip was a comedy of breakdowns and spills. As they crossed Graham Inlet[1279] on the soft rutted trail, a runner caught in a hole, and the tandem sleds careened precariously. There were breathless moments as baby Grace was extricated from under a pile of boxes, blankets and sticky snow. Martha trudged on, though, sometimes as far as 35km a day, in spite of aching limbs, and eyes swollen and burning from the glare of the sun reflecting off the snow. She kept little Grace in sight, and gave young Bill an occasional helping hand, and in the middle of April 1899, she and her family walked off the ice at Atlin City.[1280]

Eventually Matthew and Martha settled in Whitehorse where Matthew took up his old trade of tinsmithing. In 1911, Matthew Jr. established Matthew Watson's Store in Carcross which served for decades as a major community business.

Even with the convenience of rail service from Skagway to Carcross, in April 1903, a young woman from San Jose, California also discovered that just getting to Atlin could be a grand adventure. Her husband, Irving L. Ryder, was a surveyor, employed by J. H. Brownlee. They joined Brownlee for the trip north by coastal steamer, and as far as Carcross by rail. There they boarded an open sleigh, drawn by two handsome bays, and began the last part of the trip to Atlin. They were no longer pampered tourists.

Mrs. Ryder perched atop the load of trunks, grips and horse feed, while the men walked a good part of the way. Protected from the sun by her largest hat and tiny dark glasses, she viewed the lake with some mistrust. She was acutely aware of the hazards of crossing great bodies of water on spring ice, but had enough confidence in the teamster in charge of the sleigh to enjoy the trip.

That night they stopped at a log building on Taku Arm where a Swede and his partner ran an overnight stop for freighters. They were given a meal of mutton chops, boiled potatoes, corn and stewed dried apricots for 50¢ each. They paid 75¢ for pole bunks with pine bough mattresses. When it was time to turn in they strung up quilts in front of their beds for privacy, and "slept like logs in (their) clothes."

The next day, when the sleigh hit a rough section of trail on the portage between lakes, Mrs. Ryder was thrown from the sleigh into a drift, but, by thinking quickly, she clutched her husband's valuable surveying instruments in each hand, and saved them from damage.

As their party ventured out onto Atlin Lake, they encountered an obstacle about a mile from shore - a pressure crack, or "jam" as they called it. Ice was piled high along one edge of a great seam in the ice, which extended from shore to shore. Sinister open water lapped along the edge of the crack, but in some places it had frozen over again. While Brownlee drove the team, Baker, the teamster, walked along the crack testing the ice with an axe until he found a place he felt safe to cross. The ice, though, was under about 15cm of overflow. For safety the team and sleigh were taken across separately, then the passengers walked across with the water lapping at their overshoes. The ice roared and thundered ominously as it strained in the spring warmth, and Mrs. Ryder later admitted that she had been badly frightened.[1281]

Once settled in Atlin, she unpacked the silver tea service which had been hidden in her trunk expressly against the wishes of her practical husband.[1282] Mrs Ryder was ready to enjoy her summer in Atlin, and prepared for all possible social occasions.

Fourteen miles north of the barren, rocky area recognized as the summit of White Pass was a place where freighters stopped overnight to rest their teams. Log Cabin became a burgeoning settlement of tents and piled supplies as soon as gold was struck on Pine Creek. The Fantail Trail led from that point to the lakes connected to Atlin. Log Cabin, an important junction, attracted entrepreneurs who wanted to serve the thousands using the trails.

Among the business people at Log Cabin in 1898, was a heavy set, jovial woman known as Popcorn Kate. Kate had a bunkhouse squeezed next to a huge circular tent used as a barn by packers. Her back door opened on the rear entrance of the Skookum House, a hotel owned by young Richard "Dixie" Anzer, who soon found a friend in the irrepressible Kate.

Kate's sense of fun often led to riotous evenings at her place, and sometimes the customers themselves were the entertainers. One night during a dance, she had a wine drinking contest. Anyone could enter for $1.00. Participants lined up around a 50 gallon keg of wine, with an attached hose, set high on a platform. The rules called for each entrant to take a swallow from the hose and go to the back of the line, and wait for his next turn. Spectators heckled and cheered from the sideline, and when someone staggered, Kate bustled him off to his bunk, with the promise of free wine in the morning. If there was a prize for the contest, no one remembered, and the name of the participant who stayed upright to the end has been lost to history.[1283]

Popcorn Kate, whose name was probably Kate Folz,[1284] was a good businesswoman, and soon followed her clientele to Atlin, where she opened the Discovery Hotel, on Discovery Avenue. Kate herself was an attraction as she listened with friendly sympathy to the miners' problems. In 1900, when she ran into a problem trying to license her saloon, she decided it was time to move on. She gave the Discovery Hotel to Richard Anzer, and boarded the next boat leaving town, her destination unknown.

Stampeders who simply vanished during the gold rush were numerous. They left families and friends, forever wondering about their fate. Sometimes it was a woman who disappeared. Johann and Helen Lupecke were Prussian immigrants living in St. Paul, Minnesota with their three small daughters when they heard news of the astonishing gold discoveries in the Yukon. Apparently there was a serious rift in their marriage due to financial problems, and most certainly that was the issue which prompted Helen to leave her husband and children behind while she struck out for the north in a bold attempt to regain financial security. At Skagway she abandoned the Klondike Trail in favour of Atlin. She established the Pine City Laundry and Bathhouse in Discovery. During the summer her "famous German sliced potatoes" were sold at the Fraser and Co. grocery store in Atlin.[1285] In December 1900, she sent a testy letter to Johann suggesting he could do well in Discovery, but that she would not consider him a partner in her business. He could, however, work for her if he came north. When Johann replied that he was willing to join her, Helen was overjoyed and sent instructions: "Dress the children in their good clothes, so that you all look decent when arriving here. Everyone knows me here, and I should not like to see you looking poorly." She missed her small daughters terribly, and early in 1901, she sent $100 to Johann, for travel expenses. Time passed, and for some reason he never made the trip.[1286] Eventually the family lost track of their mother.[1287] It may be that, in anger and disappointment,

Helen moved on to take advantage of another business opportunity. There is no record of her after 1902.

In stark contrast to women like Helen Lupecke was Mary Hitchcock, who came to Atlin as a wealthy tourist in 1899. She thoroughly enjoyed her preeminent position in the fledgling society, and maneuvered skillfully to retain her place in the limelight.

At first she intended to follow up her successful book "Two Women in the Klondike", with another on Atlin,[1288] but other interests captured her attention. She was intrigued with the potential for mineral production in the north, and although she had not considered herself a businesswoman,[1289] she quickly involved herself in the mining industry. She dabbled in lode mining ventures in Atlin,[1290] and Dawson,[1291] and as far away as Korea,[1292] and Siberia.[1293] She had bona fide mining men in England and the Eastern United States[1294] at her beck and call. The *Atlin Claim* heralded her as "the Mining Queen of the World."[1295] Even the Premier, Sir Richard McBride, paid tribute to her contribution to mining in the Atlin District.[1296]

Mary Hitchcock invested much money, but saw little in return. Her former career as society matron had taught her little about mining, and her sense of self-importance would suffer no opposition. By 1902, interest in her mines had waned, but she continued to promote Atlin and the Yukon on her lecture tours. She was perhaps the first person to recognize fully Atlin's tourism potential, but her plan to lead parties of tourists to the north was never realized, and was perhaps an idea slightly before its time.

For a brief time in 1899, Atlin was taken by storm by Mrs. Alice Houghton, a businesswoman from Spokane, Washington, representing influential real estate and mining connections.[1297] She went into business on First Street with Archie Brownlee, the son of the surveyor, J. H. Brownlee. A flamboyant figure, she was surrounded by speculation and excitement as the rumours flew about the inflated prices she was negotiating for land and mining claims. Most Atlin businesswomen, although often astute, did not operate on such a scale.

Mrs. D. H. McDonald had never considered managing a business, but circumstances forced her to take over the family grocery store in 1910. She and her husband, Daniel, and two daughters came to Atlin in 1898 via the Chilkoot Trail. They were among the thousands of Klondike bound travellers who changed their destination to Atlin when they heard the gold strike on Pine Creek was of major importance.

Daniel McDonald was known as D. H. to everyone, and his wife became simply Mrs. D. H. Mining was never their intention. Instead they opened

AHS P-1238

*Young ladies in long skirts race to the finish line.*

a small tent grocery store in Blue Canyon on upper Spruce Creek. Later they moved the business to First Street in Atlin, where they ran a grocery and feed store.[1298]

When their third daughter, Ruth, who was born in Atlin, was only seven, D. H. became seriously ill, and had to be sent to hospital in Seattle. Mrs. D. H. knew she must go south to be with him, and in March 1910, she left Atlin with the mail carriers. Foul weather added gloom and danger to an already tense journey, and for some distance they had to cross open water in a canoe. Freezing wind whipped the lake, and water splashed over the sides. By the time they reached dry land, her sodden coat had frozen fast to the canoe, and the men had to chop the ice away, to free her so she could go ashore.

D. H. died on March 7th, and Mrs. MacDonald made the sad journey back to Atlin to face the reality of supporting her three girls alone. It was clear she would have to run the business herself, but she had another problem never encountered by her husband. She had to fight for credit to cover the twice yearly shipments of supplies that were needed to stock the store. Suppliers were unwilling to extend credit to women, and reluctant to do business with her.

She had to compete with other Atlin stores, and, when the town's boom population decreased, to make sure that her store maintained its share of the town's business. Management covered a multitude of chores, one of which was just keeping the stock from freezing on the shelves during the winter. Two wood stoves had to be stoked continuously. She also provided a store staple. Each night she mixed up a huge batch of bread which was baked the next morning, and sold warm and fresh to her customers.[1299]

Mrs D. H. provided a secure home for her daughters, a small place by the lake, a place where their friends liked to congregate. They were usually welcomed with good things from the oven, and Mrs. D. H. became famous for her whipped topping for desserts. It was a simple concoction made by whipping canned milk that had been placed out in the snow until it almost froze - a good trick when dairy products were scarce.

Mrs. D. H. lost her store in the Atlin fire of 1914, but the support of her many friends sustained her through that difficult time, and she rebuilt immediately. She continued in business until 1920 when she decided it was time to move south.[1300]

Another widow who suddenly found herself with a business on her hands was Anna Miller. In 1904, Fritz Miller and Kenneth McLaren bought the Balmoral Hotel in Discovery,[1301] just days before Miller became ill and died.[1302] Anna tried to manage the hotel, but she was unable to run an orderly house. Although she had the help and support of Kenny MacLaren, he was a shy,

*Ladies try their hands at haying at Queen's Ranch in Atlin, just south of the 60th parallel.*

AHSP-580

quiet man, and not the authoritative figure she probably needed to keep the business running properly. In 1907, the authorities forced the Balmoral to close its doors.[1303]

Nancy Simson, a young woman from Victoria, opened a store at the south end of First Street in 1899. Her stock was a reflection of her youth and included "costumes, bicycle skirts, ... shirt waists, hats (sic) and millinery".[1304] The madcap Miss Simson was also prepared to engage in more exciting ventures, which must have roused curiosity and lifted more than a few eyebrows.

Before the snow was off the ground early in 1899, she accepted an invitation to accompany a group of prospectors with Indian guides to the Pike River region to stake claims on a new creek, which the discoverers requested be named in her honour.[1305] They traveled by dog team and on snow shoes for five days, and according to the *Atlin Claim*, Nancy was not especially well outfitted for the trip, since she was wearing "French shoes." The newspaper reported that although they had to wade through waist high water, Miss Simson simply laughed heartily and nibbled chocolates along the way.[1306]

Within a decade of the gold rush, much of the retail business in Atlin was done by women, either in partnership with their husbands or independently. Margaret Mackintosh, Elizabeth Pillman, Ella Hensley, Vena Noland, Hannah Nelson, and Signe Sands were just a few of those who ran successful businesses.

Annie Smith and her four children remained in Vancouver when her husband, William, left for the north in 1899. Annie was pregnant and lonely; moreover, she missed their old home and family in Inverness, Scotland, which they had left just one year earlier.[1307] During the year William was gone, baby Constance arrived, and Annie coped with homesickness, and the frustrations of settling in a strange new country. Smith was an adventurer and an entrepreneur, and once in Atlin he built the Hotel Metropole, and became involved in a number of other business and mining ventures.

After a year Annie Smith gathered together her children and headed north to join her husband who had built a log cabin for his family. When the *Scotia* landed at the wharf one fall day, and Annie and the children stepped ashore, a delighted William was there to greet them.

As a business proposition William, who was known as "Metropole" Smith, built a 9 metre sailboat which he chartered; but more important to Annie, they used it for family excursions to the many inlets and islands on the lake.[1308] They were memorable outings, filled with exploration, hunting for the winter larder, and fishing. Annie and the children picked wild currants and other berries, which she made into jellies and sold in town, or sometimes traded for Tlingit moccasins and parkas.

Annie was happy as long as her family was together, and she was able to use her special skills in Atlin. She was an adept and economical seamstress who kept her girls supplied with pretty pinafores, made from large cotton flour sacks. She was also a midwife, who was often called upon to rush to a cabin on one of the creeks to deliver a baby.[1309] She had come a long way from her home in Inverness but life in Atlin gave her satisfaction, and an independence that she had never experienced before.

A woman of early Atlin carried out her duties as wife and mother in conditions unique to her remote location. She had to allow constantly for distances, different modes of transportation, and the weather just in the course of everyday life, but even then she might be caught in unexpected circumstances. In January 1905, Mrs. Barrie Davies found herself in a race with the stork from Boulder Creek to Atlin by dog team, a 25km run. Somewhere along the route baby Mabel was born in the sled. All went well though, and the mother and infant daughter were finally delivered into the care of the staff at St. Andrew's Hospital.[1310]

Dog team transportation was so convenient that many women drove their own or the family team. A determined team was a mighty force to control, but they learned to rule the dogs with finesse rather than the pure strength used by many men. Susie Taylor, the young daughter of a telegraph operator, handled a team as well as any man. After she married Robin McCleery and moved to Teslin, she often drove her dogs to Atlin, over the 120 km trail, to visit her parents. Janet Owen kept her own team which she used for outings with her children, to visit friends in Discovery, or on the creeks.[1311] Signe Sands could make a fast run from Discovery to Atlin, with her children in the sled, in little more than 20 minutes. She conceded that going home, up hill, took much longer.

Janet Owen, the wife of the Atlin Police Chief, sometimes had unusual help in her kitchen. The town had only one prison cell, large enough for a single inmate . If there was another arrest, the prisoner was shackled with a leg iron in the centre of Mrs. Owen's kitchen floor. She made the best of the inconvenience, by putting him to work at various chores, peeling vegetables or washing dishes.[1312]

Sometimes old conventions were ignored in Atlin, and the rule against married women continuing their careers as teachers, was shattered when Edna Miller and Charles Bourne, the druggist, were wed in 1903.[1313] Not only did

Edna continue teaching, but when she became pregnant, she remained in her classroom until just six weeks before her son was born.[1314] Perhaps her expanding figure was overlooked because a replacement was not easy to find. Later, however, Edna Bourne came up against a taboo that did stop her. In 1904, she began studying pharmacy on her own, and passed the entrance test, but was disqualified from proceeding further because she was a woman.[1315]

Edna Bourne was never a timid woman, and she could handle emergency situations, which became an asset when she later went into the acceptable profession of nursing. She would surely have occasion to recall an episode on the trail to Carcross one spring when she was going outside with her young son. While travelling on the frozen lake in the White Pass sleigh, she noticed the driver take a nip now and then from a whiskey bottle, but she thought little of it. Later as they approached the shore at Carcross, she realized he was too drunk to handle the team, which for hours had simply follwed the familiar trail. She also realized they were on thin ice over deep water and there was a wide open gap between the sleigh and dry land. With no hesitation Edna pushed the driver aside, took the reigns, and urged the horses forward over ice that broke up behind them. Then she forced the team through a stretch of shocking cold water and up on to the beach. Her luggage was thoroughly soaked, and it was still wet when they reached Victoria. Water stains on her trunks would always remind her of a very close call.[1316]

A good friend of Edna Bourne was Rosalind Watson Young, a remarkable woman for her time. She was the wife of Atlin's M. L. A., Henry Esson Young, and because of his political duties in Victoria, the couple only lived in Atlin for short periods of time, but she was highly respected in the community, and her influence was significant. She was a McGill graduate with an M.A. in Geography. She taught for 7 years at Victoria High School before becoming the first woman on the staff of Victoria College.[1317]

As a teacher Rosalind Young was interested in the children of Atlin and the standard of education they received. She was also keenly interested in mining, and established herself as an expert in the field. She wrote numerous articles for the Minister of Mines and various professional publications about mining in Atlin, and elsewhere in the province. In 1903, she was elected as an associate member of the Institute of Mining Engineers, and in 1909, was made a member of the Canadian Mining Institute.[1318]

Jenny Bender was a miner. She came to Atlin during the rush, and worked as a barber, and some believed she had been a sporting girl in Discovery. Years later she lived and maintained a barber shop in Eggert's old Discovery jewelry store, which had been moved to Pearl Avenue in Atlin. When her barbering business had dwindled to just one remaining customer, she bought claims on Wilson Creek, and made plans to work them alone. Although she was a small woman, she skillfully handled two large dogs which she used to haul her cart load of gear to the creek each spring. She remained at the claim all summer, mining with pick and shovel, underground. Her cleanups were known to be good.[1319]

In 1924, Marie Beckman, a widow in her mid-forties, came north from Washington State to marry a miner and trapper, who lived 40 km southeast of Atlin on the O'Donnel River. Nate Murphy traveled the lakes and railway to meet his future wife at Bennett Station when her train arrived from Skagway. Their simple wedding in Carcross was attended by a few Atlin friends, and they continued to Atlin, where they were honoured guests at several occasions. There was one more small celebration when, a few days later, they arrived at the O'Donnel River. Jack Prescott, a thoughtful neighbour, honoured the newlyweds with a dinner of sauerkraut and roast moose.

Marie made a cheerful home for Nate, even though amenities were few. When Marie's son, Melvin, finished school in Washington, he joined them, and became Nate's right hand man. They worked as partners in the gold mine, the trap line, hunting, gardening, tending their animals and keeping things in running order. Soon Marie was at home on the land, isolation was of little consequence to her, and her family, friends and neighbours constituted her world.

For a few years Mrs. Conroy was a neighbour and close friend. After the Conroys moved, the Murphy's only neighbours were a number of single men living on their claims. Marie's presence added a new dimension to life in the O'Donnel community and winter evenings were often spent at the Murphy's over a game of cards and Marie's homemade donuts or ice cream.

The Murphy's enjoyed a steady stream of visitors who drove to the O'Donnel in the summer. Marie welcomed them and entertained themwith meals and teas. In winter their location on the Telegraph Trail guaranteed a steady flow of traffic and visitors made up of mail drivers who plied the 375 km between Atlin and Telegraph Creek, of people travelling to and from the mines on the Taku River and to Telegraph Creek, or to Tlingit hunting camps. Often travellers used Murphy's telephone to call a taxi from Atlin to pick them up at the O'Donnel.

The telephone also kept Marie in touch with her friends in Atlin, and others living at the Warm Springs or McKee Creek. Marie learned to drive

AHS P-515

*Ethel Pillman, daughter of Edwin and Elizabeth, was a popular performer at town concerts when she was a young girl. Recitations were her speciality. She married John Cartmel, one of Atlin's Gold Commissioners.*

AHS P-1268

*Jenny Bender was a pioneer whose past is not entirely clear. In her late life, she held claims on Wilson Creek and mined them alone each summer.*

their dogs, and sometimes mushed to Atlin alone. She would harness the dogs and set out on the trail, taking tea and a break with Mrs. Adams at McKee Creek. After Nate bought a new 1927 Ford, it was only a matter of time before Marie learned to drive it, and to make solo trips to Atlin over the narrow 50 km road, which was kept in reasonable condition by the Atlin road maintenance crew.

Nate drift mined on the river banks not far from their house. When Marie needed pin money in the form of gold, she washed out a few pans from his ore dump, before heading off to Atlin to shop and visit.[1320] The land was rich in many ways, and the Murphy's lived well by using what was available. The O'Donnel valley is warm and sheltered and their gardens of flowers and vegetables produced abundantly. They planted potatoes at the Warm Springs, as did other families in the area. Marie knew where each kind of wild berry thrived, and Nate and Melvin were keen hunters and fishermen. What nature did not provide directly, they bought with money made from harvesting furs, and from gold mining. Self sufficiency was a way of life, and Marie became expert at recycling, renovating, and rejuvenating. Once when the notion for a change stirred her, she persuaded Nate to man the scissors and give her a stylish bobbed hair cut.

In 1906, a short story titled "The Unexpected," was published in *McClure's Magazine*. It was a true and chilling tale of murder and hanging, by Jack London, and it was about an Atlin woman named Hannah Nelson.

Hannah and J. H. (Hans) Nelson mined on Plum Creek, a small tributary to upper Birch.[1321] They had moved to Atlin from the Juneau area after a mining venture at Latuya Bay ended in terrible tragedy. Hans Nelson had been the foreman of the Latuya Bay Mining Company at its site on the Alaska coast west of Juneau. Hannah went to the mining camp with her husband in the summer of 1899. When fall brought an end to the work, most of the crew left the bay, but the Nelsons and three miners decided to stay to do some small scale mining of their own.

The men recovered a modest amount of gold, which was entrusted to Hannah for safe keeping. The group enjoyed a congenial relationship, until something dreadful and completely unexpected happened. On an October evening they lingered in friendly conversation around the dinner table, then one of the men, Martin Severts, went outside. When he returned, he had a Colt .45 in his hand. While his companions sat stunned, he fired point blank at Steve Fudigo, killing him on the spot. He shot again at Sam Christianson. This bullet ricocheted and grazed Christianson's neck, but knocked him to the floor. Severts then aimed the revolver straight at Hannah, but Hans seized his arm, and the diverted bullet ripped into Severts' own leg instead. Hannah leapt to the aid of her husband who was struggling with the killer. She wound a towel around Severts' neck, and twisted it until Hans was able to subdue and tie him.[1322]

In the aftermath of that sudden mayhem, the shocked survivors buried Fudigo and kept Severts heavily restrained and carefully guarded, while they tried desperately to hail a passing ship, but no help came. Eventually they hired some Indians to guard the killer in a cabin a few km away. During this time Severts begged to be hanged, knowing that this would be his fate anyway. Time and the enormity of the situation took its toll until finally the strained survivors, acting out of terror, and ahead of the law, hanged Severts.

It was early the next May when the Nelsons and Christianson reached Juneau. They recounted the grisly tale of Latuya Bay before the court, and were ultimately exonerated when the judge ruled the hanging a judicial execution. Severts' packed bag, and other clues, made it clear that he had killed for the gold.

Whether or not Hannah Nelson was the one who kicked the barrel from under Martin Severts, causing his death by strangulation, as London told it, became a small matter of controversy after the story was published. In fact the entire incident was scoffed at by a Seattle newspaper, and London had to defend the truth of his account even though it had been reported as far south as San Francisco.[1323]

Charles Gairns of Atlin had pleasant childhood memories of the Nelsons and the store they eventually established on Spruce Creek. Hannah, whom Jack London called Edith, was a kind woman who always had cookies or candies ready for children who visited. Gairns also remembered Hannah saying that she had, for a while, been required to check with Alaskan authorities periodically. She admitted to having nightmares about the Latuya Bay killings,[1324] but she went on with her life in spite of the horrors that haunted her. Moving to the Atlin district would not erase the memories, but change probably helped to ease the pain. It might have, that is, until 1906, when "The Unexpected" was published in a top selling periodical, and Hannah's story was known to a nation.

Women wove the first threads of refinement into social networks. Schools, hospitals and churches gained importance over saloons. Women promoted genteel entertainment such as study groups, recitals, lectures, and at the same time spoke out against drinking and gambling.

A group of Atlin women not accepted in the polite society of ladies, were the prostitutes, coyly termed, "sporting women" or "ladies of ultimate accessibility." Although many of their houses closed after the first rush, some remained open in both Atlin and Discovery for several decades. The police exerted some unofficial control over them, and monitored the women who worked there. Frequently they turned back unsuitable women as soon as they stepped off the boats.

In the case of one madam, who ran houses in both towns, her brothels were her livelihood, and Carrie Walker controlled her business with an iron hand. She did not shrink from dealing with the authorities directly. In 1903, she wrote to the Premier, the Hon. Richard McBride, to lay a complaint about the police, who, she claimed, were harassing her. They had closed her down because of excessive noise, and had not dealt with other "houses" as harshly.[1325]

Walker, though, was particularly adept at foiling the law, and she managed to avoid prosecution regularly when charges were filed against her. In 1903, she was accused of carrying concealed weapons, disrupting the peace, and selling liquor without a licence.[1326] In 1905, there were further charges that she was selling liquor, and none of these cases was proved.[1327] In 1908, she

AHS P-1269

*Hannah and Hans Nelson operated a store in this building on Spruce Creek for many years. Their business served the small community of miners and their families who lived near the gold workings.*

brought an 18 year girl to Atlin from Ohio for the purpose of prostitution. In order to evade the immigration laws, Walker tried to pass her off as her daughter.[1328] The girl was, in fact, her niece, but she knew nothing of her aunt's real reason for paying her way north. When Walker found out the girl was diseased, she turned her out on the street. At that point Chief Constable Conley pressed charges, but it is not known if his bid for a prosecution on this case was ever successful. It is clear, however, that he was dealing with an unscrupulous character.

Many of the working "girls", were victims of circumstances, locked into their miserable lives, and powerless to escape. Sometimes suicide was the only way out. The *Atlin Claim* reported in 1903 that a girl name Gyp had tried to take her life by swallowing antiseptic tablets.[1329] The sporting girls were seldom beautiful, as a name like "Harriet the Horse" would attest and Josie, a Discovery "girl" weighed in at over 200 lbs.[1330] Sometimes the girls became friends and confidants of homesick miners, and occasionally one would marry a local man. In time her status would change, and she would win a degree of acceptance from local society.

A few Atlin women enjoyed comfortable, protected lives. They were able to travel outside regularly to visit families and friends, and to avoid the worst winter weather. They had all the amenities Atlin could offer and their lives were well insulated against the roughness of the camp, and the wild country around them. Most of the women of Atlin, though, were true pioneers. They had left secure homes and neighbourhoods to come to the goldfields. Some lived in town, but many others kept house in simple cabins along the creeks. On the whole they adjusted well to life remote from the comforts of cities, and they appreciated many extraordinary compensations - sparkling spring mornings, glimpses of wild animals, the dazzling displays of colour in the winter sky. They enjoyed the long hours of summer sunshine that produced gardens which could not be matched in the south. In fall the hillsides were dotted with a variety of berry patches which provided for jellies and preserves in abundance. Perhaps best of all there was the warmth and generosity of neighbours, bound together by the isolation.

Common to the natures of the first Atlin women was the spirit that could cope and triumph. Whatever their role, whatever their social or moral status, they were a remarkable breed, and women of extraordinary tenacity. After the trails were hiked, homes built, friendships made, and livelihoods secured, most would not have considered returning to the predictable routine of the lives they had left behind.

CHAPTER 20

# Tourism: A Northern Destination

*"Blue as a sapphire set deep in old silver..."*

- Kathleen Keats White

The first inquisitive tourists arrived in Atlin shoulder to shoulder with rambunctious stampeders. Hamlin Garland and others like him had competed with pushy gold hunters for space on coastal steamers, bunks in mousy hotels, and meals of beans and thick coffee served in canvas huts pitched beside churned and littered trails. Mining was not for these visitors. Curiosity brought them to the noisy camp. The mesmerizing beauty of Atlin Lake, chilled by the melt of sprawling glaciers, was reward enough for the eyes of the rumpled newcomers. Any of those truly interested soon learned that meandering shorelines and high mountain nooks hid the lairs of a bewildering variety of animals, and the secret aeries of majestic birds. Adventurous visitors stumbled into a wild paradise with two towns developing in its midst. Their cameras snapped in appreciation, and no travel promoter could miss the potential.

The scenery, so awesome to early visitors, was not lost on the miners in spite of their avid preoccupation with the search for gold. Neither did they overlook the opportunity to hunt for game in the hills rising above their creek diggings. All licensed miners and prospectors were allowed by the Game Act to hunt animals for their own use in any season.[1331] That privilege was embraced out of necessity, and for the challenge and the sport.

The 1903 gold mining season brought the highest returns in the short history of the Atlin camp. That summer 200 new arrivals were quickly absorbed into the work force in just one week, and there was a cry from the creeks for more men.[1332] While there was no particular need for a second industry, another area resource was, nevertheless, recognized and promoted. From that time hunting and sport fishing were advertised as attractions that lured tourists to the area for a definite purpose.

F. G. Ashton, who had recently bought the Dawson Hotel at Taku Landing, advertised his business as the headquarters for fishing and shooting in the area.[1333] He promised good accommodations and meals, and the attentions of a genial host.[1334] In August Walter Aitken took A. W. Naylor of England on a big game hunt, establishing himself as the first game guide in Atlin.[1335] Aitken's excited client reported seeing enough game, including a flock of at least 32 sheep, to fill the steamer *Scotia*. Naylor returned from the expedition with a Fanin sheep, a black bear and one mountain goat.[1336]

In November 1903, Aitken and John Ronayne, an Irish miner, led the most ambitious hunt thus far launched. Their English hunters were Sir Brian and Lady Leighton.[1337] This was no rough neck expedition. The *Scotia* was chartered to haul horses, equipment and supplies. Despite the many conveniences and comforts provided, the trip was not an unqualified success. The party steamed to the north end of the lake, and proceeded to a place called Big Fish Lake, about 12 km east of Atlin Lake, where they established their first camp. After noticing an abundance of moose tracks in the area, they decided to move farther into the mountains, but not before a loaded pack horse fell into the water. The accident marked the end of the horse, and the loss of important supplies. One of the party was dispatched to Atlin for replacements. Later a fire consumed their second camp. With that disaster, the hunt was abandoned and the defeated hunters were forced to return to Atlin.[1338] Mishaps could be expected, but they were overlooked when hunts led by Aitken, Ronayne and others were reported in several English newspapers, with keen descriptions of the variety and abundance of game in the Atlin area.[1339] Such publicity brought a small but steady flow of English sportsmen, and in time Americans, in pursuit of trophy moose, caribou, bear, sheep, or mountain goat. Water fowl hunts were another attraction. Arctic char, also called lake trout, which weighed 12 kg and more, flashy grayling and pike drew fishermen to the lakes and unmined creeks. At Surprise City, the Victoria Hotel had a Peterborough canoe and a large canvas boat for the use of guests who were advised that grayling hit best on Red Coachman flies.[1340] Outstanding catches were duly reported in the *Claim*.

Hunting guides were not usually professionals, nor did they consider guiding a livelihood. Most were miners or merchants who were willing to lead an occasional expedition to supplement their incomes. Robert McKee had a hunting lodge at the O'Donnel River.[1341] Each fall Nate Murphy, a miner and trapper, guided a party into the country south of the O'Donnel for 4 to 6 weeks of hunting and fishing.[1342] Louis Schulz, teamster, butcher, hotel and store owner, provided equipment and guides on request.

John Noland became the best known guide in Atlin and for several years engaged in guiding and outfitting as a full time business. Much of Noland's knowledge and experience was gained while guiding for Louis Schulz. In 1924, he bought a string of horses in Hazelton and drove them to Atlin on the Telegraph Trail. They represented the start of his own business.[1343] Noland's hunters came from all over the world, and included members of the British aristocracy and American politicians and industrialists. Numerous guides, horse wranglers and cooks were employed, but all details of preparation and coordination were handled personally by Noland with the dedicated help of his wife, Vena. Hunters appreciated his smoothly run expeditions, and repeat customers proved the success of his operations.

The obvious increase of visitors to Atlin prompted a brief flurry of interest among businessmen who talked about promoting tourism in 1903.[1344] Not much happened beyond talk, though. In 1905, the White Pass Company joined with the coastal steamship companies to offer excursions to Atlin from Seattle and Vancouver. The fare, including all accommodations, and a trip to the Pine Creek gold mines, was about $120.[1345] The idea generated little response.

By 1907, most mining was in the hands of several large companies. Opportunities for small operators to make a living were limited. Dredge mining had failed, dashing all hopes for steady, large volume mining on Pine and in Blue Canyon. The inevitable slump following the boom had settled dismally over the town; many stores, offices and hotels closed, people were moving away. A business community that had exploded into prosperity on the crest of the gold rush, was floundering. Concerned merchants grasped at an obvious remedy for the sinking economy, and organized the Tourist Association with White Pass agent, James Lipscombe as President.[1346]

While local businessmen were now anxious to welcome visitors, and both accommodations and plans to entertain them were in place, the idea of turning Atlin into a tourist destination was still a dream. Promotion on a grander scale was needed, a program beyond the means of the Tourist Association. In the end it took the force of world upheaval to put Atlin on the map of popular tourist destinations.

Turmoil in Europe and the rumblings of war were curtailing the stylish practice of taking long elabourate holidays to the continent. North American tourist companies, sensitive to the trends, searched for new destinations which would keep travellers and their tourist dollars closer to home. Steamship companies, including White Pass, launched a fresh alternative to the European tour. They suggested cruises through the Inside Passage, with close views of mammoth glaciers, and stops at romantic Indian villages and towns made famous by the Klondike gold rush. They advertised the train ride from Skagway

*Nate Murphy turned to guiding each fall. A large caribou was considered a prize trophy by his hunter.*

AHS P-2231

AHS P-2318

*A reputation for well run, successful hunts brought people from around the world to Atlin for big game excursions with John Noland.*

to Whitehorse, and promoted an unusual boat trip from Carcross to Atlin. Travel brochures headlined Atlin with Alaska and the Yukon, as though Atlin were itself a territory.[1347]

White Pass publicity described Atlin as the "Switzerland of the North," a trite motto used by more than one mountain town, and hardly appropriate, but one which caught the attention of veteran European travellers. Promotions for Atlin were strikingly successful. Satisfied visitors, who were the most effective advertisers, carried home descriptions of immense blue lakes, mountains and wilderness stretching to infinity. Those who were lucky told tales of seeing the eerie aurora or sluice boxes lined with gold. White Pass had concocted a perfect blend of wild settings, adventure, and luxury service for an eager market.

Traveller's tales were reinforced and amplified by the glowing written accounts that appeared in magazines, newspapers and promotional publications. Each year the business grew, and the demand for advance bookings for 1916 was the signal to White Pass that available facilities would soon be overwhelmed. Major changes had to be made. In 1916, the British Yukon Navigation Company was commissioned to build a lake vessel to replace the aging *Scotia*, deemed too small to handle regular freight and passengers, plus the anticipated crush of summer tourists.

The other pressing problem was suitable tourist accommodation. The Royal and the Kootenay were the only hotels still operating in Atlin. Both were fine establishments with traditions of good service, but neither was designed to cater to luxury tours in the volumes anticipated. White Pass had already completed preliminary plans for a series of large hotels along a vacation route stretching through British Columbia, Yukon and Alaska.[1348] Robert C. Spencer,[1349] a Chicago architect, had designed the fashionable buildings in 1913. The first hotel would be erected in Atlin. Some construction materials arrived in the fall of 1916, and on June 10th, 1917, after a late opening of navigation, the remaining materials and 60 men were landed in Atlin, and construction started. That spring, when a fire destroyed several of the businesses on First Street, the new building was spared. On July 15th, in 35 working days, the hotel was completed[1350] at a cost of $20,000.[1351]

The imposing Atlin Inn was four stories high and it dominated the waterfront between Pearl and Rant Avenues. The hotel was a wooden building, described as "rural shooting-box style."[1352] The breakneck speed of its construction was not at all apparent to the first visitors. Large spruce-panelled smoking lounges were decorated with big game trophies which added a rustic essence to the casual decor. Luxurious leather chairs and huge fireplaces provided visitors with an opportunity for comfortable relaxation and genteel informality. Guest rooms offered splendid views of the lake and Atlin Mountain. Diners enjoyed the scenery, and chose from menus of the very best cuisine. Fresh lake trout, served regularly, was always popular with visitors interested in trying local delicacies. At the Atlin Inn no detail of service was overlooked. Guests, used to the best European resorts, found themselves comfortably content in a small isolated mining town.

On the beach near the Inn, the *M. V. Tarahne* was taking shape. She was a smart 24 meter gasoline powered vessel, a departure from the steam driven paddle wheel vessels of the era. The company apparently decided that it was less expensive to ship fuel by boat and rail, than to pay local wood cutters for the hundreds of cords needed to stoke a steam engine.

The *Tarahne* was launched the next summer on June 19, flying the blue, red and white flag of the British Yukon Navigation Company. A handsome vessel, the *M. V. Tarahne* was designed for sightseeing. Above the cargo deck was a large airy observation deck, painted in soft greens and warmed by two wood stoves. Comfortable wicker chairs were arranged next to long rows of windows. More viewing space was provided on open decks fore and aft. The new vessel made it possible for the company to add lake trips to their program of entertainment.

Cruises to Llewellyn Glacier, made two or three times weekly, followed a route through Torres Channel, a narrow waterway hidden behind Teresa Island, and past the serrated face of Cathedral Mountain.[1353] It wound past dozens of islands, large and small, covered with summer flowers and wind-bent spruce. Passengers saw snowy goats balanced on the crags of Birch Mountain, and bears, with shimmering coats of cinnamon or black, rooting for berries or grubs. Sometimes a moose swam into view. Loons cackled in the coves, and osprey swooped over precarious nests. It seemed that nature put on a show especially for the benefit of the voyagers. Slipping through Second Narrows at the end of Copper Island, the *Tarahne* cruised back into the main body of the lake, where the ancient ice of Llewellyn Glacier presented a breathtaking spectacle for all aboard.

Refreshments for excursions were catered by the Atlin Inn. On the return trip passengers enjoyed an elegant tea with sandwiches and perfect *petit-fours* served by white-coated stewards.

Mineral water had long been sought as a health tonic, and "taking the waters" was a must for discriminating tourists. White Pass wasted no time in promoting Atlin's own cure-all. A spring at the north edge of town was perfectly located, just a short stroll from the Atlin Inn. The company built wooden walkways, and a small gazebo there. Guests were encouraged to relax, enjoy the scenery and sample the water. Slightly carbonated, it tasted best with a dash of lemon juice and sugar. The mineral water was mildly laxative, but that fact was seldom mentioned.

White Pass published a small booklet, entitled "Atlin the Beautiful."[1354] It was a detailed guide to the sights and activities available to Atlin visitors. There were hikes to Como Lake, Fourth of July Creek, the Indian Cemetery, and up Monarch Mountain. The company maintained a tennis court in town, and established a golf course about three kilometres away on the Warm Springs Road. At the Atlin Inn a fleet of 4 metre skiffs with outboard motors were ready for hopeful fishermen. Throughout each season, dances, and an occasional grand ball, complete with an accomplished orchestra, were held at the Inn. Invitations were often extended to include local people who blended with the tourists at these gala events.[1355]

*The M.V. Tarahne was a sleek 36 meters long after being lengthened in 1928.*

*The Atlinto, moored in front of the White Pass resort hotel, the Atlin Inn, was used to take tourists on midnight excursions to the Llewellyn Glacier.*

AHS P-1271

*Munro Mountain stood as a backdrop behind the town of Atlin as it was seen by tourists arriving aboard the M.V. Tarahne in the busy tourist seasons of the 1920's.*

AHS P-1274

*Atlin attracted many distinguished visitors. In 1922 the Governor General of Canada, Lord Byng of Vimy, and Lady Byng were in town. They drew a crowd of fascinated bystanders as they posed graciously for photographers at the White Pass Wharf.*

Beyond the natural sights, sports and planned activities, tourists found another attraction. They were fascinated with the Atlin people, whom they considered somewhat exotic, for the very fact that they lived in the remote wilderness.

A few local characters unequivocally enjoyed their roles as colourful individuals and story tellers. Taku Jack, the local Tlingit Chief, liked to greet passengers arriving on the *Tarahne*, and he was always willing to have his picture taken with them. An accomplished silversmith, he would solicit a silver dollar and take it home to shape a bracelet, napkin ring or a brooch, and decorated with traditional Tlingit designs. The delighted tourist was usually willing to pay him well for his trouble, and thus his work found its way to many corners of the globe.[1356]

Captain "Scotia Mac" MacDonald, skipper of the *Scotia* and Captain Coughlan of the *Tutshi* regaled tourists with romantic tales of the paddle wheelers. Strolling around town on the boardwalks was a way for tourists to meet the people they found so interesting. They might have had chance conversations with the town constable, a nurse or schoolteacher, and perhaps the busy Gold Commissioner, who represented the many arms of the provincial bureaucracy. Grinning school children, Tlingit hunters, merchants and miners told favourite stories of the stampede,[1357] or reflected on winter weather so extreme it was scarcely imaginable, and in doing so added spice that flavoured an Atlin holiday.[1358]

White Pass was at the centre of the Atlin tourist trade, but local businesses were also involved, and their ventures were seen by the White Pass as complementary.[1359] Merchants catered to tourists by stocking their shelves with gifts and souvenirs with a local flavour. Jules Eggert, a master watchmaker and jeweller from Switzerland, landed in Atlin in 1899.[1360] In a short time he had lured clientele from his two competitors, and working from his Atlin store, on the corner of First and Pearl, became the sole jeweller in the area. When his son Paul became part of the family business, he was put in charge of a Discovery store. Gold nugget jewellery was Jules Eggert's speciality. He designed original pieces for discriminating tourists, and fashioned unique gifts and awards given locally. Showy gold nugget watch chains with distinctive fobs, and tie pins were favourites with most men. Necklaces, bracelets, and brooches featuring gold from specific creeks were highly prized by the ladies.[1361]

Moosehide moccasins made by the Tlingit women, rich with the smell of wood smoke, and bright with intricate beading, were striking souvenirs.

Luxurious furs from Clarence Sands' store were fashioned into neck pieces, and worn as elegant reminders of a summer trip to Atlin.

The thriving tourist economy generated other enterprises. Louis Schulz began operating several taxis in conjunction with his Royal Hotel. A fleet of polished McLaughlan Buick touring cars, owned by the Roxborough Brothers, William and Robert, stood ready at the Atlin Inn to take guests on long breezy rides to Spruce Creek, Discovery, Surprise Lake, and the placer mines. If a clean up was in progress, tourists saw gold lying in thick gleaming patterns against sluice box riffles. Visitors sometimes experienced a thrill closely related to gold fever when they had the opportunity to hold the heavy nuggets, and gasp at the estimated value. The miner would elevate the excitement by boasting about the rich potential of his claims, and his great prospects for "next year."

Visitors also ventured the 25 km south to the Warm Springs. Over a large area near the lake, 20° C. thermal water seeped from subterranean springs. Some was captured in a small pool in a clearing of meadow and garden beside a two-story log house. Here visitors were received and entertained by Mr. and Mrs. Bruce Morton, who served refreshments.[1362]

The opportunity to explore an ancient, yet active, ice mass was among the many choices for adventure offered an Atlin visitor. And in the summer of 1919, a spectacular feature drew international attention, and throngs of curious tourists, who travelled to Llewellyn Glacier aboard the *Tarahne*. During the previous winter, movement and pressure had formed a massive arch of ice, a solid bridge, that stood over 9 metres high. It formed a magnificent bow, and was so strong that it easily supported all who wanted to walk across or pose at the very top for pictures. The gigantic ice bridge was named the King's Triumphal Arch, but sadly, its reign was short. By 1920, after a winter of grinding glacial activity, the phenomenon had all but disappeared, and only the abutments remained.[1364]

Still, the glacier's irresistible allure was present, and a trip to Llewellyn, that also had a dash of romantic appeal, was the midnight cruise of a small gas launched named the *Atlinto*. It took eager explorers to the glacier long after the famous arch was gone.

Sailing time was 10 p.m., and with red ensign flying, the *Atlinto* cruised through the bewitching hours of a northern summer's soft half light. At 1 a.m., passengers disembarked at a camp in Glacier Bay. [1363] A large tent, furnished with a few cots for guests overcome by fresh air and the late hour, and some picnic tables stood ready on the beach.

AHSP-2238

*The Mountain King's Triumphal Arch was a magnificent natural attraction for tourists visiting Llewellyn Glacier in the summer of 1919.*

*Morris Taylor, photographer and ardent outdoorsman, hiked glacier ice with his camera and his faithful pal, Prince.*

AHSP-1275

After coffee and sandwiches, all interested hikers, usually from 10 to 18 people, took the trail to the glacier. Stewart James and Mickey Eggert, led the way over the moraine misty with tiny wild flowers, and through ice cold streams where the ladies were often carried. As the sun rose over the mountains, they arrived at the broken face of the glacier, and the lake thick with icebergs.

After exploring the mysteries of the ice, the hikers returned to the beach camp at about 7 a.m. to tackle plates of sizzling bacon and eggs with the appetites of miners. The *Atlinto* returned to port through First Narrows and down Torres Channel. She set a smart pace of about 12 knots, so with time allowed for stops, she arrived back at the Atlin dock about noon. It was a memorable 14 hour tour, and well worth the $10 fare.[1365]

The *Atlinto* was launched in 1911 by Jules Eggert who had spent many hundreds of hours building her over a period of two years. Her performance was, at first, disappointing. The engine persistently overheated, and no amount of tinkering could solve the problem. There were no other gas boats in the area, so there was a lack of local expertise to tap for a remedy. Eggert's project was finally abandoned, and for years the pretty little vessel sat on the shore. It was not until after Jules Eggert died in 1922, that his son, Paul and his grandson, Mickey were able to solve the engine problems and launch their glacier tour business.

John Noland bought the *Atlinto* in 1927, and for a number of years he offered similar midnight excursions.[1366] The boat also became part of his hunting outfit, used to carry hunters and their equipment.

In the years during and immediately after World War I, when L. C. Read was a robust, white-haired gentleman in his 70's, he led ambitious hiking expeditions into the Llewellyn Glacier area. His clients booked trips 6 to 10 days in length, at a cost of $6 per day per person. They stayed at Read's well-appointed camp at Glacier Bay, and from that point they explored not only the glacier and its treasures, but also surrounding mountains, and lakes to the south.[1367] Sloko Lake, glacial-fed, with remarkably milky waters was a favourite destination for the veteran naturalist. Read was a dedicated outdoorsman who had explored the area with the curiosity of a scientist. He had an impressive knowledge of the flora and fauna, the intimidating character of the glacier, and the geological history of the Atlin area.[1368]

Read was more than a naturalist, he was a professional photographer, who kept a studio on Pearl Avenue across from the Royal Hotel. Dedication to his craft, coupled with his enthusiasm for hiking the back country, resulted in a pictorial record of places rarely seen before the advent of the airplane. His fine photographs were widely distributed and often used for promotional purposes.[1369]

Following the footsteps of L. C. Read was a young photographer named Morris Taylor, who shared Read's fascination with the glacier. Taylor also had a photo shop on Pearl Avenue. The importance of lighting to black and white photography was always on Taylor's mind. It was not unusual for him to abruptly leave a social gathering, after a quick scan of the sky and his watch. Gathering his camera and his gear, he was on the run to some precarious vantage point to catch the instant of perfect lighting. Taylor's constant companion on hikes along dangerous glacial crevasses or on rough mountain slopes was a big black and white dog he called Prince. When tourists purchased Taylor's scenic photographs, Prince was often part of the composition.

Closely connected with the tourist industry in Atlin is the romantic story of Ben-My-Chree. Tourists visited this idyllic spot aboard the *S. S. Tutshi*, as a side trip, from Carcross, Yukon, but Ben-My-Chree was in British Columbia, and had close associations with the people and spirit of Atlin. Tucked up against a mountain at the southwest tip of Taku Arm, Ben-My-Chree was the wilderness home of an English couple, Otto and Kate Partridge. Otto had grown up on the Isle of Man, and the name Ben-My-Chree, Manx for "Girl of My Heart," was a tribute to his wife.

The *Tutshi* made twice weekly trips to the Partridge's retreat, usually carrying more than 100 passengers. The cruise down Tagish Lake was spectacular, but in no way prepared the voyagers for the sight of Ben-My-Chree. The boat nosed to a stop at a steep shoreline where tourists were greeted by Otto Partridge, silver-haired and gentlemanly. They followed him along catwalks and paths to the garden gate where Mrs Partridge, quaint in lace collar and dark dress, welcomed them to the beautiful grounds of their home. The paradox of riotous English gardens and tidy greenhouses against a setting of mountain spires and glaciers, entranced visitors. That flowers could grow in such profusion in a place deeply frozen for nearly 8 months of the year, was believable only in observing, but the truth of it created the magic of Ben-My-Chree. At the Partridges' log house, Mrs Partridge and her companion, Miss Emily Dalton, served tea, rhubarb wine, and cakes. Then Mrs. Partridge would seat herself at the harmonium to lead a sing song. Out of respect for the predominant nationalities present it was traditional to sing both "God Save the King" and the American national anthem. After a pleasant hour or two the guests would return to the *Tutshi* for a wistful return to Carcross, deeply touched by their experience.[1370]

Otto Partridge was a gold rush stampeder of 1898, who slipped by Soapy Smith's gang in Skagway, but stopped short on his way to the Klondike. After a brief time working for the Bennett Lake and Klondyke Navigation Company, he took up mining on Tagish Lake. His long association with White Pass tourism ended with his death in 1930. His wife survived him by just 5 months. White Pass took over Ben-My-Chree and operated tours until 1955. Each spring, before break up, a staff of three was flown to the homestead. They planted gardens and greenhouses, and brewed the famous Ben-My-Chree wine before the first visitor stepped ashore.[1371]

Tourism became a force in the Atlin economy as soon as White Pass had it firmly in hand. Five new guest rooms were added to the original 35 at the Atlin Inn the year after it opened.[1372] By 1926, tourism was entering its peak years, and White Pass found more expansion was necessary. A third wing with 18 more guest rooms, and five staff rooms was added to the Atlin Inn.[1373]

At the close of navigation in 1927, the *Tarahne* was hauled up on the beach in Atlin, instead of being left to freeze in the ice at Safety Cove, as usual. Drastic alterations were planned for the boat, and on April Fools Day, 1928, a crowd gathered to see the *Tarahne* being cut into two separate pieces.[1374] What appeared to be a bad joke was the first step in a major overhaul. Carpenters added a 9 metre extension amidships. The *M. V. Tarahne* had a new overall length of 36.4 metres. Larger engines and new propellers were installed to increase her speed to 12 knots per hour.

White Pass understood the nature of its clientele, and provided a high standard of accommodation and transportation. There is no doubt the northern tours were geared primarily to wealthier travellers. Atlin tourists included politicians, figures from the entertainment world, industrialists, British aristocracy, and, in 1922, the Governor General, Lord & Lady Byng of Vimy.[1375]

The stock market crash of 1929 had an effect on northern tourism, but the impact was slow in coming. Seven years passed before White Pass cut back their operations. When coastal steamer companies reduced the number of ships on the Alaska run, train schedules were changed, and it became impossible to make suitable connections with tourists at Carcross for the side trip to Atlin. Citizens watched in bitter resignation as the company withdrew, after selling off furnishings from the Atlin Inn, and boarding up the windows.[1376] They particularly resented the fact that the Ben-My-Chree trips continued, as it turned out, for nearly 20 more years.

The future of the *Tarahne* was not as clear. Gold mining gained momentum when a higher gold price was set in Washington D.C. Men were clamouring for jobs, and they found them on the Atlin creeks. There was considerable traffic in and out of the community, and residents hoped the *Tarahne* would continue to run. When the ice went out in the spring of 1937 the vessel remained on the ways at the foot of Trainor Street, and a smaller vessel, the *Norgold*, was brought in to take her place.

Tourism was gone as a major component in the Atlin economy, and summers would never be quite the same again. The *Tarahne* grew gray on the beach, and some of the wharves buckled and sagged. The Atlin Inn was silent. The busy taxis disappeared. On the creeks, though, gold mining regained a good share of its former importance, and in Atlin no one felt the real misery and heartbreak of the Depression. There was no unemployment, there were no hungry families, and none of the despair which was prevalent in southern communities. Again chaotic change that spelled crisis for most of the world, came with the opposite effect on Atlin. Miners replaced the vanished tourists; Atlin survived, and all the natural attractions remained for visitors of another era.

AHSP-43

*At Scotia Bay, travellers leaving Atlin Lake disembarked from the M.V. Tarahne and boarded coach cars for the short rail trip to Taku Landing.*

# Epilogue

The Second World War brought an end to the gold mining that flourished in Atlin during the Depression. Valleys scarred by the large mining concerns began to heal as only a few small operators worked each summer. Jobs became scarce and young adults left Atlin to find work elsewhere. The population fell off drastically, and by the early 1960's only about 100 citizens remained in a subdued town. Yet, when it seemed Atlin might be teetering on the brink of extinction, the philosophy of the sixties youth, the abandonment of materialism, and the quest for a simple life, seeped into the thinking of the over-thirty crowd as well. Young families pulled up stakes, loaded possessions and struck out to find a different way to live. Peace, quiet, and freedom from rat race anxiety were the idyllic goals of the legions who fled the cities. Those who found Atlin, were sure they had stumbled upon the place of their foggy dreams, and many stayed.

During the school term of 1966-67 there were only eight children in the Atlin School and one teacher, but the next year there were two teachers and twenty-nine children. Soon a new school was needed, and, not many years later, another. The town, still referred to by old timers as the gold camp, was experiencing a rebirth. It was not a resurrection from ashes of disaster, but instead a slow steady growth due to new arrivals who, for the most part, created their own jobs, and revelled in a complete departure from their former lives. The population slowly climbed to a count of five hundred spirited, busy, mostly new, citizens, who were providing services and amenities which had for a time disappeared or never existed; a new Atlin Inn, attractive bed and breakfasts, a second grocery store, an expanded general store where Clarence Sands once sold boots and shovels. A modern gas station replaced the old pump with the glass top and a handle. In time there was also an efficient library, fire department and ambulance corps. Streets were lighted and surfaced.

As during the gold rush, the work people did usually bore little relation to their previous occupations, but a strong undercurrent of vitality was again detectable. Mining continued to play a part in the economy as gold prices rose. Tourists began to arrive in growing numbers. The Atlin story continued to unfold. And present, as they had always been while Atlin's fortunes ebbed and flowed, were the magnificent lake and the mountains at its shoreline.

# Glossary

| | |
|---|---|
| BEDROCK | The impenetrable layer of rock on which gold settles to form a pay streak. |
| BENCH CLAIMS | Are in the area beyond the banks of a creek and away from the gold bearing creek itself. Sometimes bench areas are the sites of ancient creek channels and may be rich with gold. |
| DERRICKS | Are high booms, hinged at the base. They may be rigged with cables which pull and maneuver a large bucket used to scoop up and deposit earth. This would be called a dragline. |
| DRAGLINE | This is also part of a dredge. It is the cable on which the dredge buckets move when it is operating. |
| DREDGES | These are machines which operate on floating barges. They scoop up gravels with a "ladder" of buckets conveyed by a dragline, thus moving great amounts of earth quickly. The gold is separated from the gravel within the dredge complex. Dredges are self-contained plants which perform several functions. They are powered electrically. |
| DRIFTING | Means tunneling horizontally into a creek bank, hillside, or mountain to find a pay streak or to follow a gold bearing lead. |
| FLUMES | These are ditches or wooden trough-like structures used to conduct water from a source, such as a creek or lake, to a placer mining site. |
| HYDRAULIC | This type of mining is done by using water under great pressure to cut into banks and wash away overburden from bedrock. |

MONITORS These are the nozzles used to direct water for hydraulic mining.

OVERBURDEN This is the ground covering bedrock, on or near which gold is found. Often gold in small amounts is recovered from this gravel covering, but the gold is usually concentrated at bedrock.

PLACER This is a Spanish word meaning marine plane or sandy bank. Placer mining generally means mining for gold deposits mixed with sand or gravel in creek beds.

RIFFLES These are the baffles set in sluice boxes to catch the gold, while other lighter weight gravel and rock wash through the boxes. They are usually made of wood, but sometimes steel is used. Manganese steel plates once used by some Atlin miner were much more durable than wooden riffles and they were easily cleaned.

SLUICE BOXES These are shallow troughs often made of wood, but sometimes of steel. Riffles are laid in the bottom of a sluice and ore washed through with fast moving water. Gold, because of its high specific gravity, sinks to the bottom and catches in the riffles. After a period of sluicing, the gold is recovered from around the riffles.

TAILINGS These are the gravel and boulders left after gold has been extracted in a mining operation. Since the ground is thoroughly loosened and churned at a mining site, the residue from placer mining is far moe voluminous than was originally packed at the site. Finding room to deposit tailings in narrow creek valleys was a serious problem for miners.

TROY WEIGHT

| | | |
|---|---|---|
| 24 grains | = | 1 pennyweight (dwt) |
| 20 dwt | = | 1 Troy ounce |
| 12 Troy ounces | = | 1 Troy pound |

# References

## UNPUBLISHED SOURCES

Abbott, Doris Nichols, MS, A. H. S.

Allard, J., *Atlin Mission's History Resumed*, MS, Oblate Archives, Ottawa.

Ashby, Gwynneth, Letter, MS, A. H. S.

Bilsland, W. W., *Additional MS after 1910*, copy, A. H. S.

Bourne, Clayton, MS, A. H. S.

Bruce, Charles, MS, A. H. S.

Buckpitt, Matthew, MS, G. Smedley Andrews private collection.

Burke, Brian, MS, copy, A. H. S.

Burry, Ralph, R. *Louis Schulz and His Tokens of Atlin, B.C.*, nd., MS, A. H. S.

Carmichael, Alfred, *Diary*, MS, P. A. B. C.

Cartmel, J. E., Letter to Christine Dickinson, 1977, A. H. S.

Connolly, Hugh, MS, A. H. S.

de Armond, R. MS, A. H. S.

Ely, Charles, Letter to Diane Smith, 1 Dec., 1994, A. H. S.

Gairns, Jean, MS, A. H. S.

Godfrey, Alex, MS, Vancouver City Archives.

Hartshorn, Hazel, MS, A. H. S.

Hicks, Roger Hayes, *Diary*, MS, Alberta Glenbow Institute.

Horral, S. W., MS, A. H. S.

James, T. MS, Bourne Collection, A. H. S.

Jardine, Hugh, *Diary*, MS, A. H. S.

Kelly, Pat, Letter, Jan. 1983, MS, A. H. S.

Kendall, *Cattle Drive to Alaska, 1896-1898*, MS, P. A. B. C.

Kipling, Rudyard T., MS, A. H. S.

Kruty, Paul, Letter to A. H. S., 1982, MS, A. H. S.

Legat, Allice, MA Thesis, University of Alberta.

Lupecke, Helen, Letters to Johann, Dec. 1900 - Feb. 1901, copies, A. H. S.

Mackenrot, Dorothy A., Trek of W. J. Smith and family from Inverness, Scotland to Atlin, B. C., MS, A. H. S., Letter, 2 April, 1979, MS, A. H. S.

McCandless, R., *Yukon Wildlife: A Social History*, 1900-1950, unpublished manuscript, Yukon Archives.

McKinnon, Ruth, *Notes on the D. H. McDonald Family*, MS, A. H. S.

McLaren, Leola, Letter, A. H. S.

Murphy, Marie, *Diaries 1924-1958*, MS, A. H. S.

O'Bannon, Marcella, Letter, 27 Feb. 1978, A. H. S.

Olive, W. H. T., *Trail of '98: Memoirs*, MS, P. A. B. C.

Reed, Thomas Harper, *A History of the Engineer Mine*, MS A. H. S.

Royal Canadian Mounted Police, Atlin Detachment, Inspection Report, 12 July, 1921, MS, A. H. S.

Ryder, Mrs. Irving, *Diary*, copy A. H. S.

St. Martin's Anglican Church Records, MS, A. H. S.

Sproule, A. F. *The Role of Patrick Burns in the Development of Western Canada*, MA Thesis, University of Alberta, 1962.

Stephenson, F. L., MS, A. H. S.

Tapp, Lila, Letter to Sue Morhun, 6 Sept.1975, A. H. S.

Tervo, *Diary*, MS, P. A. B. C.

Walker, Carrie, Letter to Hon. Richard McBride, 21 Dec. 1903, Lottie Bowron Papers, MS, P. A. B. C.

Ward, James, The Tlingit Tribe History, Part 1, 1974, MS, Yukon Archives.

Watson, Martha Grace, *Diary*, MS, Yukon Archives.

White Pass Records, River Division, Yukon Archives.

## INTERVIEWS

Acheson, Winnie, with Christine Dickinson, Atlin, c. 1984.

Bennett, Charles, with Christine Dickinson, Atlin, B.C. circa 1980.

Brooks, Reginald, with Sue Morhun, Atlin, B.C, 1975.

Coleman, Tom and Irene, with Sue Morhun, Atlin, 1975.

Eggert, E.J. "Mickey", with Sue Morhun, Sechelt, B.C., 10 Nov. 1978

Fisher, Norman, with Sue Morhun & Shirley Connolly, Atlin, B.C., Oct., 1974, Jan. 1975 & March, 1975.

Hensley, Bill, with Sue Morhun, Aug., 1976.

Husselbee, Wm, with Sue Morhun at Logger Bay, B.C., 25 Mar. 1975.

Jack, John Taku, with Christine Dickinson, Atlin, Jan., 1978.

James, Cyril, with Sue Morhun & Shirley Connolly, Atlin, B.C., Jan. 1975 & March, 1975.

James, Cyril, with Shirley Connolly & Nora Smith, Atlin, 31 Jan, 1973.

MacDonald, Mabel, with Christine Dickinson, Atlin, July, 1981.

McKinnon, J. S.,

Nelson, Bill, with Christine Dickinson, Atlin, 11 Dec., 1986 & 15 April, 1987.

Nelson, Bill, with Diane Smith, Atlin, 1 May, 1984.

Owen, Walter, with Sue Morhun, Victoria, 12 Feb. 1975.

Roxborough, Jack

## GOVERNMENT DOCUMENTS

Aitken Report, c1950, Gold Commissioner's Office, Atlin.

Bennett-Atlin Commission of Inquiry, 1899, Miscellaneous Exhibits and Shorthand Notes of Cases, MS. P. A. B. C.

Black, J. M., *Report on the Atlin Placer Mining Camp*, Ministry of Mines and Petroleum Resources, 1953.

British Columbia Department of Lands and Works, Correspondence. P. A. B. C.

British Columbia Ministry of Lands Records. Atlin Hospital Lot File, copy, A. H. S.

British Columbia Gazette, Victoria, Dec. 1898.

British Columbia Journals, Victoria, 1902.

British Columbia Provincial Police, Correspondance Inward 1891-1910, P. A. B. C.

British Columbia Sessional Papers, Victoria, 1899-1942.

Cairnes, D. D., *Report on Portions of the Atlin District With Special Reference to Lode Mining*, Geological Survey of Canada, Memoir #37, 1913..

Complaints Against Officials, Atlin District, British Columbia, *Sessional Papers*, Ottawa, 1900, p.643.

Correspondence of the Government Agent, Atlin, P.A.B.C.

Gauvreau, Narcisse B., Exploration of the North-western Portion of the Province, British Columbia, *Crown Land Surveys for the Year Ending 31st December, 1892*, Victoria, 1893.

Government Records (GR) Series 440, P. A. B. C.

Graham, J. D., Gold Commissioner's Report, British Columbia *Sessional Papers*, Victoria, 1900.

McKenna-McBride Commission, Indian Affairs Records, GR 1342, P. A. B. C.

Minister of Mines Annual Reports, Victoria, 1898 - 1950.

Privy Council Records, R.G. 2, Series 3, Vol. 128, File 2800.

Proudlock, P. J. & W. J., *Stratigraphy of the Placers in Atlin Placer Mining Camp*, Victoria; Ministry of Mines and Resources, 1976.

Shipping Registers, various, A. H. S.

Snyder, Supt., Annual Report, Sessional Papers, Ottawa, 1906.

Statutes of British Columbia, Victoria, 1899, 1900, 1903, 1904.

Steele, Supt. S. B., Annual Report, Sessional Papers, Ottawa, 1898.

Wood, Supt. Z. T., Annual Report, Sessional Papers, Ottawa, 1898.

NEWSPAPERS AND MAGAZINES

*Alaska and Northwest Quarterly*, 1899

*Alaska Journal*, 1975-1979

*Alaska Mining Record*, 1898-1900

*Alaska Miner*, 1898

*Alaska Record*, 1898

*Alaska Searchlight*, 1896.

*Alaska Sportsman*, 1958-1960

*Atlin Claim*, April 1899 - Oct. 1907.

*Atlin Globe*, 28 Sept. 1899, 19 Oct., 1899.

*Atlin News Miner*, 1938-1976

*Atlin Nugget*, 1936.

*Bennett Sun*, 1899.

*British Columbia Mining Exchange*, 1904-05

*British Columbia Review of London*, 1903.

*Canadian Churchman*, 1941.

*Dawson Daily News*, 1899-1900

*Dawson Weekly News*, 1907-

*Douglas Island News*, 1898 -1900.

*Dyea Trail Special Edition*, Aug. 1898

*Glenora News*, Sept. 1898

*Revelstoke Kootenay Mail*, 1897-1902

*Skagway News*, 1898-1899

*Vancouver Province*, 1898 - 1930

*Vancouver Sun Magazine*, 13 Jan. 1951.

*Victoria Colonist*, Aug, 1898-1962.

*Western Miner*, 1946

*Whitehorse Star*, 1914-1978

*Yukon Midnight Sun*, 15 Aug. 1898

PUBLISHED SOURCES

Akrigg, G. P. V. & Helen B., *1001 British Columbia Place Names*, Vancouver, 1969.

Albee, Bill & Ruth, *Alaska Challenge*, New York; Dodd, Mead & Co., 1943.

Allard, J., "Rapport sur La Mission d'Atlin," *Missions de la Congregation des Missionaires Oblats de Marie Immaculee*, Vol 46.

Anzer, Richard, *Klondike Gold Rush*, New York, 1959.

Atlin District Board of Trade, *The Atlin Gold Fields*, Pamphlet, n.d., P. A. B. C.

Atwood, Fred N. *Alaska-Yukon Gold Book*, Seattle, 1930.

Bancroft, H. H., *History of British Columbia, 1792-1887*, San Francisco, 1887.

Barkdull, Calvin, "I Saw Soapy Smith Killed," *Alaska Sportsman*, June, 1952.

Barlee, N. L., "Historic Atlin," *Canada West*, Vol. III, No. 4, Winter, 1974.

Berton, Pierre, *The Great Railroad*, Toronto; McClelland & Stewart, 1972.

, *Klondike*, Toronto; McClelland & Stewart, 1972.

Bilsland, W. W., *Atlin 1898-1910*, Atlin Centennial Committee, 1971.

Bock, Billy, *The Book of Skeletons*, c 1960.

Bordynuk, Ernest, "The Mail Driver," The *Yukon News*, June-July 1967.

Brooks, Alfred Hulse, "Blazing Alaska's Trails," *The Century Magazine*, March, 1898.

Caldwell, Francis E., *Land of the Ocean Mists*, Edmonds, WA; Northwest, 1986.

Cameron, Charlotte, *A Cheechako in Alaska and Yukon*

Carter, Margaret, "Bennett - Town or Illusion," *Alaska Journal*, Vol. 8, No. 1, 1978.

Clifford, Howard, *Rails North*, Seattle; Superior.

Cole, Dermot, Frank Barr, Edmonds, WA.; Alaska Northwest, 1986.

Cross, W. K. et al. (eds.), *British Columbia Land Bulletin Area Source Book*, Victoria, 1966, p. 68.

Curtis, Edward S., "The Rush to the Kondike Over the Mountain Passes" *Century Magazine*, March, 1898.

Davidson, J. N. *The Arctic Brotherhood*, Seattle, 1909.

Dawson, G. M., *Report on an Exploration in the Yukon District, N. W. T., and Adjacent Portion of British Columbia 1887*, Montreal, 1887.

DeArmond, R. N. "Down Through the Years," *Alaska Journal*, Spring, 1979.

Edwards, William Seymour, *In to the Yukon*, Cincinnati; Robert Clarke, 1904.

Ellis, Frank H., "Ordeal on the Liard," Art Downs, (ed.), *Pioneer Days in British Columbia*, Vancouver, 1975.

"Yukon's Flying Heritage," *Whitehorse Star*, 11 Aug. 1969.

Emmons, George, *The Tlingit Indian*, edited with additions by Frederick de Laguna, Seattle; University of Washington, 1991.

Freemasons, British Columbia, *History of the Grand Lodge of British Columbia, 1871-1970.* Victoria; Colonist Printers, 1971.

Garland, Hamlin, *The Trail of the Goldseekers,* New York, 1899.

George, Melvin, *Post Offices of British Columbia,* Vancouver, 1972.

Gray, L. H., "Atlin Goldfields and the White Pass Railway," *Alaska and Northwest Quarterly,* Vol. 1, No 1 (Jan. 1899).

Hamilton, Lord Ernest, *Forty Years On,* New York, n.d.

Hamilton, W. R., *The Yukon Story,* Vancouver; Mitchell, 1972.

Harbottle, Jean, "Clyde Wann, Father of Yukon Aviation," *Alaska Journal,* Autumn, 1973.

Harrington, Reba, *Cinderella Takes a Holiday in Northland*

Harvey, Judge, J. T. "Meditations in the Atlin Vault," *The Advocate,* Vol.22, Part 4, 1964.

Hibben's B. C. Directory, 1877-78, Vancouver, 1878.

Hitchcock, Mary E., *Two Women in the Klondike,* New York; G. P. Putnam, 1899.

Howay, F. W., *British Columbia From the Earliest Times To the Present,* Vol III, Vancouver, 1914.

Ince, Clifford H., *Royal Bank of Canada. A Chronology: 1864-1969.*

Johnson, Kirke E., "Skagway-Atlin Letters," *Oregon Historical Quarterly,* Vol. 81, No 2 (Summer, 1980).

Krause, Aurel, *The Tlingit Indians,* Seattle; University of Washington, 1956.

Lawrence, Guy, "Atlin Miners Ignored Riot Act," *Vancouver Sun Magazine,* Supplement, 31 July 1948.

"The Atlin Placer Camp," *Cariboo and Northwest Digest,* Dec. 1950, p. 38.

*Forty Years on the Yukon Telegraph,* Vancouver; Mitchell, 1965.

"The Sluice Box Robbery," *The Shoulder Strap,* June 1950.

"The Stikine and Teslin Trails," Parts 1-4, *Alaska Sportsman,* March, 1943.

"The Stikine Trail," Part 3, *Alaska Sportsman,* July, 1957.

"When the Riot Act Was Read in Atlin," *The Shoulder Strap,* Dec. 1949, pp. 12-13.

"The Yukon Telegraph Service," Part I-IV, *Alaska Sportsman,* Oct, 1958-Jan. 1959.

Lee, Norman, *Klondike Cattle Drive,* Vancouver; Mitchell, 1960.

Le Fevre, H. B., "A Quartz Discovery," *Alaskan Magazine,* Vol 1, No. 1, 1899, p. 19-23.

Leinweber, Martin, "Stumbling Stampeders," *Alaska Sportsman,* March, 1944.

London, Jack, Letter to the *Seattle Post Intelligencer,* 2 Aug. 1906.

M. S. P., *The Early History of St. Andrew's Hospital, Atlin, B.C.,* Woman's Homen's Home Missionary Society of the Presbyterian Church in Canada, 1907.

McBride, W. S., "Railroad Built to Serve Miners," *Whitehorse Star,* 9 Aug. 1956.

"Saga of the River Boats", *Caribou and Northwest Digest,* 1948-49.

"The Taku Tram," *Cariboo and Northwest Digest,* April 1950.

"Trails of Ninety-Eight," *Farthest North Collegian,* 8 Oct. 1947.

McClellan, Catherine, "The Inland Tlingit," *American Antiquity,* Vol. 8, No. 3, Pt. 2 Jan., 1953.

Mills, Thora McIlroy, "Bala Sent Nurses to the Klondike Too!" *Bracebridge Herald-Gazette,* 27 Jan. 1972.

*The Church and the Klondike,* Toronto, 1977

"The Hanna Family in the Early Life of Muskoka," Parts 4 and 5, The *Muskoka Sun,* 25 July & 1 Aug. 1974

Minter, Roy, *The White Pass: Gateway to the Klondike,* University of Alaska Press, 1987.

Morritt, Hope, *Land of the Fireweed,* Edmonds, WA; Alaska Northwest, 1985.

Ogilvie, William, *Early Days on the Yukon,* Ottawa; Thorburn & Abbott, 1913.

Okun, S. B. The Russian-American Company, Mass; Cambridge, 1951.

Oliphant, John, "The Goldtimers", *Vancouver Sun Magazine*, 14 Feb. 1979.

Ormsby, M, *British Columbia: A History*, Vancouver; MacMillan, 1958.

*Pioneer Breweries of British Columbia*, Nanaimo, 1974.

Presbyterian Church of Canada, *Women's Home Missionary Society: Our Jubilee Story*, 1924.

Pugsley, Ed E., "The Duchess Retires," *Family Herald*, 16 Jan. 1899.

Richeson, David R., "The Yukon Telegraph," *British Columbia Historical News*, Winter, 1982.

Ridley, W., "New Mission to Miners on Atlin and Bennett Goldfields," Parts I & II, *The Gospel Missionary*, Nov-Dec 1899.

*Snapshots of the North Pacific*, London; C. M. S., 1904.

Roppell, Patricia, "Porcupine," *The Alaska Journal*, Vol. 5, No 1, Winter, 1975.

St. Andrew's Session, *In Memoriam, Rev. John Pringle*, Sydney, N.S.

Schwatka, Frederick, *A Summer in Alaska*, St. Louis, MO; J. W. Henry, 1894.

Scholefield, E. O. S. & F. W. Howay, *British Columbia From the Earliest Times to the Present*, Vol II, Vancouver, 1914.

Sinclair, James M., *Mission: Klondike*, Vancouver; Mitchell, 1978.

Stephenson, F. L., "Atlin Mining Camp," *Vancouver Star*, 26 Aug. 1931.

"The Church in the Frozen North," *The Mission Field*, 1 May, 1901.

Swarth, Harry S., "Mammals of the Atlin Region, Northwestern British Columbia," *Journal of Mammalogy* #17, 1936

Tuomaala, Eileen, "Warm Memories of Icy Atlin Winters", *White Rock Sun*, 6 Jan. 1972.

"Veteran of the Force," *Shoulder Strap*, Oct. 1936

Walker, F. T., *Reminiscences of F. T. Walker, 1899-1912.*

Warner, Iris, "Taylor & Drury Ltd.", *Alaska Journal*, Spring 1975.

Wells, Hulet M., "Backdoor to the Klondike," Parts 1-6, *Alaska Magazine*, April-Sept. 1960.

Wheeler, Keith, *The Old West - The Railroaders*, New York; Time-Life, 1973.

White Pass Co., Atlin the Beautiful, Pamphlet, nd. *A Handbook of Vacation Trips in Alaska, Atlin and the Yukon*, Pamphlet, nd, A. H. S.

*The Story of Ben-My-Chree*, n.d.

Wilcoxen, I. N., *History of the Arctic Brotherhood*, 1906.

Wilkes, L. C., "Packers on the Dyea Trail," *Alaska Sportsman*, June 1948.

Williams, Bryan, *Game Trails in B. C.*, London; John Murray, 1925.

Williams, Glyndwr (ed.) *London Correspondance Inward from George Simpson, 1841-1842*, London; Hudson Bay Company Record Society, 1973.

Williams, Mostyn, "Recent Operations in the Atlin District," *The Mining Record*, Vol 9, No. 2, Feb. 1902.

Wright, Allen, *Prelude to Bonanza*, Victoria; Gray, 1976.

Young, Rosalind Watson, "Hydraulic and Placer Mining in the Atlin District," *Mining and Metallurgy*, 1 Oct.,1901.

"Mining and Smelting in the Province", in Flumerfelt, Alfred (ed.), *Potential Riches of British Columbia*, Vancouver, 1907.

"Mining Operations in Atlin, B. C.," Parts I & II, *Mines and Minerals*, 1 Dec. 1901, & Jan. 1902.

# Notes

## Chapter 1

[1] Emmons, George, The Tlingit Indian, edited with additions by Frederick de Laguna, Seattle; University of Washington, 1991, p. 7.

[2] McClellan, Catherine, "The Inland Tlingit," American Antiquty, Vol. 8, No. 3, Pt. 2, Jan. 1953, pp. 47-51.

[3] McClellan, The Inland Tlingit.

[4] Krause, Aurel, The Tlingit Indians, Seattle; University of Washington, 1956, p. 53.

[5] Ward, James, The Tlingit Tribe History, Part I, 1974, MS, Yukon Archives.

[6] Krause, The Tlingit Indians, p. 69.

[7] McClellan, The Inland Tlingit.

[8] Krause, The Tlingit Indians, p. 136.

[9] McClellan, The Inland Tlingit.

[10] Bancroft, H. H., History of British Columbia, 1792-1887, San Francisco, 1887, p.71.

[11] Williams, Glyndwr (ed.) London Correspondence Inward from George Simpson, 1841-1842. London; Hudson Bay Company Record Society, 1973. p. xxxii.

[12] Emmons, The Tlingit Indian, p.7.

[13] McClellan, The Inland Tlingit.

[14] McCandless, R., Yukon Wildlife: A social history, 1900-1950, unpublished MS, Yukon Archives.

[15] Jack, John Taku. Interview with Christine Dickinson, Jan., 1978.

[16] Gauvreau, Narcisse B., Exploration of the North-Western Portion of the Province, British Columbia, Crown Land Surveys for the Year Ending 31st December, 1892, Victoria, 1893, p. 486.

[17] Albee, Bill & Ruth, Alaska Challenge, New York; Dodd, Mead & Co., 1943, p. 127.

[18] Wright, Allen, Prelude to Bonanza, Victoria; Gray, 1976, p. 93.

[19] Scholefield, E. O. S. & F. W. Howay, British Columbia From the Earliest Times to the Present, Vol II, Vancouver 1914, p. 264

[20] Akrigg, G.P.V. & Helen S., 1001 British Columbia Place Names, Vancouver 1969

[21] Schwatka, Frederick, A Summer in Alaska. St. Louis, Mo; J. W. Henry, pp. 117-9 & Dawson, G.M., Report on an Exploration in the Yukon Dist., NWT, and Adjacent Northern Portion of BC 1887, Montreal, 1887.

[22] Bilsland, W.W., Atlin 1998-1910, Atlin Centennial Committee, 1971, p.7.

[23] Brooks, Alfred Hulse. "Blazing Alaska's Trails," The Century Magazine, March, 1898.

[24] Dawson, Report on an Exploration in the Yukon District, 168B-170B.

[25] Schwatka, A Summer in Alaska, pp. 117-18.

[26] Schwatka, A Summer in Alaska, p.9.

[27] Wright, Prelude to Bonanza, pp. 143-48.

[28] Dawson, Report on an Exploration in the Yukon District, p. 1.

[29] Dawson, Report on an Exploration in the Yukon District, p. 168.

[30] Dawson, Report on an Exploration in the Yukon District, p.168.

[31] Gauvreau, Exploration of the North-Western Portion of the Province, p.496.

[32] *Alaska Mining Record*, 21 Dec. 1898.

[33] Okun, S. B., The Russian-American Company, Mass; Cambridge, 1951, pp. 245-8.

[34] de Armond, R. MSS, nd., A. H. S.

[35] Wright, Prelude to Bonanza, pp 122-3.

[36] Hibben's B.C. Directory, 1877-78, Vancouver, 1878, p.92.

[37] *Al?ska Searchlight*, v.2, no. 38, 29 Aug. 1896.

[38] *Alaska Searchlight*, 29 Aug. 1896.

[39] *Alaska Searchlight*, 29 Aug. 1896.

[40] Young, Rosalind Watson, Mining and Smelting in the Province, in Flumerfelt, Alfred (ed.), Potential Riches of B.C., Vancouver, 1907.

# Chapter 2

[41] Bennett-Atlin Commission of Inquiry, 1899, Miscellaneous Exhibits and Shorthand Notes of Cases, MS. P.A.B.C.

[42] *Victoria Colonist*, 14 Aug. 1898.

[43] *Vancouver Sun Magazine*, 13 Jan. 1951.

[44] *Victoria Colonist*, 23 Oct. 1898.

[45] Minister of Mines, Annual Report, 1936, p. B39; Wm. Husselbee, Interview with Sue Morhun, 25 Mar. 1975, A. H. S.; *Alaska Mining Record*, 9 Aug. 1899, & 4 Sept. 1898; *Alaska Miner*, 13 Sept. 1898.

[46] *Alaska Mining Record*, 9 Aug. 1899 & *Alaska Miner*, 17 Sept. 1898.

[47] *Alaska Mining Record*, 4 Sept. 1898.

[48] Olive, W. H. T., Trail of '98: Memoirs, MS, P. A. B. C. pp. 30-31.

[49] *Atlin Claim*, 4 Nov. 1899. This fragment printed in the *Atlin Claim* is all of this document known to exist.

[50] The diary, as reprinted, actually cites 1897 as the year of this trip, but this is possible an error made in transcription. All other accounts name the year as 1898.

[51] Minister of Mines, Annual Report, 1901, p. 6B.

[52] *Atlin Claim*, 4 Nov. 1899. February 1st is the date given by C. D. Knapp in *Alaska Miner*, 17 Sept. 1898.

[53] *Victoria Colonist*, 28 Sept. 1899.

[54] *Atlin Claim*, 3 Sept. 1904.

[55] McLaren, Leola, Letter, A. H. S.

[56] McKinnon, J. S., Interview with Sue Morhun at Juneau, June 1976.

[57] *Victoria Colonist*, 28 Sept. 1898.

[58] *Atlin Claim*, 17 June, 1899.

[59] *Atlin Claim*, 6 April 1901.

[60] Barkdull, Calvin, "I Saw Soapy Smith Killed" *Alaska Sportsman*, June 1952. Barkdull gives credit for the Pine Creek discovery to Jim Greenslate, Al Short, Fritz Miller and McLean (McLaren?).

[61] *Alaska Mining Record*, 10 Aug. 1898.

[62] *Atlin News Miner*, 17 Dec. 1938.

[63] Knapp, C.D. in *Alaska Miner*, 17 Sept. 1898 also names George Eisenburg, & Mr. & Mrs Al Short as Miller and MacLaren's associates on Pine Creek before their claims were recorded.

[64] *Alaska Miner*, 13 Aug. 1898, & Dyea Trail, Aug. 1898 - Special Edition.

[65] Bennett-Atlin Commission.

[66] *Yukon Midnight Sun*, 15 Aug. 1898, & *Alaska Miner*, 30 Aug. 1898.

[67] Bennett-Atlin Commission.

[68] *Victoria Colonist*, 5 Oct. 1898.

[69] There may be some error in the dates given since Strickland was at Tagish to record Miller's claims on July 30, and by July 31st, or August 1 was returning from Atlin lake after staking his own claim.

[70] Supt. S.B. Steele, Annual Report, Sessional Papers, 1898. p 6.

[71] Rant, W. J., Letter to F. S. Hussey, 10 Aug. 1898, British Columbia Provincial Police, Correspondence Inward, 1891-1910, MS, P. A. B. C.

[72] Rant, W. J. Letter to F. S. Hussey, 10 Aug. 1898, British Columbia Provincial Police. Constable Clifford Little had resigned and left for Pine Creek.

[73] Rant, W. J., Letter to F. S. Hussey, 10 Aug. 1898, British Columbia Provincial Police.

[74] Steele, Annual Report, 1898, p.6.

[75] Olive, Trail of '98, p. 6.

[76] *Victoria Colonist*, 14 Aug. 1898.

[77] *Victoria Colonist*, 14 Aug. 1898.

[78] Johnson Kirke E., "Skagway-Atlin Letters" *Oregon Historical Quarterly* Vol.81, No.2 (Summer, 1980), p.134.

[79] L.H. Gray, "Atlin Goldfields and the White Pass Railway," *Alaska and Northwest Quarterly*, Vol 1. No. 1 (Jan. 1899), p.73.

[80] Minter, Roy, The White Pass: Gateway to the Klondike, Toronto; McClelland and Stewart, 1987, p. 229.

[81] Knapp, C. D. *Alaska Mining Record*, 14 Sept. 1898.

[82] Olive, Trail of '98, p 254-55.

[83] *Glenora News*, 9 Sept. 1898.

[84] Garland, Hamlin, The Trail of the Goldseekers, New York, 1899. p. 217.

[85] Garland, The Trail of the Goldseekers, p.218.

[86] *Victoria Colonist*, 3 Sept. 1898.

[87] *Victoria Colonist*, 3 Sept. 1898.

[88] Mike King of the Victoria and Yukon Trading Company, Bennett City, as quoted in Victoria Colonist, 8 Nov. 1899.

[89] *Victoria Colonist*, 16 Aug. 1898.

[90] *Victoria Colonist*, 16 Aug. 1898.

[91] It is worth noting that the Province's mining laws, the cost of transportation, a shortage of water, and the depth of the bedrock later became major problems for both individual miners and companies.

[92] *Victoria Colonist*, 25 Aug., 31 Aug., 13 Sept. 1898.

[93] *Victoria Colonist*, 23 Oct. 1898.

[94] Atlin City, Pine City, Taku City, Surprise City, and the Blanco Site (at the foot of Atlin Mountain). Victoria Colonist, 23 Oct. 1898, (map) & 5 Oct. 1898.

[95] *Victoria Colonist*, 23 Oct. 1898.

[96] Minister of Mines, Annual Report, 1936, p.985.

[97] Minister of Mines, Annual Report, 1936, p. 227.

[98] Minister of Mines, Annual Report, 1936, p. B85.

[99] Hussellbee, Interview.

# Chapter 3

[100] Leinweber, Martin, "Stumbling Stampeders", *Alaska Sportsman*, March 1944, pp. 11 & 27.

[101] Berton, Pierre, Klondike, Toronto: McClelland & Stewart, 1972, p.333.

[102] Carter, Margaret, "Bennett - Town or Illusion," *Alaska Journal*, Vol. 8, No. 1, 1978, p.56.

[103] Curtis, Edward S., "The Rush to the Klondike Over the Mountain Passes,"*The Century Magazine*, March, 1898.

[104] Wilkes, L. C., "Packers on the Dyea Trail," *Alaska Sportsman*, June 1948 p. 10.

[105] Brooks, Alfred Hulse, "Blazing Alaska's Trails," *The Century Magazine*, March 1898. p 360.

[106] Wilkes, "Packers on the Dyea Trail", p.9. Actual loads added up to some 900 Kg which included food, clothing, tools and other necessities.

[107] Minister of Mines, Annual Report, 1898. p. 990.

[108] Hicks, Roger Hayes, Diary, MS, Alberta Glenbow Institute.

[109] Now known as Graham Inlet.

[110] Hicks, Diary.

[111] Hicks, Diary.

[112] *Atlin Claim*, 3 June 1899.

[113] Minister of Mines, Annual Report, 1898, p.990.

[114] *Victoria Colonist*, 22 March 1899.

[115] Minter, Roy, The White Pass: Gateway to the Klondike, Toronto; McClelland & Stewart, 1987, p.312.

[116] *Atlin Claim*, 12 Aug. 1899.

[117] *Atlin Claim*, 12 Aug. 1899.

[118] *Atlin Globe*, 28 Sept. 1899.

[119] Also known as Taku Landing.

[120] *Atlin Claim*, 12 Aug. 1899.

[121] *Douglas Island News*, 1 July 1899.

[122] Minister of Mines, Annual Report, 1898. p.990.

[123] *Douglas Island News*, 11 Jan. 1899.

[124] *Douglas Island News*, 22 Feb. 1899.

[125] Wright, Prelude to Bonanza, p. 258

[126] Ogilvie, William, Early Days on the Yukon, Ottawa; Thorburn & Abbott, 1913.

[127] Wells, Hulet, "Backdoor to the Klondike," Part I, *Alaska Magazine*, April, 1960. p.13.

[128] Hamilton, W. R., The Yukon Story, Vancouver: Mitchell Press, 1972, p.8.

[129] Tervo, Diary, MS, P. A. B. C.

[130] Lawrence, Guy, "The Stikine and Teslin Trails," *Alaska Sportsman*, March 1943.

[131] Lee, Norman, Klondike Cattle Drive, Vancouver; Mitchell, 1960, p. 24.

[132] Wells, "Back Door to the Klondike," Part 3, p. 29.

[133] Lawrence, "The Stikine and Teslin Trails", Part 1.

[134] Lawrence, "The Stikine and Teslin Trails", Part 1, p. 22, & Leinweber, Stumbling Stampeders, p.11.

[135] Kendall, Cattle Drive to Alaska 1896 - 1898. MS. P. A. B. C.

[136] Lawrence, "The Stikine and Teslin Trails", Part 1.

[137] Wells, "Backdoor to the Klondike", Part I, p.42.

[138] Wells, "Backdoor to the Klondike", Part 4, p.43.

[139] Wells, "Backdoor to the Klondike", Part 4, p. 27.

[140] Guy Lawrence, Forty Years on the Yukon Telegraph, Vancouver: Mitchell, 1965, p.16. In an earlier account (The Stikine and Teslin Trails) Lawrence tells this incident as happening to "one traveller."

[141] Jardine, Hugh, Diary, MS. A. H. S.

[142] Leinweber, "Stumbling Stampeders", p.31.

[143] Lienweber, "Stumbling Stampeders", p. 32.

[144] Drummond. A. D., Letter to F. S. Hussey, nd, British Columbia Provincial Police.

[145] Wells, "Backdoor to the Klondike", Part 5, p.54.

[146] Hamilton, The Yukon Story, p. 23.

[147] Wells, "Backdoor to the Klondike", Part 5, p.54.

[148] Kendall, Cattle Drive to Alaska.

[149] *Atlin Claim*, 12 Aug. 1899.

[150] Hamilton, The Yukon Story, p. 50.

## Chapter 4

[151] *Vancouver Province*, 15 Aug. 1898.

[152] *Victoria Colonist*, 5 Oct. 1898.

[153] *Skagway News*, 23 Sept. 1898.

[154] *Victoria Colonist*, 5 Oct. 1898.

[155] *Victoria Colonist*, 1 Aug. 1898.

[156] Minister of Mines, Annual Report, 1902, p 985. In *Alaska Mining Record*, 25 Jan. 1899, p.1, Rev. G. H. Gurr places the figure at around 450.

[157] *Skagway News*, 12 Jan. 1899.

[158] *Victoria Colonist*, 3 Sept. 1898.

[159] Anzer, Richard, Klondike Gold Rush, New York, 1959, p. 167.

[160] *Skagway News*, 23 Dec. 1899.

[161] "Discovery" acquired acceptance over "Pine City" after the first couple of years.

[162] *Atlin Claim*, 6 May, 1899.

[163] *Atlin Globe*, 19 Oct. 1899.

[164] *Atlin Globe*, 19 Oct. 1899.

[165] *Atlin Claim*, 10 June 1899.

[166] *Atlin Claim*, 6 May 1899 & *Victoria Colonist*, 5 Jan. 1899.

[167] Anzer, Klondike Gold Rush, p.170.

[168] *Victoria Colonist*, 27 May 1899.

[169] Hicks, Diary.

[170] *Atlin Claim*, 20 May 1899.

[171] *Victoria Colonist*, 29 April 1899.

[172] *Atlin Claim*, 24 June 1899.

[173] *Atlin Claim*, 29 July 1899.

[174] Anzer, Klondike Goldrush, p. 168.

[175] *Atlin Claim*, 28 Oct. 1899.

[176] *Atlin Claim*, 13 May 1899.

[177] *Atlin Claim*, 15 July 1899.

[178] *Atlin Claim*, 15 July 1899.

[179] Burns File, A.H.S. Archives.

[180] Sproule, A. F., The Role of Patrick Burns in the Development of Western Canada. M.A. Thesis, University of Alberta, 1962, pp 80-83.

[181] *Atlin Claim*, 1 July 1899.

[182] Hicks, Diary. Hicks' claim that there was one doctor for every 10 residents may have been an exaggeration, but it does appear that there were many who were unable to practice for lack of a B.C. licence

[183] *Atlin Claim*, 13 April 1899.

[184] *Atlin Claim*, 3 June 1899.

[185] *Atlin Claim*, 11 Nov. 1899.

[186] *Atlin Claim*, 3 Feb. 1899.

[187] *Atlin Claim*, 18 Nov. 1899.

[188] Wells, "Backdoor to the Klondike", Part 6, p.33.

[189] Wells, "Backdoor to the Klondike", Part 6, p.43.

[190] *Atlin Claim*, 5 Aug. 1899.

[191] Wells, "Backdoor to the Klondike", Part 6, p.57.

[192] Wells, "Backdoor to the Klondike", Part 6, p. 57.

[193] *Atlin Claim*, 27 May 1899.

[194] *Atlin Claim*, 29 April 1899.

[195] *Atlin Claim*, 22 April 1899.

[196] *Atlin Claim*, 13 May 1899; *Alaska Sportsman*, Sept. 1958, p.29.

[197] *Atlin Claim*, 27 May 1899.

[198] Hitchcock, Mary E, Two Women in the Klondike, New York; G. P. Putnam, 1899.

[199] *Atlin Claim*, 14 Oct. 1899.

[200] *Atlin Claim*, 14 Oct. 1899.

[201] *Atlin Claim*, 14 Oct. 1899.

[202] Hamilton, Lord Ernest, Forty Years On, New York, n.d.

[203] *Atlin Claim*, 8 & 22 July 1899.

[204] *Atlin Claim*, 5 July 1899.

[205] Hicks, Diary.

[206] Hicks, Diary.

[207] *Atlin Claim*, 27 May 1899.

[208] *Atlin Claim*, 20 May 1899.

[209] *Atlin Claim*, 20 May, 1899.

[210] *Atlin Claim*, 27 May 1899.

[211] *Atlin Claim*, 6 May 1899.

[212] *Atlin Claim*, 29 April 1899.

[213] *Atlin Claim*, 16 July 1899.

[214] *Atlin Claim*, 9 Sept. 1899.

[215] Hicks, Diary.

[216] *Atlin Claim*, 26 Aug. 1899.

[217] *Atlin Claim*, 13 May 1899.

[218] *Atlin Claim*, 15 July 1899.

[219] *Atlin Claim*, 1 Sept. 1900.

[220] Hicks, Diary.

[221] *Atlin Claim*, 30 Dec. 1899.

[222] *Atlin Claim*, 6 Jan. 1900.

[223] Stephenson, F. L., "Atlin Mining Camp", *Vancouver Star*, 26 Aug. 1931.

[224] Bilsland, Atlin 1898-1910, pp 44-45.

[225] Atlin District Board of Trade. The Atlin Gold Fields, Pamphlet, n.d. P. A. B. C.

## Chapter 5

[226] *Victoria Colonist*, 5 Oct. 1898.

[227] Graham, J. D., Gold Commissioner's Report, British Columbia Sessional Papers, 1900, p.643.

[228] Jardine, Diary.

[229] Complaints Against Officials Atlin District, British Columbia Sessional papers, 1900, p.

[230] Complaints Against Officials, Atlin District, p.991.

[231] Complaints Against Officials, Atlin District, p. 991.

[232] Stables, James, Letter to Fred Hume, Minister of Mines, 5 June, 1899, Complaints Against Government Officials, Atlin District, p. 994.

[233] *Atlin Claim*, June 24, 1899.

[234] *Victoria Colonist*, 5 Jan. & 6 Feb. 1899.

[235] *Victoria Colonist*, 5 Oct. 1898.

[236] Carmichael, Alfred, Diary, MS, P. A. B. C.

[237] *Atlin Claim*, 1 Mar. 1899.

[238] Graham, Gold Commissioner's Report, 1900, p.643.

[239] *Victoria Colonist*, 3 Dec. 1898, & *Revelstoke Kootenay Mail*, 3 Dec. 1898.

[240] *Victoria Colonist*, 28 Oct. 1898.

[241] *Victoria Colonist*, 8 March 1899.

[242] Graham, Gold Commissioner's Report, 1900, p. 643, & Graham, J. D., Letter to the Minister of Mines, 16 Jan.1899, Correspondence of the Government Agent, Atlin, B. C. P. P.

[243] Graham, Gold Commissioner's Report, 1900, p. 643.

[244] Graham, Gold Commissioner's Report, 1900, p. 643.

[245] *Revelstoke Kootenay Mail*, 27 Nov. 1897.

[246] *Victoria Colonist*, 3 Dec. 1898.

[247] *Revelstoke Kootenay Mail*, 25 Mar. 1899.

[248] Statutes of British Columbia, 1899, Chapter 50, p.155.

[249] Graham, Gold Commissioner's Report, 1900, p. 643.

[250] *Atlin Claim*, 3 June, 1899.

[251] *Victoria Colonist*, 25 Mar. 1899.

[252] *Atlin Globe*, 28 Sept. 1899.

[253] *Victoria Colonist*, 18 Feb. 1899.

[254] *Victoria Colonist*, 11 Feb., 1899.

[255] *Atlin Globe*, 19 Oct. 1899.

[256] *Atlin Claim*, 10 June 1899.

[257] *Alaska and Northwest Quarterly*, Vol. I, Jan. 1899, p.75.

[258] *Atlin Claim*, 23 Sept. 1899 & 10 Feb. 1900.

[259] British Columbia Sessional Papers, 1900, p. 485.

[260] Roppell, Patricia, "Porcupine", *The Alaska Journal*, Vol. 5, No. 1, Winter, 1975.

[261] *Atlin Claim*, 21 April 1900.

[262] *Douglas Island News*, 11 Oct. 1900.

[263] *Victoria Colonist*, 28 Oct. 1898.

[264] British Columbia Gazette, 15 Dec. 1898.

[265] *Victoria Colonist*, 5 Jan. 1899.

[266] *Skagway News*, 20 Aug. 1898

[267] Semlin, C. C., Chief Commissioner of Works, Letter to W. Davis Esq., Bennett, 11 Dec. 1898, British Columbia Department of Lands and Works, Correspondence, P. A. B. C.

[268] *Glenora News*, 9 Sept. 1898.

[269] *Victoria Colonist*, 8 Mar. 1899.

[270] *Victoria Colonist*, 8 Mar. 1899.

[271] Semlin, C. A., Letter to J.D. Graham, 26 January, 1899, Correspondence of the Government Agent, Atlin.

[272] *Atlin Claim*, 20 May 1899.

[273] *Atlin Claim*, 20 May, 1899.

[274] *Victoria Colonist*, 28 June 1899.

[275] Statutes of British Columbia, 1899, p.155-6.

[276] Bennett-Atlin Commission. p 495.

[277] *Victoria Colonist*, 1 July 1899.

[278] *Atlin Claim*, 24 June 1899.

[279] *Atlin Claim*, 10 June 1899.

[280] *Atlin Claim*, 10 June 1899.

[281] Hume, Hon. J., Letter to James Stables, 27 June 1899, Complaints Against Officials, Atlin, p. 994.

[282] *Victoria Colonist*, 29 July 1899.

[283] Bennett-Atlin Commission.

[284] Pine Creek Discovery Confirmation Act, Statutes of British Columbia, 1900, op. cit., Chapter 29, p. 113.

[285] Graham, Gold Commissioner's Report, 1900, p. 644.

[286] Complaints Against Officials, Atlin District, p. 1001.

[287] Complaints Against Officials, Atlin District, p.

[288] *Victoria Colonist*, 22 April 1902.

[289] Complaints Against Officials, Atlin District, p. 1001.

[290] Hitchcock, Mary, Letter to J. D. Graham, 1 Sept. 1900, Correspondence of the Government Agent.

[291] Complaints Against Officials, Atlin District, pp 997-9

[292] *Victoria Colonist*, 22 April 1902.

[293] *Atlin Claim*, 20 July 1901.

[294] *Victoria Colonist*, 24 April 1902.

[295] *Atlin Claim*, 28 Jan. 1902.

[296] British Columbia Journals, 1902, p 55.

[297] British Columbia Journals, 1902, p 141.

[298] British Columbia Journals, 1902, pp 141-2.

[299] British Columbia Journals, 1902, pp. 141-2.
[300] *Victoria Colonist*, 10 May 1902.
[301] *Atlin Claim*, 7 June 1902.
[302] *Atlin Claim*, 7 June, 1902.

## Chapter 6

[303] British Columbia Mining Exchange, Vol. III, Oct. 1905, & Skagway News, 23 Sept. 1898.
[304] Minister of Mines, 1899, p. 653, & *Atlin Claim*, 26 Aug. 1899.
[305] Minister of Mines, 1936. p. B42.
[306] *Atlin Claim*, 27 May 1899.
[307] *Alaska Record*, 14 Sept. 1898.
[308] Minister of Mines, 1936, p B40. The figures reported by banks and express companies indicate a discrepancy between the gold mined, and the gold reported to the Minister of Mines.
[309] Minister of Mines, 1900, p 755.
[310] Minister of Mines, 1904, p 694.
[311] Minister of Mines, 1900, p 754.
[312] *Atlin Claim*, 21 Oct. 1899.
[313] Minister of Mines, 1899, p 645, & *Atlin Claim*, 19 Aug. 1899.
[314] *Atlin Claim*, 27 May 1899.
[315] Minister of Mines, 1901, p 981.
[316] *Atlin Claim*, 14 May 1900.
[317] Young, Rosalind Watson, "Mining Operations in Atlin, B.C." Part I, Mines and Minerals, 1 Dec. 1901, p 195.
[318] *Atlin Claim*, 23 June 1900.
[319] Young, Mining Operations in Atlin, p. 195.
[320] Young, Mining Operations in Atlin, p. 195.
[321] Minister of Mines, 1905, p. 66-67.
[322] Minister of Mines, 1905, pp 66-67.
[323] Minister of Mines, 1904, pp. G66-69.
[324] Minister of Mines, 1902, p H31.
[325] *Atlin Claim*, 3rd May 1902.
[326] Minister of Mines, 1904, pp. G63-64.
[327] Minister of Mines, 1904, pp. G63-64.
[328] Minister of Mines, 1936, p. B43.
[329] Minister of Mines, 1904, p. G66.
[330] Minister of Mines, 1902, p.H32.
[331] Young, Rosalind Watson, Mining and Smelting in the Province, p. 38.
[332] Minister of Mines, 1903, p. H40.
[333] Minister of Mines, 1936, p. B43.
[334] Minister of Mines, 1909, p. K50.
[335] Minister of Mines, 1909, p. B43.
[336] Minister of Mines, 1908, p. H50.
[337] Young, Mining Operations in Atlin Part II, p. 273.

[338] *Atlin Claim*, 9 Feb. 1901.
[339] *Atlin Claim*, 2 May 1901.
[340] Minister of Mines, 1906, p. H50-1.
[341] *Atlin Claim*, 29 Sept. 1906.
[342] Minister of Mines, 1936, p. B43.
[343] Minister of Mines, 1908, p. J47.
[344] Minister of Mines, 1936, p. B43.
[345] Minister of Mines, 1936, p. B43.
[346] Minister of Mines, 1936, p. B44.
[347] Minister of Mines, 1925, p. 117.
[348] Minister of Mines, 1925, p. 117.
[349] Minister of Mines, 1930, p. 124.
[350] Complaints Against Officials, Atlin, p. 991.
[351] Lawrence, Guy. Forty Years On the Yukon Telegraph, p.18.
[352] *Douglas Island* News, 31 May 1899.
[353] *Atlin Claim*, 12 Aug. 1899.
[354] *Atlin Claim*, 2 Sept. 1899.
[355] *Atlin Claim*, 26 July 1899.
[356] *Atlin Claim*, 26 Aug. 1899.
[357] *Atlin Claim*, 2 Sept. 1899.
[358] *Atlin Claim*, 26 July 1902.
[359] *Atlin Claim*, 26 July 1902.
[360] Minister of Mines, 1906, p. H50.
[361] Minister of Mines, 1936, p. B50.
[362] Minister of Mines, 1899, p. 643.
[363] Minister of Mines, 1900, p. 757.
[364] *Atlin Claim*, 3 Dec. 1904.
[365] *Atlin Claim*, 3 Dec. 1904 & 25 Feb. 1905.
[366] *Atlin Claim*, 22 Jan. 1907.
[367] *Atlin Claim*, 28 March 1908.
[368] Barlee, N. L., "Historic Atlin", *Canada West*, Winter, 1974, Vol. III, #4.
[369] *Victoria Colonist*, 24 April 1902.
[370] Minister of Mines, 1901, p. 782.
[371] Minister of Mines, 1905, p.67.
[372] *Atlin Claim*, 29 Dec. 1906.
[373] Minister of Mines, 1904, p. G73.
[374] Minister of Mines, 1906, p. H50.
[375] *Atlin Claim*, 27 April 1904.
[376] Minister of Mines, 1936, p. B50.
[377] Minister of Mines, 1936, p. B51.
[378] Minister of Mines, 1936, p. B51.
[379] Minister of Mines, 1906, p. H40.
[380] *Atlin Claim*, 17 June 1899.
[381] *Atlin Claim*, 29 Sept. 1900.

382 *Atlin Claim*, 30 Sept. 1899.
383 *Atlin Claim*, 17 March 1899.
384 *Atlin Claim*, 22 Sept. 1900.
385 *Atlin Claim*, 22 Sept. 1900.
386 *Atlin Claim*, 13 April 1901.
387 Minister of Mines, 1902, p. H37.
388 *Atlin Claim*, 29 Sept. 1900.
389 *Atlin Claim*, 28 Oct. 1899 & 29 Sept. 1900.
390 Minister of Mines, 1904, p. G73.
391 Minister of Mines, 1902, p. H36.
392 *Atlin Claim*, 7 June 1899.
393 *Atlin Claim*, 29 Sept. 1900.
394 Minister of Mines, 1904, p. G74.
395 Minister of Mines, 1904, p. G74.
396 *Atlin Claim*, 11 Mar. 1905.
397 Minister of Mines, 1904, p. G74.
398 Minister of Mines, 1905, p. J69.
399 Minister of Mines, 1908, p. J45.
400 Minister of Mines, 1909, p. K49.
401 Minister of Mines, 1910, p. K51.
402 Minister of Mines, 1912, p. K55.
403 Minister of Mines, 1914, p. K74.
404 Minister of Mines, 1909, p. K49.
405 Minister of Mines, 1919, p. N88.
406 Minister of Mines, 1924, p. 81.
407 *Atlin Claim*, 26 April 1902.
408 Minister of Mines, 1901, p. 983.
409 Minister of Mines, 1902, P. H35.
410 Young, Rosalind Watson, Hydraulic and Placer Mining in the Atlin District, Mining and Metallurgy, Oct 1 1901, p. 58.
411 Minister of Mines, 1900, p. 757.
412 Young, Mining Operations in Atlin, Part I, p. 194.
413 *Atlin Claim*, 21 May 1900.
414 *Atlin Claim*, 9 March 1901.
415 Minister of Mines, 1906, p. H52.
416 Minister of Mines, 1903, p. H41.
417 *Atlin Claim*, 28 Jan. 1905.
418 Minister of Mines, 1904, p. G69.
419 Minister of Mines 1911, p K57
420 Minister of Mines, 1911, P. K55.
421 Minister of Mines, 1913, p. K69.
422 Minister of Mines, 1902, p. H35.
423 Minister of Mines, 1917, p. F78.
424 Minister of Mines, 1917, p. F78.

[425] Minister of Mines, 1920, p. N72.

[426] Minister of Mines, 1921, p. N72.

[427] Minister of Mines, 1927, p. C115.

[428] *Atlin Claim*, 5 May 1900.

[429] Minister of Mines, 1927, p. C115.

[430] Minister of Mines, 1930, p. 124.

[431] Black, J. M., Report on the Atlin Placer Mining Camp, Ministry of Mines and Petroleum Resources, 1953, p. 66.

[432] Minister of Mines, 1936, p. B59.

[433] Minister of Mines, 1899, p. 645.

[434] Minister of Mines, 1899, p. 645.

[435] Scholefield, E. O. S., & F.W. Howay, British Columbia From the Earliest Times to the Present, Vol II, Vancouver, 1914, p. 500.

[436] Minister of Mines, 1900, p. 756.

[437] Minister of Mines, 1901, p. 982.

[438] Minister of Mines, 1902, p. H35.

[439] Minister of Mines, 1905, p. J71.

[440] Minister of Mines, 1905, p. J72.

[441] Minister of Mines, 1906 p. H40.

[442] Minister of Mines, 1908, p. J45.

[443] Minister of Mines, 1914, p. K77.

[444] Minister of Mines, 1913, p. K69.

[445] Minister of Mines, 1915, p. K61.

[446] Minister of Mines, 1916, p. K45.

[447] *Atlin Claim*, 28 Oct. 1899.

[448] Carmichael, Diary, pp. 117-120.

[449] *Atlin Claim*, 2 March 1901.

[450] Minister of Mines, 1914, p. K78.

[451] Minister of Mines, 1903, p. H43.

[452] Minister of Mines, 1904, p. G90.

[453] Carmichael, Diary, pp. 161-171.

[454] Minister of Mines, 1906, p. H53.

[455] Carmichael, Diary, p. 176.

[456] Minister of Mines, 1914, p. K78.

[457] Minister of Mines, 1907, p. L51.

[458] Minister of Mines, 1908, p. J48.

[459] Minister of Mines, 1910, p. K54.

[460] Minister of Mines, 1911, p. K58.

[461] Minister of Mines, 1912, p. K59.

[462] Minister of Mines, 1914, p. K78.

[463] Minister of Mines, 1915, p. K62.

[464] Minister of Mines, 1916, p. K45.

[465] Minister of Mines, 1917, p. F75.

[466] Minister of Mines, 1918, p. K100.

[467] Minister of MInes, 1927, p. C115.

[468] Minister of Mines, 1918, p. K100.

[469] Nelson, Bill, Interview, 11 Dec. 1986.

[470] Minister of Mines, 1928, p. C122.

[471] Nelson, Interview, 11 Dec. 1986.

[472] Minister of Mines, 1929, p. C121.

[473] Nelson, Interview, 11 Dec. 1986.

[474] Minister of Mines, 1930, p. A125.

[475] Minister of Mines, 1931, p. A66.

[476] Minister of Mines, 1931, p. A66.

[477] Minister of Mines, 1936, p. B59.

[478] Aitken Report, c1950, Gold Commissioner's Office, Atlin, p. 76.

[479] *Atlin Claim*, 7 Feb. 1903.

[480] Proudlock, P. J. & W. J., Stratigraphy of the Placers in the Atlin Placer Mining Camp, Victoria; Ministry of Mines and Resources, 1976, p. 42.

[481] Minister of Mines, 1901, p. 983.

[482] *Atlin Claim*, 7 Feb. 1903 & 10 Oct. 1903.

[483] Minister of Mines, 1904, p. G89.

[484] Minister of Mines, 1908, p. J48.

[485] Minister of Mines, 1909, p. K51.

[486] Minister of Mines, 1911, p. K58.

[487] Minister of Mines, 1914, p. K77.

[488] Minister of Mines, 1915, p. K61.

[489] Minister of Mines, 1917, p. F78.

[490] Minister of Mines, 1917, p. F78.

[491] Minister of Mines, 1919, p. N87.

[492] Minister of Mines, 1922, p. N88.

[493] Minister of Mines, 1924, p. B80.

[494] Minister of Mines, 1927, p. C115.

[495] Minister of Mines, 1928, p. C122.

[496] Minister of Mines, 1930, p. A124.

[497] Minister of Mines, 1932, p. A70.

[498] *Atlin Claim*, 16 Sept. 1899.

[499] Minister of Mines, 1899, p 646..

[500] Minister of Mines, 1899, p. 649.

[501] *Atlin Globe*, 28 Sept. 1899.

[502] Minister of Mines, 1901, p. 983.

[503] Minister of Mines, 1900, p. 757.

[504] *Atlin Claim*, 31 Mar. 1900.

[505] Williams, Mostyn, "Recent Operations in the Atlin District", The Mining Record, Vol 9, #2, Feb. 1902.

[506] *Atlin Claim*, 23 June 1900.

[507] Minister of Mines, 1902, p. H36.

[508] Minister of Mines, 1905, p. J72.

[509] Minister of Mines, 1906, p. H53.
[510] Minister of Mines, 1916, p. 45.
[511] Minister of Mines, 1920, p. 72.
[512] Carmichael, Diary, pp. 140-46.
[513] Minister of Mines, 1905, p. J73.
[514] Minister of Mines, 1906, p. H53.
[515] Minister of Mines, 1906, p. H54.
[516] Minister of Mines, 1911, p. K59.
[517] Minister of Mines, 1912, p. K59.
[518] Minister of Mines, 1913, p. K71.
[519] Minister of Mines, 1914, p. K79.
[520] Minister of Mines, 1917, p. F75.
[521] *Atlin Claim*, 28 April 1906.
[522] *Atlin Claim*, 5 May 1905.
[523] Minister of Mines, 1906, p. H54.
[524] Minister of Mines, 1908, p. J49.
[525] Minister of Mines, 1912, p. K59.
[526] Minister of Mines, 1912, p. K60.
[527] Minister of Mines, 1913, p. K72.
[528] Minister of MInes, 1914, p. K79.
[529] Minister of Mines, 1929, p. C123 & 1930, p. A127.
[530] Minister of MInes, 1932, p. A65.
[531] Minister of Mines, 1934, p. 36.

## Chapter 7

[532] Wheeler, Keith, The Old West - The Railroaders, New York; Time-Life, 1973, pp. 94-95 & 98.
[533] Berton, Pierre, The Great Railroad, Toronto; McClelland & Stewart, 1972, p. 202.
[534] Bancroft, History of British Columbia, p. 710.
[535] Berton, The Great Railroad, p. 202.
[536] *Atlin Claim*, 19 Oct. 1901.
[537] *Atlin Claim*, 8 March 1902.
[538] *Victoria Colonist*, 2 April 1902.
[539] *Victoria Colonist*, 2 April 1902.
[540] Lawrence, Guy, "The Atlin Placer Camp," Cariboo and Northwest Digest, Dec. 1950, p.38.
[541] Lawrence, Guy, "When the Riot Act Was Read in Atlin," *The Shoulder Strap*, Dec. 1949, pp 12-13.
[542] Lawrence, "When the Riot Act Was Read in Atlin", p. 39.
[543] *Atlin Claim*, 29 Mar. 1902.
[544] Lawrence, Guy, "The Atlin Placer Camp", p. 39.
[545] *Atlin Claim*, 29 Mar. 1902.
[546] British Columbia Sessional Papers, 1902, p.843.
[547] *Atlin Claim*, 21 June 1902.
[548] *Atlin Claim*, 1 Oct. 1902.

[549] *Atlin Claim*, 10 April 1904.
[550] *Atlin Claim*, 27 April 1907.
[551] *Atlin Claim*, 4 May 1907
[552] Minister of Mines, 1902, p. H13.
[553] Minister of Mines, 1927, p. C115

## Chapter 8

[554] *Atlin Claim*, 9 Sept. 1899.
[555] *Atlin Claim*, 9 Sept. 1899 & *Victoria Colonist*, 28 Sept. 1899.
[556] *Victoria Colonist*, 5 & 23 Oct. 1898.
[557] Minister of Mines, 1899, p.648.
[558] *Atlin Claim*, 28 Oct. 1899.
[559] *Atlin Claim*, 2 Sept. 1899.
[560] *Atlin Claim*, 2 Sept. 1899.
[561] Minister of Mines, 1899, p. 648.
[562] Hamilton, Forty Years On, p.226.
[563] Hamilton, Forty Years On, p. 226.
[564] *Atlin Globe*, 28 Sept. 1899.
[565] Hamilton, Forty Years On, p. 245.
[566] *Atlin Claim*, 2 Sept. 1899.
[567] *Atlin Claim*, 30 Sept. 1899 & 19 May 1900.
[568] *Atlin Claim*, 23 Sept. 1899.
[569] *Atlin Claim*, 19 May 1900.
[570] *Atlin Claim*, 29 Sept. 1900.
[571] *Atlin Claim*, 22 Dec. 1900.
[572] *Atlin Claim*, 29 Sept. 1900.
[573] *Atlin Claim*, 3 May & 14 June 1902.
[574] *Atlin Claim*, 8 June 1901.
[575] *Atlin Claim*, 11 Feb. 1905, & 14 Dec. 1907.
[576] *Atlin Claim*, 18 Nov. 1899.
[577] *Atlin Claim*, 7 April 1900.
[578] *Atlin Claim*, 16 March 1901.
[579] Minister of Mines, 1902, p. H38.
[580] *Atlin Claim*, 2 Aug. 1902.
[581] *Atlin Claim*, 22 Nov. 1902.
[582] Minister of Mines, 1902, p. H37.
[583] Minister of Mines, 1903, p. H44.
[584] Minister of Mines, 1904, p. G78.
[585] Cairnes, D.D., Report on Portions of the Atlin District with Special Reference to Lode Mining, Geological Survey of Canada, Memoir #37, 1913, p. 103.
[586] Minister of Mines, 1899, p. 648.
[587] *Atlin Claim*, 16 Sept. 1899 & 11 Feb. 1905.

[588] *Atlin Claim*, 11 Jan. 1908.
[589] Minister of Mines, 1908, p. J50.
[590] Minister of Mines, 1917, p. F80.
[591] Minister of Mines, 1917, p. F80.
[592] Minister of Mines, 1904, p. G 52.
[593] *Atlin Claim*, 4 Feb. 1905.
[594] Young, Mining and Smelting in the Province, p. 38.
[595] B.C. Mining Exchange, Vol VII. No 2.
[596] Minister of Mines, 1910, p. K36.
[597] Minister of Mines, 1919, p. N92.
[598] Minister of Mines, 1920, p. N74.
[599] Le Fevre, H.B. "A Quartz Discovery," *Alaskan Magazine*, Vol I No.1, 1899, pp. 19-23.
[600] Minister of Mines, 1900, p. 760.
[601] Le Fevre, A Quartz Discovery, p. 24.
[602] *Atlin Claim*, 14 July 1900.
[603] *Atlin Claim*, 5 May 1900.
[604] Minister of Mines, 1900, p. 761.
[605] Minister of Mines, 1902, p. 11.
[606] Minister of Mines, 1904, p. G91.
[607] Reed, Harper, A History of the Engineer Mine, MS, A. H. S.
[608] Minister of Mines, 1910, p.K58.
[609] Minister of Mines, 1910, p. 55.
[610] Minister of Mines, 1911,p. K60.
[611] Minister of Mines, 1912, p. K61.
[612] Minister of Mines, 1912, p. 61; 1913, p. 72; 1914, p. 79.
[613] Minister of Mines, 1915, p. 64.
[614] Minister of Mines, 1916, p. 46.
[615] Minister of Mines, 1918, p. 92.
[616] Minister of Mines, 1924, p. 79.
[617] Minister of Mines, 1925, p. 115.
[618] Minister of Mines, 1932, p. 249.
[619] Cross, W.K. et al (eds.) British Columbia Land Bulletin Areas Source Book, Victoria, 1966, p. 68.
[620] Reed, A History of the Engineer Mine.
[621] Minister of Mines, 1924, p.77.
[622] Baker Lake is not found on modern maps, but it is thought to be situated north of the Engineer Mine.
[623] *Atlin Claim*, 9 June 1900.
[624] *Atlin Claim*, 22 Dec. 1900 & 5 Jan. 1901.
[625] Minister of Mines, 1913, p. K73. & Atlin Globe, 28 Sept. 1899.
[626] *Atlin Claim*, 1 Mar. 1902.
[627] Cairnes, Report on Portions of the Atlin District, p. 96.
[628] Minister of Mines, 1914, p. K94.
[629] Brooks, Reginald, Interview with Sue Morhun, 1975.
[630] Minister of Mines, 1934, p. B34.

[631] White Pass & Yukon Route, The Story of Ben-My-Chree, n.d., p. 10.
[632] Minister of Mines, 1911, p. K60.
[633] Minister of Mines, 1913, pp. 72-73.
[634] Minister of Mines, 1915, p. K64.
[635] Minister of Mines, 1919, p. N74.
[636] Edwards, Into the Yukon, Cincinnati; Robt. Clarke, 1904, p. 94.
[637] Minister of Mines, 1921, p. N 74.
[638] Minister of Mines, 1927, p.110.
[639] Minister of Mines, 1926, p. 107
[640] Minister of Mines, 1930, p. 123.
[641] Minister of Mines, 1934, p. B.36.
[642] Minister of Mines, 1930. p. 120.
[643] Minister of Mines, 1936, p. B21.
[644] *Atlin News Miner*, 28 Oct. 1938.

## Chapter 9

[645] *Atlin Claim*, 29 April 1899.
[646] Hamilton, The Yukon Story, p. 27.
[647] *Atlin Claim*, 20 May 1899.
[648] *Atlin Claim*, 24 June 1899.
[649] *Atlin Claim*, 4 & 25 Jan.& 15 Feb. 1902.
[650] Hamilton, The Yukon Story, p. 28.
[651] *Atlin Claim*, 9 June 1900.
[652] *Atlin Claim*, 28 May 1900.
[653] Atlin Claim, 7 April 1900.
[654] Godfrey, Alex. MS, Vancouver City Archives.
[655] Godfrey, MS.
[656] *Atlin Claim*, 5 April 1902.
[657] *Atlin Claim*, 19th Sept. 1903.
[658] *Atlin Claim*, 15 Sept. 1900.
[659] *Atlin Claim*, 15 Aug. 1903.
[660] *Atlin Claim*, 15 & 22 Aug. 1903.
[661] *Atlin Claim*, 10 Oct. 1903.
[662] i, 7 Nov. 1903.
[663] Ormsby, M. British Columbia: A History, Vancouver; MacMillan, 1958. p. 356.
[664] *Atlin Claim*, 12 Jan. 1907.
[665] *Victoria Colonist*, 28 Sept. 1899.
[666] Howay, F. W., British Columbia From the Earliest Times To the Present, Vol 3, Vancouver, 1914, p. 77.
[667] Laurence, Guy, "Yukon Telegraph Service", Part IV in *The AlaskaSportsman*, January, 1959, p. 42. Laurence actually places this election in 1906, but it is likely he is referring to 1907.
[668] *Atlin Claim*, 7 Feb. 1903.
[669] *Atlin Claim*, 23 Nov. 1901; 18 July 1903; 7th Dec. 1907.

[670] Atwood, Fred N. Alaska-Yukon Gold Book, Seattle, 1930, np.
[671] Wilcoxen, I.N. History of the Arctic Brotherhood, 1906, np.
[672] Davidson, J.N. The Arctic Brotherhood, Seattle, 1909, p. 41.
[673] *Atlin Claim*, 22 & 29 Sept. 1900.
[674] *Atlin Claim*, 10 June 1905.
[675] Davidson, The Arctic Brotherhood, p.95.
[676] *Atlin Claim*, 11 May 1901.
[677] *Atlin Claim*, 15 July 1899.
[678] Freemasons, British Columbia, History of the Grand Lodge of British Columbia, 1871-1970. Victoria: Colonist Printers, 1971. p. 203-5.
[679] *Atlin Claim*, 11 Feb. 1905.
[680] *Atlin Claim*, 18 April 1901.
[681] *Atlin Claim*, 23 May, 1903.
[682] *Atlin Claim*, 24 & 31 Jan. 1903.
[683] *Atlin Claim*, 18 April, 1904.

## Chapter 10

[684] *Victoria Colonist*, 23 Oct. 1898.
[685] *Atlin Claim*, 16 Feb. 1901.
[686] *Atlin Claim*, 31 March 1900.
[687] *Atlin Claim*, 10 March 1900.
[688] *Atlin Claim*, 21 July 1900.
[689] *Atlin Claim*, 16 Feb. 1901 & 10 Aug. 1901.
[690] *Atlin Claim*, 11 Jan. 1902.
[691] *Atlin Claim*, 7 June 1902.
[692] *Atlin Claim*, 10 Jan. 1903.
[693] *Atlin Claim*, 11 July 1903.
[694] *Atlin Claim*, 24 Sept. 1904.
[695] *Whitehorse Star*, 19 Feb. 1943.
[696] Ince, Clifford H. Royal Bank of Canada. A Chronology: 1864 -1969, p.13
[697] Ince, Royal Bank of Canada, p. 13.
[698] Atlin District Board of Trade, The Atlin Gold Fields.
[699] Walker, F.T. Reminiscences.
[700] *Atlin Claim*, 7 Jan. 1905.
[701] *Atlin Claim*, 12 Aug. 1899. For many years a small building in Discovery was known as the bank, and it seems likely that one of the banks did open a branch or an agency there at some later time.
[702] *Atlin Claim*, 2 Feb. 1901.
[703] *Atlin Claim*, 16 Mar. 1907.
[704] *Atlin Claim*, 6 April 1901.
[705] *Atlin Claim*, 6 April 1901 & *Victoria Daily Colonist*, 5 Jan. 1899.
[706] *Victoria Daily Colonist*, 18 Feb. 1899.
[707] *Atlin Claim*, 6 April 1901 & *Victoria Daily Colonist*, 22 Mar. 1899.
[708] *Atlin Claim*, 29 April 1899 & Tapp, Lila, Letter to Sue Morhun, 6 September, 1975, A. H. S.

[709] *Atlin Claim*, 9 Dec. 1899.
[710] *Atlin Claim*, 30 Aug. 1899.
[711] *Atlin Claim*, 23 Mar. 1901.
[712] *Atlin Claim*, 26 April 1905.
[713] *Atlin Claim*, 28 Oct. 1899.
[714] *Atlin Claim*, 23 June 1900.
[715] *Atlin Claim*, 15 Aug. 1900.
[716] *Atlin Claim*, 22 Dec. 1900.
[717] *Atlin Claim*, 22 Sept. 1900.
[718] *Atlin Claim*, 29 Dec. 1900.
[719] *Atlin Claim*, 6 April 1904.
[720] *Atlin Claim*, 15 June 1901.
[721] *Atlin Claim*, 28 Mar. 1903.
[722] *Atlin Claim*, 3 Oct. 1903.
[723] *Atlin Claim*, 30 April 1904 & 29 April, 1905.
[724] Watson, George & Ilene, Pioneer Breweries of British Columbia, Nanaimo, 1974.
[725] *Atlin Claim*, 19 Sept. 1903.
[726] *Atlin Claim*, 6 June 1903.
[727] *Atlin Claim*, 12 Mar. & 14 Nov. 1903.
[728] *Atlin Claim*, 7 Jan. 1905.
[729] *Atlin Claim*, 12 Mar. 1903.
[730] *Atlin Claim*, 19 Nov. 1904.
[731] *Atlin Claim*, 7 Jan. 1905.
[732] *Atlin Claim*, 3 June 1905.
[733] *Atlin Claim*, 26 May 1900.
[734] *Atlin Claim*, 6 April 1901.
[735] *Atlin Claim*, 20 April 1904.
[736] *Atlin Claim*, 13 June 1903.
[737] *Vancouver Province*, 1 April 1899.
[738] *Atlin Claim*, 29 April 1899.
[739] *Atlin Claim*, 29 April 1899.
[740] *Atlin Claim*, 3 June 1899.
[741] *Atlin Claim*, 15 Sept. 1900.
[742] *Atlin Claim*, 29 Sept. 1900.
[743] *Atlin Claim*, 7 Sept. 1901.
[744] Carmichael, Diary, p. 159.
[745] *Atlin Claim*, 8 Oct. 1904.
[746] *Atlin Claim*, 7 Sept. 1901.
[747] *Atlin Claim*, 17 Jan. 1903.
[748] *Atlin Claim*, 6 May 1905.
[749] *Atlin Claim*, 14 Mar. 1908.
[750] *Atlin Claim*, 2 Sept. 1899 & 14 May 1900.
[751] *Atlin Globe*, 28 Sept. 1899.
[752] *Atlin Claim*, 14 June 1900.

[753] James, Cyril, Interview with Sue Morhun, Atlin, 1975.
[754] *Atlin News Miner*, 13 Dec. 1941.
[755] *Atlin News Miner*, 13 Dec. 1941.
[756] *Atlin Claim*, 13 May 1899.
[757] *Victoria Colonist*, 5 Jan 1899.
[758] These figures were calculated from hotels mentioned in the *Atlin Claim* for 1899.
[759] *Atlin Claim*, 12 May 1900.
[760] *Atlin Claim*, 6 May 1899.
[761] *Atlin Claim*, 1 July 1899.
[762] *Atlin Claim*, 13 May 1899.
[763] Anzer, Klondike Gold Rush, p. 167.
[764] Fisher, Norman, Interview with Sue Morhun and Shirley Connolly, 1975.
[765] Lawrence, Guy, "The Stikine Trail" Part III, *Alaska Sportsman*, July, 1957.
[766] *Atlin Claim*, 7 Oct. 1899.
[767] *Atlin Claim*, 22 Dec. 1900.
[768] Fisher, Norman, Interview.
[769] *Atlin Claim*, 14 Oct. 1899 & 28 May 1900.
[770] *Atlin Claim*, 23 April 1903.
[771] Edwards, William Seymour, In to the Yukon, p. 88.
[772] *Atlin Claim*, 25 July 1903.
[773] *Atlin Claim*, 5 Sept. 1903.
[774] *Alaska Mining Record*, 28 Dec. 1898.
[775] *Alaska Mining Record*, 27 Jan. 1900.
[776] Godfrey, Alexander, MS.
[777] *Atlin Claim*, 6 Sept. 1902.
[778] *Atlin Claim*, 24 May & 22 Nov. 1902.
[779] *Atlin Claim*, 28 Mar. 1903.
[780] Ryder, Mrs. Irving, Diary, copy, A. H. S.
[781] *Atlin Claim*, 17 Jan. 1903.
[782] *Atlin Claim*, 11 Oct. 1902.
[783] *Atlin Claim*, 15 Oct. 1904 & 8 July 1905.
[784] Ryder, Diary.
[785] *Atlin Claim*, 11 Nov. 1905.
[786] *Atlin Claim*, 7 Oct. 1899.
[787] *Atlin Claim*, 29 Dec. 1900.
[788] *Atlin Claim*, 6 April 1901.
[789] *Atlin Claim*, 15 Nov. 1902 & 17 Jan. 1903.
[790] *Atlin Claim*, 28 May 1904.
[791] *Atlin Claim*, 12 Mar. 1904.
[792] Fraser, J. A., Letter to F. S. Hussey, 17 June 1910, British Columbia Provincial Police.
[793] *Atlin Claim*, 22 Dec. 1906.
[794] *Victoria Colonist*, 24 Sept. 1898.
[795] *Atlin Claim*, 23 Sept. 1899.
[796] *Atlin Claim*, 17 Jan. 1903.

[797] Fisher, Norman, Interview.
[798] *Atlin Claim*, 12 May, 1900.
[799] *Atlin Claim*, 24 Mar. 1900.
[800] Godfrey, MS.
[801] Warner, Iris, "Taylor & Drury, Ltd.", Alaska Journal, Spring, 1975, p. 74.
[802] Sproule, The Role of Patrick Burn. pp. 80-83.
[803] *Atlin Claim*, 5 Oct. 1901.
[804] Burry, Ralph R., Louis Schulz and his tokens of Atlin, B.C., n.d., MS, A. H. S.
[805] *Atlin Claim*, 4 June 1904.
[806] *Atlin Nugget*, 8 Aug. 1936.
[807] Timbers used in underground mining.
[808] Coleman, Tom & Irene, Interview with Sue Morhun, Atlin, 1975
[809] Burry, Louis Schulz.
[810] *Skagway News*, 14 Oct. 1898.
[811] *Atlin Claim*, 29 April 1899.
[812] Cartmel, J.E., Letter to the Christine Dickinson, 1977, A. H. S.
[813] *Atlin Claim*, 14 April 1904.
[814] Owen, Walter, Interview with Sue Morhun, Victoria, 1975.
[815] Carmichael, Diary, p. 157.
[816] *Atlin Claim*, 28 May 1900.
[817] Cartmel, Letter.
[818] James, Cyril, Interview.
[819] *Atlin Claim*, 16 July, 1904.
[820] *Atlin Claim*, 6 April, 1904.

## Chapter 11

[821] *Glenora News*, 9 Sept, 1898.
[822] Berton, Pierre, Klondike, p.227.
[823] Shipping Registration, copy, A. H. S.
[824] McBride, W. S., "Saga of the River Boats," *Caribou and Northwest Digest*, 1948-49, p. 108.
[825] Shipping Registration, copy, A. H. S.
[826] *Atlin Claim*, 10 June 1899.
[827] McBride, W. D. Saga of the River Boats.
[828] *Atlin Claim*, 10 June 1899.
[829] McBride, Saga of the River Boats, p. 106. Although McBride had a thorough knowledge of the history of northern transportation, he may have relied on his memory for some details.
[830] *Atlin Claim*, 10 June 1899.
[831] *Atlin Claim*, 10 June, 1899.
[832] *Atlin Claim*, 10 June 1899.
[833] MacBride, W. D. "Railroad Built to Serve Miners," *Whitehorse Star*, 9 Aug. 1956.
[834] *Atlin Claim*, 19 Aug. 1899.
[835] *Atlin Claim*, 12 Aug. 1899.

[836] *Atlin Claim*, 9 Aug. 1899.
[837] *Atlin Claim*, 19 Aug. 1899.
[838] Shipping Registration, copy, A. H. S.
[839] *Atlin Claim*, 19 Aug. 1899.
[840] *Atlin Claim*, 9 Sept. 1899.
[841] *Atlin Claim*, 19 Aug. 1899.
[842] *Atlin Claim*, 12 Aug. 1899.
[843] *Atlin Claim*, 16 Sept. 1899.
[844] Statutes of British Columbia, 1899, Chapt. 79, p. 343.
[845] Clifford, Howard, Rails North, Seattle; Superior, p. 58.
[846] Clifford, Rails North, p. 58.
[847] Pugsley, Ed E., "The Duchess Retires," *Family Herald*, 16 Jan. 1899.
[848] *Atlin Claim*, 26 Aug. 1899.
[849] Carter Cotton, F., Letters to J. D. Graham, 29 April & 7 Sept. 1899, G. R. 440, Vol 74, P. A. B. C.
[850] *Atlin Claim*, 24 Mar. 1900.
[851] *Atlin Claim*, 31 Mar. 1900.
[852] Pugsley, "The Duchess Retires".
[853] *Atlin Claim*, 5 May, 1900.
[854] *Atlin Claim*, 23 June 1900.
[855] *Atlin Claim*, 21 July 1900.
[856] McBride, W. D. "The Taku Tram," *Cariboo and Northwest Digest*, April 1950.
[857] *Atlin Claim*, 24 Aug. 1900.
[858] *Atlin Claim*, 27 Aug. 1904.
[859] Shipping Registration, copy, A. H. S.
[860] *Atlin Claim*, 13 Aug. & 8 Oct. 1904.
[861] *Atlin Claim*, 21 July 1906.
[862] *Atlin Claim*, 22 Oct. 1904 & 1 Feb. 1908.
[863] White Pass Records, River Division, Scotia Log Book, 1918, Yukon Archives.
[864] White Pass Records, River Division, Annual Report, 1917, Yukon Archives
[865] Pugsley, "The Duchess Retires".
[866] Pugsley, "The Duchess Retires".
[867] Gore, W. S., Letter to J. D. Graham, 9 Feb. 1899, MS, G.R. #440, Vol. 74, P.A.B.C.
[868] *Atlin Claim*, 10 May 1902.
[869] Ryder, Diary.
[870] McBride, W.D. "Trails of 'Ninety-Eight," Farthest North Collegian, 8 Oct. 1947.
[871] Ryder, Diary.
[872] Abbott, Doris Nichols. MS. A. H. S.
[873] *Alaska Journal*, Spring 1979, p.56.
[874] *Alaska Journal*, Spring 1979, p. 56.
[875] *Atlin Claim*, 30 Mar. 1904.
[876] Oliphant, John, The Goldtimers, *Vancouver Sun Magazine*, 14 Feb. 1979.
[877] *Atlin Claim*, 6 Dec. 1902.
[878] *Atlin Claim*, 6 Dec. 1902.

[879] *Atlin Claim*, 2 May & 16 May 1903.
[880] Oliphant, "The Goldtimers," & Fisher, Norman, Interview, 1974.
[881] *Dawson Weekly News*, 14 June 1907.
[882] Oliphant, The Goldtimers.
[883] *Atlin News Miner*, Mar. 1976.
[884] Lawrence, Yukon Telegraph Service, Part IV, p. 30.
[885] Fisher, Norman, Interview, March 1975.
[886] *Atlin Claim*, 2 Dec. 1899.
[887] *Atlin Claim*, 2 Nov. 1901.
[888] *Atlin Claim*, 12 Sept. 1903.
[889] *Atlin Claim*, 12 May 1906.
[890] *Atlin Claim*, 5 Oct. 1907.
[891] Bordynuk, Ernest, "The Mail Driver," *The Yukon News*, June-July 1967, p.25.
[892] *Victoria Colonist*, 27 April 1900.
[893] *Alaska Sportsman*, Sept. 1958, p. 29.
[894] *Alaska Sportsman*, Sept. 1958, p. 29.
[895] *Dawson Daily News*, 31 May 1900.
[896] *Atlin Claim*, 23 July 1904.
[897] Bruce, Charles, MS, A. H. S.
[898] *Atlin Claim*, 29 July & 19 Aug. 1899.
[899] *Atlin Claim*, 17 Feb. 1900.
[900] *Atlin Claim*, 14 May 1900.
[901] Murphy, Marie, Diaries, MS, A. H. S.
[902] *Atlin Claim*, 7 Oct. 1899.
[903] *Atlin Globe*, 28 Sept. 1899.
[904] Richeson, David R., "The Yukon Telegraph," in *British Columbia Historical News*, Winter 1982, p. 6.
[905] *Atlin Claim*, 29 July 1899.
[906] *Atlin Claim*, 24 Mar. 1900.
[907] Richeson, The Yukon Telegraph, p. 6.
[908] Lawrence, Yukon Telegraph Service, Part I, p. 14, & *Atlin Claim*, 16 Mar. 1901.
[909] Richeson, The Yukon Telegraph, p. 9.
[910] Lawrence, Forty Years on the Yukon Telegraph, p. 39.
[911] Lawrence, Yukon Telegraph Service, Part IV. p. 30.
[912] *Atlin Claim*, 7 Oct. 1899.
[913] Richeson, The Yukon Telegraph, p. 15.
[914] Nelson, Bill, Interview, 15 April 1987.
[915] *Atlin Claim*, 5 Aug. 1899.
[916] George, Melvin, Post Offices of British Columbia, Vancouver, 1972.
[917] *Atlin Claim*, 13 May 1899.
[918] *Atlin Claim*, 4 July 1899 & 30 Dec. 1905.
[919] George, Post Offices of British Columbia.
[920] *Atlin Claim*, 2 Sept. 1899.
[921] Harbottle, Jean, "Clyde Wann, Father of Yukon Aviation," in *Alaska Journal*, Autumn 1973, p. 242.
[922] Ellis, Frank H., "Yukon's Flying Heritage" Whitehorse Star, 11 Aug. 1969.

[923] Harbottle, Clyde Wann, p. 242.
[924] Bilsland, W. W. Additional MS after 1910, copy A. H. S.
[925] Burke, Brian, MS, copy A. H. S.
[926] Burke, MS.
[927] Ellis, F.H. "Ordeal on the Liard", Art Downs (ed.), Pioneer Days in British Columbia, Vol II, Vancouver, 1975, p. 55-59.
[928] Cole, Dermot, Frank Barr, Edmonds, WA; Alaska Northwest, 1986, pp 33-38.
[929] Cole, Frank Barr, pp 39-55.
[930] Bilsland, Additional MS.
[931] Bilsland, Additional MS.
[932] Nelson, Interview, 15 April, 1987.
[933] *Atlin News Miner*, 15 April 1939.
[934] Nelson, Interview, 15 April, 1987.
[935] Statutes, of British Columbia, 1899, pp. 347-356.
[936] Statutes of British Columbia, 1900, pp. 247-253.
[937] Statutes of British Columbia, 1903-04, pp. 387-391.
[938] *Atlin Claim*, 4 May 1901.
[939] *Atlin Claim*, 2 Mar. 1901.
[940] Minister of Mines, 1914, p. K87.
[941] *Atlin Claim*, 4 & 18 Feb. 1905.
[942] *Atlin Claim*, 1 April 1904.
[943] *Atlin Claim*, 24 June 1905.
[944] *Atlin Claim*, 8 July 1905.
[945] *Atlin Claim*, 6 Jan. 1906.
[946] *Atlin Claim*, 10 Dec. 1904; 23 Sept. 1905; 14 Dec. 1908.
[947] *Atlin Claim*, 22 Feb. 1908.
[948] *Atlin News Miner*, 22 July 1939.
[949] Coleman, Tom, Interview with Sue Morhun, 1975.
[950] Morritt, Hope, Land of the Fireweed, Edmonds, WA; *Alaska Northwest*, 1985, p. 172.
[951] Morritt, Land of the Fireweed, pp. 171-2.
[952] Morritt, Land of the Fireweed, p. 186.
[953] Morritt, Land of the Fireweed, p. 186.

## Chapter 12

[954] Steele, Annual Report, 1898, p.6.
[955] Wood, Supt. Z. T., Annual Report, Sessional Papers, 1898, Ottawa, p. 2.
[956] Wood, Annual Report, 1898, pp 35-6.
[957] Wood, Annual Report, 1898, p.39. & Annual Report, 1899, p.9.
[958] Privy Council Records, R.G. 2, Series 3, Vol. 128, File 2800.
[959] See Chapter 2.
[960] See Chapter 5.
[961] See Chapter 5.

[962] Privy Council Records.

[963] *Victoria Colonist*, 7 Oct. 1962.

[964] See Chapter 5.

[965] Graham, J. D., Letter to F. S. Hussey, 12 Aug. 1899, British Columbia Provincial Police.

[966] *Atlin Claim*, 13 Jan. 1906, & Owen, W.S. Interview.

[967] Graham, J. D., Letter to F. S. Hussey, 2 Jan 1899, British Columbia Provincial Police.

[968] DesBrisay, T. D., Letter to F. S. Hussey, 15 July 1899, British Columbia Provincial Police.

[969] Bickle had resigned to take up a position in J. D. Graham's office. *Shoulder Strap*, Oct. 1942.

[970] Wood, Annual Report, 1898, p.2.

[971] DesBrisay, T. D., Letter to F. S. Hussey, 7 July, 1899, British Columbia Provincial Police.

[972] DesBrisay, T. D., Letter to F. S. Hussey, 10 Aug. 1899, British Columbia Provincial Police.

[973] Graham, J. D., Letter to F. S. Hussey, 12 Aug. 1899, British Columbia Provincial Police.

[974] DesBrisay, T. D., Letter to F. S. Hussey, 2 Nov. 1899, British Columbia Provincial Police.

[975] "Veteran of the Force," *The Shoulder Strap*, Oct. 1936.

[976] This son, also Walter Owen, was to become Lt. Governor of British Columbia.

[977] "Veteran of the Force", p. 24.

[978] *Atlin Claim*, 9 Mar. 1909.

[979] Harvey, Judge J.T., "Meditations in the Atlin Vault," *The Advocate*, Vol 22, Part 4, 1964, pp 133-4.

[980] Harvey, "Meditations in the Atlin Vault".

[981] *Atlin Claim*, 10 Aug. 1901.

[982] Harvey, "Meditations in the Atlin Vault".

[983] *Atlin Claim*, 23 Aug. 1899.

[984] Wood, Annual Report, 1899, p.9.

[985] *Atlin Claim*, 6 May 1899.

[986] Owen, W., Letter to F.S. Hussey, 13 Feb., 1905, British Columbia Provincial Police.

[987] Lawrence, Guy, "The Sluice Box Robbery", *The Shoulder Strap*, June 1950, pp 48-49 & *Atlin Claim*, 29 Aug. 1903.

[988] British Columbia Provincial Police, unidentified newspaper clipping.

[989] *Atlin Claim*, 29 Aug. 1903.

[990] *Atlin Claim*, 5 Sept. 1905.

[991] Owen, Interview, & British Columbia, Sessional Papers, Reports of the Superinterindent of Provincial Police, 1927-42.

[992] *Atlin Claim*, 24 June, 1899.

[993] *Atlin Claim*, 12 Oct. 1901 & 21 & 28 June 1902.

[994] Owen, W., Letter to F. S. Hussey, 29 Mar. 1906, British Columbia Provincial Police.

[995] Snyder, Supt., Annual Report, Sessional Papers, 1906, p. 23.

[996] Horral, S. W., MS, A. H. S.

[997] Royal Canadian Mounted Police, Atlin Detachment, Inspection Report, 12 July 1921, MS, A. H. S.

[998] Bennett, Charles, Interview with Christine Dickinson, circa 1980.

## Chapter 13

[999] Sinclair, James M., Mission: Klondike, Vancouver; Mitchell, 1978, p. 25.

[1000] Sinclair, Mission: Klondike, pp 63-66.

[1001] Hamilton, The Yukon Story, p.164.
[1002] St. Andrew's Session, In Memoriam, Rev. John Pringle, Sydney, NS p. 23.
[1003] Sinclair, Mission: Klondike, p. 78.
[1004] Sinclair, Mission: Klondike, p. 78.
[1005] Leinweber, Stumbling Stampeders.
[1006] Sinclair, Mission: Klondike, pp 11-12.
[1007] Sinclair, Mission: Klondike, pp. 11-12.
[1008] Sinclair, Mission: Klondike, p. 192.
[1009] Hicks, Diary.
[1010] Kendall, Cattle Drive to Alaska.
[1011] Sinclair, Mission: Klondike, p. 192.
[1012] Sinclair, Mission: Klondike, pp. 200-1.
[1013] Sinclair, Mission: Klondike, p. 197.
[1014] Bennett Sun, 24 June 1899.
[1015] St. Andrew's Session, In Memoriam, p. 7.
[1016] Sinclair, Mission: Klondike, p. 200.
[1017] Hartshorn, Hazel, MS, A. H. S.
[1018] Lawrence, Forty Years on the Yukon Telegraph, p.13.
[1019] Canadian Churchman, 16 Jan. 1941. Carmichael claims that Stephenson also worked at Metlakatla.
[1020] Ridley, W. Snapshots of the North Pacific, C.M.S., London, 1904, p. 154
[1021] Stephenson, F.L., "The Church in the Frozen North," The Mission Field, 1 May 1901.
[1022] Stephenson, F.L., MS, A. H. S.
[1023] Ridley, W., "New Mission to the Miners on Atlin and Bennett Goldfields," Part I, The Gospel Missionary, Nov.1899, p.125.
[1024] Ridley, "New Mission to Miners", Part II, pp 137-8.
[1025] Stephenson, MS.
[1026] *Atlin Claim*, 12 Aug. 1899.
[1027] Carmichael, Diary, p.135.
[1028] Carmichael, Diary.
[1029] *Atlin Claim*, 19 July 1902.
[1030] *Atlin Claim*, 1 Feb. 1902.
[1031] Carmichael, Diary, p. 137.
[1032] Carmichael, Diary.
[1033] St. Martin's Anglican Church Records, MS, A. H. S.
[1034] St. Martin's Records.
[1035] Allard, J. "Atlin Mission's History Resumed", MS, Oblate Archives, Ottawa.
[1036] Allard, Atlin Mission's History Resumed.
[1037] Allard, Atlin Mission's History Resumed.
[1038] *Atlin Claim*, 4 June 1904.
[1039] *Atlin Claim*, 13 April 1907.
[1040] Allard, Atlin's Mission History Resumed.
[1041] Atlin Claim, 14 Dec. 1907.
[1042] Allard, J. "Rapport sur La Mission d'Atlin," Missions de la Congregation des Missionaires Oblats de Marie Immaculee, Vol. 46, p. 242.

[1043] Allard, J. "Atlin Mission's History Resumed.

[1044] *Atlin Claim*, 5 Nov. 1904.

[1045] M. S. P., The Early History of St. Andrew's Hospital, Atlin, B.C., Women's Home Missionary Society of the Presbyterian Church in Canada, 1907, p.2.

## Chapter 14

[1046] *Atlin Claim*, 6 May 1899.

[1047] Hicks, Diary.

[1048] *Atlin Claim*, 6 May 1899.

[1049] Gore, W. S., Letter to J. D. Graham, 15 April 1899, GR #440, Vol. 73 File #719, P.A.B.C.

[1050] *Atlin Claim*, 6 May 1899.

[1051] *Atlin Claim*, 29 April 1899.

[1052] *Atlin Claim*, 1 July 1899

[1053] *Atlin Claim*, 6 May 1899.

[1054] *Atlin Claim*, 20 May 1899.

[1055] *Atlin Claim*, 20 May, 1899.

[1056] Sinclair, Mission: Klondike, pp. 69-74.

[1057] M. S. P., The Early History of St. Andrew's Hospital, p. 2.

[1058] Presbyterian Church of Canada, Women's Missionary Society: Our Jubilee Story, 1924.

[1059] M. S. P. Early History of St. Andrew's Hospital, p. 3.

[1060] Mills, Thora McIlroy, The Church and the Klondike, Toronto, 1977. p. 53.

[1061] Sinclair, Mission: Klondike, p. 193.

[1062] Mills, The Church and the Klondike, p. 53.

[1063] Mills, The Church and the Klondike, p. 53.

[1064] M. S. P. The Early History of St. Andrew's Hospital, pp 4-5.

[1065] M. S. P., The Early History of St. Andrew's Hospital, p. 5.

[1066] Mills, The Church and the Klondike, p. 53.

[1067] Sinclair, Mission: Klondike, pp 193-4.

[1068] Sinclair, Mission: Klondike, p. 193.

[1069] M. S. P., The Early History of St. Andrew's Hospital, p. 5.

[1070] *Atlin Claim*, 3 Feb. 1900.

[1071] *Atlin Claim*, 13 Jan. 1900.

[1072] M. S. P., The Early History of St. Andrew's Hospital, p. 6.

[1073] Wells, Backdoor to the Klondike, Part 4, p. 35.

[1074] *Atlin Claim*, 13 Jan. & 24 Feb. 1900.

[1075] *Atlin Claim*, 4 May 1901.

[1076] See Chapter 5.

[1077] *Atlin Claim*, 2 Aug. 1902.

[1078] *Atlin Claim*, 2 Aug. 1902.

[1079] Mills, The Church and the Klondike, p.54.

[1080] M. S. P., The Early History of St. Andrew's Hospital, p. 9.

[1081] M. S. P., The Early History of St. Andrew's Hospital, p. 8.

[1082] M. S. P., The Early History of St. Andrew's Hospital, p. 6.
[1083] M. S. P., The Early History of St. Andrew's Hospital, p. 10.
[1084] Mills, Thora McIlroy. "Bala Sent Nurses to the Gold Rush Too!", *Bracebridge Herald-Gazette*, 27 Jan. 1972.
[1085] Mills, The Church and the Klondike, p. 55.
[1086] *Atlin Claim*, 4 Aug. 1906.
[1087] *Atlin Claim*, 4 Aug., 1906.
[1088] Sinclair, Mission: Klondike, p. 70.
[1089] Sinclair, Mission: Klondike, pp. 70-4.
[1090] Mills, Thora McIlroy, "The Hanna Family in the Early Life of Muskoka", Parts V & VI, *The Muskoka Sun*, 25 July & 1 Aug. 1974.
[1091] Mills, The Church and the Klondike, p. 67.
[1092] Connolly, Hugh, MS, A. H. S.
[1093] Mills, The Church and the Klondike, p. 86.
[1094] Kipling, R. MS., A. H. S.
[1095] *Atlin Claim*, 9 Dec. 1905.
[1096] *Atlin Claim*, 28 Nov. 1903.
[1097] *Atlin Claim*, 19 July 1902.
[1098] *Atlin Claim*, 1 Nov. 1902.
[1099] *Atlin Claim*, 3 Sept. 1904.
[1100] Records of St. Martin's Church.
[1101] Records of St. Martin's Church.Ibid.
[1102] Abbott, MS.
[1103] Mills, The Church and the Klondike, p.55.
[1104] British Columbia Ministry of Lands Records. Atlin Hospital Lot File, copy A. H. S.

## Chapter 15

[1105] *Atlin Claim*, 1 Sept. 1900.
[1106] Godfrey, MS.
[1107] *Atlin Claim*, 1 Sept. 1900.
[1108] *Alaska Sportsman*, Sept. 1960, picture, p.34.
[1109] *Atlin Claim*, 1 Sept. 1900.
[1110] *Atlin Claim*, 1 Sept. 1900.
[1111] Godfrey, MS.
[1112] *Atlin Claim*, 1 Sept. 1900.
[1113] *Atlin Claim*, 1 Sept., 1900.Ibid.
[1114] *Atlin Claim*, 1 Sept., 1900.
[1115] *Atlin Claim*, 1 Sept., 1900.
[1116] Lawrence, Guy, "Atlin Miners Ignored Riot Act", *Vancouver Sun Magazine*, Supplement, 31 July 1948.
[1117] Lawrence, "Atlin Miners Ignored Riot Act"
[1118] *Vancouver Province*, 8 Jan. 1964.
[1119] *Atlin Claim*, 20 & 27 May 1899.

[1120] *Atlin Claim*, 15 July 1899.
[1121] *Atlin Claim*, 19 Aug. 1899.
[1122] *Atlin Claim*, 16 Sept. 1899.
[1123] *Atlin Claim*, 16 Sept., 1899.
[1124] *Atlin Claim*, 5 Jan. 1901.
[1125] *Atlin Claim*, 20 April 1901.
[1126] *Atlin Claim*, 29 June 1901.
[1127] *Atlin Claim*, 13 July 1901.
[1128] *Atlin Claim*, 14 Nov. 1903.
[1129] *Atlin Claim*, 3 Nov. & 1 Dec. 1906.
[1130] *Atlin Claim*, 3 June 1905.
[1131] James, T, MS, Bourne Collection, A. H. S.
[1132] James, T. MS.
[1133] *Whitehorse Star*, 29 May 1914.
[1134] *Whitehorse Star*, 29 May 1914.
[1135] James, Cyril, MS. A. H. S.
[1136] *Whitehorse Star*, 5 April 1929.
[1137] *Atlin Claim*, 21 Jan. 1905.

## Chapter 16

[1138] *Atlin Claim*, 19 Aug. 1899.
[1139] *Atlin Claim*, 19 Aug. 1899.
[1140] *Atlin Globe*, 28 Sept. 1899.
[1141] *Atlin Globe*, 28 Sept., 1899.
[1142] *Atlin Claim*, 28 Oct. & 16 Dec. 1899.
[1143] *Atlin Claim*, 11 Nov. 1899.
[1144] Wells, "Back Door to the Klondike", Part 6, p. 35.
[1145] *Atlin Claim*, 28 Oct. & 11 Nov. 1899.
[1146] *Atlin Claim*, 1 Sept. 1900.
[1147] British Columbia, Sessional Papers, 1901.
[1148] Bruce, MS.
[1149] *Atlin Claim*, 3 Aug. 1901.
[1150] Wells, "Backdoor to the Klondike", Part 6, p. 35.
[1151] *Atlin Claim*, 30 Aug. 1902.
[1152] *Atlin Claim*, 11 Jan. 1902.
[1153] *Atlin Claim*, 23 Aug. 1902.
[1154] British Columbia Sessional Papers, 1901.
[1155] *Atlin Claim*, 31 Nov. 1903.
[1156] *Atlin Claim*, 12 Mar. 1904.
[1157] *Atlin Claim*, 27 Aug. 1904.
[1158] *Atlin Claim*, 22 July 1905.
[1159] *Atlin Claim*, 21 & 27 July 1906.

[1160] *Atlin Claim*, 11 Nov. 1905.
[1161] *Atlin Claim*, 1 April 1905, & 17 Feb. 1906.
[1162] *Atlin Claim*, 1st July 1905.
[1163] *Atlin Claim*, 17 June 1905.
[1164] *Atlin Claim*, 11 Nov. 1905.
[1165] British Columbia Sessional Papers, 1902.
[1166] Bruce, MS.
[1167] British Columbia Sessional Papers, 1901.
[1168] British Columbia Sessional Papers, 1904.
[1169] Bruce, MS.
[1170] Buckpitt, Matthew, MS, G. Smedley Andrews private collection.
[1171] *Atlin Claim*, 29 Aug. 1905.
[1172] Gairns, Jean, MS, A.H.S.
[1173] Kipling, R.T. MS, A.H.S.
[1174] British Columbia Sessional Papers, 1925.

## Chapter 17

[1175] Wells, "Backdoor to the Klondike", Part 6, p. 57.
[1176] *Atlin Claim*, 26 May 1900.
[1177] Coleman, Irene, Interview with Sue Morhun, 1975.
[1178] Ely, Charles, Letter to Diane Smith, 1 Dec. 1994, A. H. S.
[1179] *Atlin Claim*, 1 Aug. 1903.
[1180] *Atlin Claim*, 21 May 1900 & 29 July 1905.
[1181] *Atlin News Miner*, 8 July & 12 Aug. 1939.
[1182] *Atlin News Miner*, 2 Sept. 1939.
[1183] *Atlin Claim*, 2 Mar. & 23 Mar. 1901.
[1184] *Atlin Claim*, 27 Jan. 1900 & 2 Feb. 1901.
[1185] *Atlin Claim*, 17 Feb. 1900.
[1186] *Atlin Claim*, 16 Dec. 1899.
[1187] *Atlin Claim*, 5 Jan. 1901.
[1188] *Atlin Claim*, 2 Dec. 1899.
[1189] *Atlin Claim*, 7 Oct. 1899.
[1190] *Atlin Claim*, 22 Sept. & 30 Sept. 1900.
[1191] Fisher, Norman & James, Cyril, Interview, 4 Oct. 1974.
[1192] *Atlin Claim*, 28 May 1904.
[1193] *Atlin Claim*, 9 Dec. 1899.
[1194] *Atlin Claim*. 25 May, 1901.
[1195] *Atlin Claim*, 2 May 1902.
[1196] Owen, Interview.
[1197] Ashby, Gwynneth, Letter, MS, A. H. S.
[1198] *Atlin Claim*, 2 Dec. 1899.
[1199] *Atlin Claim*, 6 Dec. 1902 & 17 Jan. 1903.

[1200] *Atlin Claim*, 31 Jan. 1903.
[1201] *Atlin Claim*, 14 Nov. 1903.
[1202] Lawrence, Forty Years on the Yukon Telegraph, p. 13.
[1203] *Atlin Claim*, 9 Dec. 1899.
[1204] Fisher, Norman & James, Cyril, Interview, 4 Oct. 1974.
[1205] *Atlin Claim*, 2 Dec. 1905.
[1206] *Atlin Claim*, 10 June 1905.
[1207] *Atlin Claim*, 24 Oct. 1903.
[1208] *British Columbia Mining Exchange*, Vol VI, July, 1904, pp. 13-14.
[1209] *Atlin Claim*, 13 Jan. 1900.
[1210] *British Columbia Mining Exchange*, Vol. VII, October, 1905, p. 6.
[1211] *British Columbia Mining Exchange*, Vol. VI, July, 1904, pp. 13-14.
[1212] *Atlin Claim*, 13 Aug. 1904.
[1213] *British Columbia Mining Exchange*, Vol VI, July, 1904, p. 13.
[1214] *British Columbia Mining Exchange*, Vol VII, October, 1905, p. 6.
[1215] *Atlin Claim*, 13 Aug. 1904.
[1216] *Atlin Claim*, 25 Feb. 1905.
[1217] *Atlin Claim*, 29 May 1905.
[1218] *Atlin Claim*, 6 July 1907.
[1219] *Atlin Claim*, 3 Mar. 1900.
[1220] *Atlin Claim*, 22 Dec. 1900.
[1221] *Atlin Claim*, 14 Dec. 1901.
[1222] Outcroppings of hydro magnesite, locally called "alkali" covered large areas east of town, between Third and Sixth Streets, and several other nearby locations.
[1223] *Atlin Claim*, 25 Jan. 1902.
[1224] *Atlin Claim*, 25 Jan. 1902.
[1225] *Atlin Claim*, 8 Nov. 1902.
[1226] *Atlin Claim*, 18 Jan. 1902.
[1227] *Atlin Claim*, 17 Dec. 1904.
[1228] *Atlin Claim*, 15 Feb. 1902.
[1229] *Atlin Claim*, 8 Mar. 1902.
[1230] *Atlin Claim*, 8 Nov. 1902.
[1231] *Atlin Claim*, 29 Oct. & 19 Nov. 1904.
[1232] *Atlin Claim*, 15 Feb. 1908.
[1233] *Atlin Claim*, 10 Dec. 1904 & 1 & 10 April, 1905.
[1234] Anon, *Atlin Claim*, 18 Nov. 1899.
[1235] *Atlin Claim*, 25 Nov. 1899.
[1236] *Atlin News Miner*, 4 Feb. 1939.
[1237] *Atlin News Miner*, 28 Dec. 1940.

## Chapter 18

[1238] *Atlin Claim*, 2 June 1900.

[1239] *Atlin Claim*, 22 Nov. 1902.
[1240] *Atlin Claim*, 3 Aug. 1907.
[1241] *Atlin Claim*, 30 April 1904.
[1242] *Atlin Claim*, 10 Oct. 1902.
[1243] *Atlin Claim*, 10 Oct. 1905.
[1244] *Atlin Claim*, 10 Oct. 1902.
[1245] Gore, W. S., Letter to J.A. Fraser, 13 Feb. 1903, Correspondence of the Government Agent, Atlin.
[1246] *Atlin Claim*, 4 April 1904.
[1247] *Atlin Claim*, 13 Dec. 1902.
[1248] James, Cyril, Interview with Shirley Connolly and Nora Smith, 31 Jan.1973.
[1249] Nelson, Bill, Interview with Diane Smith, 1 May, 1984.
[1250] Minister of Mines, 1899, p 989.
[1251] *Atlin Claim*, 31 May, 1902.
[1252] Swarth, Harry S., "Mammals of the Atlin Region, Northwestern British Columbia," Journal of Mammalogy #17, 1936, p. 404.
[1253] Williams, A. Bryan, Game Trails in B.C., London; John Murray, 1925, p. 162. & *Atlin Claim*, 5 Aug. 1905.
[1254] *Atlin Claim*, 11 May 1901.
[1255] *Atlin Claim*, 18 Nov. 1899 & 15 Sept. 1900.
[1256] *Alaska Mining Record*, 28 Dec. 1898.
[1257] *Atlin Claim*, 15 June 1901.
[1258] Hensley, Bill, Interview with Sue Morhun, Aug. 1976.
[1259] Husselbee, Wm., Interview, A. H. S.
[1260] *Atlin Claim*, 3 Mar. 1906 & 24 Nov. 1906.
[1261] McCandless, R. Yukon Wildlife: A Social History 1900 - 1950.
[1262] Nelson, Bill, Interview, 1984.
[1263] Nelson, Interview, 1984.
[1264] McKenna - McBride Commission, Indian Affairs Records, GR 1342, P. A. B. C.
[1265] McKenna- McBride Commission.
[1266] McKenna - McBride Commission.
[1267] McKenna - McBride Commission.

## Chapter 19

[1268] Bennett-Atlin Commission.
[1269] See Chapter 2
[1270] *Alaska Mining Record*, 14 Sept. 1898.
[1271] *Alaska Mining Record*, 14 Sept. 1898. Other claimants were Mrs. J. Merrifield, Douglas Island News, 8th Jan. 1899; Mrs. M. Turner, *Atlin Claim*, 29 April, 1899; Mrs. L. White - *Atlin Claim*, 29 Sept. 1900.
[1272] Bennett-Atlin Commission.
[1273] *Atlin Claim*, 29 April 1899.
[1274] *Atlin Claim*, 17 June 1899.
[1275] Douglas Island News, 8 Jan. 1899.

[1276] Watson, Martha Grace, Diary, MS, Yukon Archives.

[1277] Anzer, Klondike Gold Rush, p. 131.

[1278] Watson, Diary.

[1279] Mrs. Watson's diary actually locates this event at Otter Lake, but since this would place the party far from the trail, it is probaby safe to assume that she actually meant Graham Inlet.

[1280] Watson, Diary.

[1281] Ryder, Diary.

[1282] O'Bannon, Marcella, Letter, 27 Feb. 1978, A. H. S.

[1283] Anzer, Klondike Gold Rush.

[1284] Graham, J. D., Letter to F. S. Hussey, 3 Aug. 1899, British Columbia Provincial Police.

[1285] *Atlin Claim*, 15 June 1901.

[1286] Lupecke, Helen, Letters written to Johann, Dec. 1900 - Feb. 1901, copies, A. H. S.

[1287] Kelly, Pat, letter, Jan. 1983, MS, A. H. S.

[1288] *Victoria Colonist*, 3 Aug. 1899.

[1289] *Atlin Claim*, 6 May 1899.

[1290] *Atlin Claim*, 14 Oct. 1899.

[1291] *Atlin Claim*, 7th April 1900.

[1292] *Atlin Claim*, 7 Sept. 1901.

[1293] *Atlin Claim*, 1 Sept. 1900.

[1294] *Atlin Claim*, 7 April 1900 & 15 Sept. 1900.

[1295] *Atlin Claim*, 15 Sept. 1900.

[1296] Hitchcock, M. E., Letter to J. D. Graham, 5 Oct., 1900, Correspondence of the Government Agent, Atlin.

[1297] *Atlin Claim*, 3 June 1899.

[1298] McKinnon, Ruth, Notes on D. H. McDonald and Family, MS, A. H. S.

[1299] Tuomaala, Eileen, "Warm Memories of Icy Atlin Winters" in *White Rock Sun*, 6 Jan. 1972.

[1300] McKinnon, Notes on D. H. McDonald.

[1301] *Atlin Claim*, 13 Aug. 1904.

[1302] *Atlin Claim*, 3 Sept. 1904.

[1303] *Atlin Claim*, 29 June 1907.

[1304] *Atlin Claim*, 29 April 1899.

[1305] Godfrey, Letter, 26 April 1899.

[1306] *Atlin Claim*, 29 April 1899.

[1307] Mackenrot, Dorothy A., Trek of W.J. Smith and Family from Inverness, Scotland to Atlin, B.C., MS, A. H. S.

[1308] Mackenrot, Trek of W. J. Smith.

[1309] Mackenrot, Trek of W.J. Smith.

[1310] MacDonald, Mabel, Interview with Christine Dickinson, July, 1981.

[1311] Owen, Interview.

[1312] Owen, Interview.

[1313] Bourne, Clayton, MS. A. H. S.

[1314] *Atlin Claim*, 12 Mar. & 30 April 1904.

[1315] Bourne, MS.

[1316] Bourne, MS.

[1317] Howay, British Columbia From the Earliest Times to the Present, Vol. 3, p.917.

[1318] Howay, British Columbia From the Earliest Times to the Present, Vol 3, p. 917.

[1319] Nelson, Interview, 1984.

[1320] Murphy, Marie, Diaries, 1924- 58. MS, A. H. S.

[1321] Gairns, Charlie, as told to Diane Smith, c.1968.

[1322] Caldwell, Francis E. Land of the Ocean Mists. Edmonds, WA, 1986, pp. 150-51.

[1323] London, Jack. Letter to the *Seattle Post Intelligencer*, 2 Aug. 1906.

[1324] Gairns, Charlie, as told to Diane Smith, c1968.

[1325] Walker, Carrie, Letter to Hon. Richard McBride, 21 Dec. 1903, Lottie Bowron Papers, MS, P. A. B. C.

[1326] *Atlin Claim*, 9 May 1903.

[1327] *Atlin Claim*, 8 July 1905.

[1328] Conley, O., Letter to his supt., Nov. 7, 1908, British Columbia Provincial Police.

[1329] *Atlin Claim*, 26 Sept. 1903.

[1330] Bock, Billy, The Book of Skeletons, c. 1960, p.92.

## Chapter 20

[1331] *Atlin Claim*, 1 April 1904.

[1332] *Atlin Claim*, 27 June 1903.

[1333] *Atlin Claim*, 29 Aug. 1903

[1334] *Atlin Claim*, 1 Aug. 1903.

[1335] *Atlin Claim*, 1 Aug. 1903.

[1336] *Atlin Claim*, 17 Oct. 1903.

[1337] *Atlin Claim*, 26 Sept. 1903.

[1338] *Atlin Claim*, 17 Oct. 1903.

[1339] B. C. Review of London, 1903.

[1340] *Atlin Claim*, 17 June 1899.

[1341] Carmichael, Diary.

[1342] Murphy, Marie, Diaries.

[1343] James, Cyril, Interview with Sue Morhun, 1975.

[1344] *Atlin Claim*, 7th November, 1903

[1345] *Atlin Claim*, 29 April, 1904.

[1346] *Atlin Claim*, 24 Aug. 1907.

[1347] White Pass, "A Handbook of Vacation Trips in Alaska, Atlin and the Yukon", Pamphlet, n.d. A. H. S.

[1348] *Atlin Claim*, 7 Nov. 1903 & *Seattle Times*, 1 June, 1913.

[1349] Kruty, Paul, Letter to A.H.S., 1982, MS, A. H. S.

[1350] White Pass Records, River Division, Superintendent's Report, 1916, Yukon Archives.

[1351] White Pass & Yukon Co., Superintendent's Report, 1916.

[1352] Cameron, Charlotte, A Cheechako in Alaska and Yukon, London, 1920, p. 83

[1353] White Pass Records, River Division, Tarahne Log Book, 1926

[1354] White Pass, Atlin the Beautiful, Pamphlet, n.d., A. H. S.

[1355] Acheson, Winnie, Interview with Christine Dickinson, c1984.

[1356] Eggert, Mickey, Interview with Sue Morhun, Sechelt, B.C., 10 Nov. 1978

[1357] Cameron, Charlotte, A Cheechako in Alaska and Yukon.

[1358] Harrington, Reba, Cinderella Takes a Holiday in the Northland, p. 136. & Cameron, A Cheechako in Alaska and Yukon, p. 79-94.

[1359] White Pass, Atlin the Beautiful.

[1360] *Atlin Claim*, 5 Aug. 1899.

[1361] Eggert, Interview.

[1362] White Pass, Atlin the Beautiful. p. 5-7

[1363] Glacier Bay is now known as Llewellyn Inlet.

[1364] Natural History, May-June, 1921, p. 252.

[1365] Eggert, Interview.

[1366] *Whitehorse Star*, 4th Aug. 1978.

[1367] Canadian Alpine Journal, 1918, p. 139.

[1368] Cameron, A Cheechako in Alaska and Yukon, p. 92.

[1369] Canadian Alpine Journal, 1918, pp. 135-139 & Natural History, 1921, pp. 251-254

[1370] White, Kathleen Keats, Vancouver Province, 13 April, 1930.

[1371] Temple, L.G. Victoria Colonist, 28 Oct. 1951.

[1372] White Pass Records, Report of the Superintendent, 1916.

[1373] White Pass Records, Superintendent's Report, 1926.

[1374] Abbott, MS.

[1375] St Martin's Anglican Church Records.

[1376] Coleman, Irene & Tom, Interview.

# INDEX

B

## D

# Q



# R

S

## T

## U



## V



## W

## Y